TRANSFER
PROCESSES

TRANSFER PROCESSES

*An Introduction to Diffusion,
Convection, and Radiation*

D. K. EDWARDS
V. E. DENNY
A. F. MILLS
University of California, Los Angeles

● HEMISPHERE PUBLISHING CORPORATION
Washington   London

McGRAW-HILL BOOK COMPANY

New York   St. Louis   San Francisco   Auckland   Bogotá
Düsseldorf   Johannesburg   London   Madrid   Mexico
Montreal   New Delhi   Panama   Paris   São Paulo
Singapore   Sydney   Tokyo   Toronto

TRANSFER PROCESSES

ISBN 0-07-019040-2

Library of Congress Catalog Card Number: 76-28628

1234567890   KPKP   785432109876

PREFACE

This textbook is intended for use in a first course in transfer processes. Heat and mass transfer and heat exchanger and mass transfer units are described to the extent permitted by *one-dimensional* ordinary differential equations. Formulation of transfer processes in partial differential equations is introduced but only transient diffusion in the semi-infinite solid and off-shoots of this problem (slug flow and the falling film with short contact time) are developed. Momentum transfer is treated to a lesser extent, and neutron transport is introduced very briefly.

The restriction to one-dimensional processes permits teaching a course to an undergraduate student who is still rather unsteady in exercising his newly acquired mathematical skills. Such a student is too easily bewildered by a great deal of mathematical complexity which often serves not to elucidate the physical nature of the phenomena in question but only to obscure it. The undergraduate student has had basic courses in chemistry, physics, and mathematics but has had only limited experience in applying all of it at one time to solving engineering problems. He is led to do so in a series of easy steps in this text.

The viewpoint taken in the presentation is first macroscopic, then microscopic, then macroscopic on a systems scale. This sequence of viewpoints was arrived at after some thought and experimentation. The micro-macro sequence has appeal in that it permits a highly ordered development using deductive logic. However, there is difficulty in sustaining student interest and motivation through the rather tedious early phases, and this approach was shunned. In Chapters 2 through 5, Appendix A, and the first half of Chapter 6 the macroscopic view is used to introduce the student to heat conduction, mass diffusion, convection, and the macroscopic concepts used in radiation and free molecule transfer. Starting in the second half of Chapter 6 and continuing through the first half of Chapter 8 use is made of the idea of microscopic carriers of heat, mass, and momentum moving in straight lines between collisions or interactions with walls or molecules. The idea of a transport mean free path is introduced, and this idea is used to develop transport properties: mass diffusivity, viscosity, thermal conductivity, and neutron diffusivity. In the first half of Chapter 8 the phenomenon of turbulence is introduced using the mean free path concept. The latter half of Chapter 8 and Chapters 9 and 10 return to the macroscopic world by considering systems of ducts, heat exchangers, and mass transfer units.

Both the instructor and the student are urged not to get bogged down in Chapters 2 or 3. It is true that many institutions teaching engineering and/or applied science devote approximately a full course to conduction, a full course to convective heat transfer, a full course to convective mass transfer, a full course to radiation, a full course to free molecule flow, one or two full courses to neutron transport, and often additional courses in specialized topics. It is so at our own institution. This text and the course intended to be offered from it are not to replace all these specialized courses. Rather the text is to introduce the subjects of these courses so that they may start at a somewhat higher level and may proceed at a somewhat faster pace. It is intended that the student who does not specialize in heat and mass transfer, fluid mechanics, or neutron transport obtain a broader view of these subjects than a first course in conduction.

The instructor is also urged not to neglect the engineering content of a course in transfer processes. Differential equations are certainly important in engineering and, furthermore, are enjoyable to teach. But the engineering student needs to be introduced to the practice of his profession, a profession in which algebra and logic are important too, and in which judgment in selecting the approach to a given problem is of paramount importance. It is our experience that appreciable lecture time should be devoted to the presentation of examples and applications which both reflect the subject matter of the text and relate to current engineering problems.

As mentioned earlier, momentum transfer is treated to a lesser extent than heat and mass transfer. For curricula in which fluid mechanics is not taught apart from heat and mass transfer, the instructor should augment this text with a fluid mechanics book of his choice. To cover the present text thoroughly, including Appendix A, and to augment the fluid mechanics would require 90 lecture hours. It is feasible to offer a 40 lecture-hour course by emphasizing only a portion of the text. For example, courses in (essentially)

1. convective heat and mass transfer, or
2. heat transfer, or
3. transport phenomena

could be taught by emphasizing Chapters 2, 3, 4, 5, 9, and 10; Chapters 2, 4, 5, the first half of 6, and 8 and 9; and Chapters 2 through 8 and Appendix A (the latter in conjunction with Chapter 5), respectively. Alternatively or additionally, the instructor may choose to emphasize fundamental macroscopic concepts treated in early sections of a given chapter and pass over the later sections. In this case, the instructor might omit in a first course Sections 2.11, 2.12.3–4, 2.13; 3.7–3.12; the whole of Chapter 4; Sections 6.5–6.9; 7.4–7.7; 8.2–8.3; 9.5–9.9; 10.5–10.7; and the whole of Appendix A.

We acknowledge the assistance of I. Catton, H. Buchberg, and D. N. Bennion, colleagues who contributed to Chapters 8, 9, and 10, respectively. Section 1.2.4 was based in part on notes prepared by Mr. Anil Vasudeva. Messrs. Richard Clever, John Rauscher, and Robert Turner assisted in the collection of the property data in Appendix B. Grateful acknowledgment is made to Mrs. Phyllis Gilbert, who typed the manuscript.

D. K. EDWARDS
V. E. DENNY
A. F. MILLS

March 1973

CONTENTS

Chapter **10**   MASS EXCHANGERS   262

Appendix **A**   CONSERVATION EQUATIONS   295

Appendix **B**   SELECTED VALUES OF CONSTANTS,
CONVERSION FACTORS, AND PROPERTIES   326

TRANSFER
PROCESSES

CHAPTER 1

TRANSFER PROCESSES

1.1 INTRODUCTION

Transfer of heat, momentum, and species of matter pose recurring problems to engineers and scientists. Unless a uniform condition exists — that is, unless thermodynamic equilibrium exists — such transfer processes will take place. The rates at which these transfers occur are important considerations in the design of technological systems as diverse as aerospace vehicles, chemical process plants, nuclear power stations, sewage treatment units, and thermal control devices for electronic gear. In planning such systems, the engineer analyzes and designs heat shields and exchangers, mass exchangers, chemical and nuclear reactors, settling tanks, filters, separators, and other such devices in which heat, momentum, and/or mass are transferred. The scientist, too, is concerned with transfer processes, for he frequently must understand and solve transfer problems in order to plan an experiment and interpret laboratory data. Experiments investigating superconductivity at cryogenic temperatures or nuclear fusion in high temperature gas plasmas or chemical syntheses involve transfer processes.

The physical mechanisms by which transfer processes occur are complex. Three broad categories are treated in this text: **diffusion**, **convection**, and **radiation**. In the *diffusion* category are such processes as heat conduction, viscous transfer of momentum, and mass diffusion. *Convection* is the transport of heat, momentum, or mass by a moving fluid. In the *radiation* category are such processes as heat radiation, neutron transport, and free molecule flow.

Diffusion occurs in dense matter. Slow neutrons or molecules in a gas at normal density are able to go but very short distances before colliding with matter. Radiation, on the other hand, occurs in low density matter. Fast neutrons travel vast open spaces between tiny atomic nuclei. Photons emitted by the sun are unimpeded on their path to the earth or to a spacecraft. Molecules in a vacuum chamber traverse from one wall to another without a collision. Diffusion is thus a short mean free path process while radiation is a long mean free path process.

The term convection, strictly interpreted, means transport by bulk motion of a moving medium. A fluid globule, by virtue of its mass and velocity, will transport momentum. Likewise, by virtue of its temperature and composition, it will transport energy and mass species. However, it is common practice to use the term convection more broadly and describe transfer processes involving moving fluids as convective, even though diffusion (or radiation) plays a dominant role in the transfer from a wall to the fluid. In this broad sense, convection is *both* convective and diffusive transport in a moving fluid.

1.2   EXAMPLES OF TRANSFER PROCESSES

1.2.1   Scope

One can cite examples of transfer processes within the human body itself or in the atmosphere and ocean. But of particular interest to engineers are the applications within the industrial complex which man has erected to satisfy his material needs. Properly used, transfer processes help man to reduce famine, disease, and stupefying labor. Examples could be chosen from such areas as power production, food processing, refining, and so on. The processes which are described in the following sections have been chosen to convey an idea of the variety of physical situations in which diffusion, convection, and radiation play key roles, individually or in concert.

1.2.2   Desalination by Multistage Flash Evaporation

One method for large-scale production of fresh water from sea water is multistage flash evaporation. Dual-purpose plants are proposed in which a nuclear reactor supplies steam for power generation, and low temperature exhaust steam from the turbines is used to provide heating required for the distillation process. A plant typical of those being proposed will produce 150 million gallons per day (MGD) and provide an electrical power output of perhaps 750 megawatts (Mw). The design consists of three parallel 50 MGD units, each having 53 stages. A stage is in the form of a large chamber. Hot brine enters through an orifice from the previous, higher pressure stage and a small fraction immediately evaporates (flashes). The remainder collects in a pool covering the base of the chamber before flowing on to the next, lower pressure chamber. The top of the chamber is filled with a tube bundle through which cooler brine flows and upon which flashed vapor condenses. Trays under the tube bundle collect the fresh water distillate, which is removed from the chamber. The 53 stages are connected in series, their pressures ranging from about 0.4 psia at one end, where cold sea water enters, to 25 psia at the other end, where brine at 235°F emerges from the condenser tubes. This brine is heated by power plant exhaust heat to 250°F and then returns to the highest pressure stage as feed, to cascade from one stage to the next, a small fraction flashing into vapor in each stage.

We see that each stage is simply a large boiler-condenser in which there are a number of transfer processes to be understood and controlled by the design engineer. The evaporation poses few problems. The pressure in the stage is less than the saturation pressure of the entering brine; the liquid thus boils vigorously, agitating the brine pool and facilitating transport of latent heat to bubbles of vapor and the pool surface. In fact, the agitation is sometimes so violent that baffles must be provided to prevent brine spray from contaminating the distillate. (A pilot plant at Point Loma, California, gave a design guarantee of less than 50 parts per million (ppm) salinity in the distillate but was able to achieve less than 4 ppm in practice.)

The vapor condenses as a film or as drops on the condenser tubes. The rate of condensation depends on the available temperature difference, the amount of air or other trace gas present, and the number and size of drops or the film thickness, since the latent heat of condensation must be conducted through the liquid. The latent heat must also be conducted across the tube walls and through deposits on the tube surfaces before being convected away by the brine coolant stream. The engineer's task is to maximize the condensation rate for a given temperature drop, because this will minimize the number of condenser tubes required. A 150 MGD plant will have a condensing surface area of 10.5 million sq ft; thus even a small gain in the effectiveness of the condensation process will be worthwhile.

1.2.3   Blood Oxygenation

Artificial blood oxygenation is required during open heart surgery. It may be necessary to bypass the lungs for several hours, during which time a machine is used to provide the patient with freshly oxygenated blood. Engineers have collaborated with surgeons in the development of devices known as *heart-lung* machines. The ideal machine must oxygenate venous blood up to the levels found in the pulmonary arteries, and remove CO_2, without altering the physico-chemical properties of the blood. In particular, blood may be damaged by shear stress levels resulting from excessive agitation. Another important design constraint is that the blood volume in the system must be kept small. A large blood volume is costly and may lead to problems of blood matching during surgery.

Of the various types of oxygenators in use, the rotating disk is perhaps the most versatile. A schematic of a typical layout is shown in Figure 1.1. As the

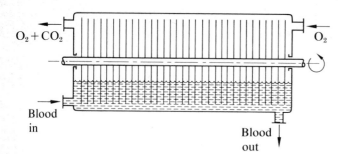

$O_2 + CO_2$

O_2

Blood
in

Blood
out

Figure 1.1   A   rotating
disk   blood
oxygenator.

disks rotate, they dip into a pool of blood and pick up a thin film which is exposed
to the oxygen. The rate of oxygen transfer depends on the number and size of the
disks, the depth to which the disks dip into the blood, and the rotational speed of the
disks. The engineer's objective is to optimize mass transfer without damage to the
blood. In addition the blood must be maintained close to body temperature to
avoid thermal shock to the patient. Thus the engineer also has a heat transfer
problem to solve.

1.2.4   Diffusion in a Semiconductor Diode

Another example of a transport process is the diffusion of carriers near
the junction of a semiconductor diode. A silicon atom, having only 4 electrons in
the M-shell (which is *complete*, that is, highly stable, with 8 electrons), arranges
itself in the solid state to share electrons with 4 neighbors. If silicon is *doped*
with atoms having only 3 valence electrons, such as boron, gallium, or indium, the
foreign atoms leave *holes* for electrons in the shells of shared electrons. This doping makes the material a p-type semiconductor. On the other hand, doping silicon
with atoms having 5 valence electrons, such as phosphorus, arsenic, or antimony,
creates an n-type semiconductor with loosely bound electrons which are readily
mobile.

When two such semiconductors are first joined to form a p-n junction,
holes, which are mobile in the p-material, diffuse toward the junction since there
are no holes there. Electrons from the n-material fill the holes reaching the junction. Very soon a negative charge is built up on the p-side of the junction, because
the protons in the foreign nuclei are insufficient to balance the electrons in "com-
pleted" shared shells. Similarly, a region of positive charge arises on the n-side,
because free electrons are lost.

The spatial integral of the charge density distribution gives rise to an
electric field (in the same way that a distributed load on a beam gives rise to a shear
force). Figure 1.2 illustrates this field. The field in turn gives rise to a potential (in the
same way that a shear on a beam gives rise to a moment). The gradient of the
potential causes a drift current which counteracts the current set up by the diffusion, so that the net current sums to zero when the junction is open circuit.

When a forward bias is applied, that is, when a positive voltage is applied
to the p-side of the junction, the potential gradient causing backward drift is re-

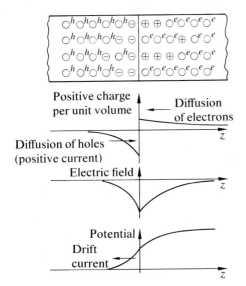

Figure 1.2 Transport phenomena at a semiconductor junction.

duced and may be reversed, if the bias is sufficiently large. The junction then readily conducts. When a backward bias is applied, the backward drift current is increased, but this increase causes a further depletion of holes on the p-side near the junction so that the number of carriers is reduced. This reduction in carriers causes the electrical conductivity to become very small and the resistivity very large. However, there will be a very slight increase in diffusion, because the concentration gradient steepens slightly. There is, as a result, a very slight negative current (unless too large a back bias is applied and a new phenomenon arises). The junction therefore acts as a diode conducting well in one direction and poorly in the other.

The diffusion and drift phenomena described very briefly above are examples of transfer processes governed by the type of equations and tractable to the methods of analysis to be developed in this text.

1.2.5 Spacecraft Thermal Control

High altitude earth satellites have already improved communications so that people in their homes may watch events as they occur on other continents, on the moon, and, one day, on nearby planets. Low altitude satellites make it possible to discover the location and extent of earth resources and to learn of the activities of the other inhabitants of the earth. To ensure smooth operation of a space probe or satellite, a detailed analysis is made of the heat transfer to each thermally sensitive component on the craft. Batteries must not freeze; transistors and diodes must not overheat; telescope optics must not be thrown out of alignment by thermal expansions and contractions; photographic records must be processed at just the right temperature to preserve high resolution.

On the exterior of a spacecraft one transfer process dominates, namely thermal radiation. Electromagnetic radiation flows from the sun and impinges upon the craft, imparting to it a small impulse and an appreciable amount of energy. Under the influence of these warming rays and internal heating from electrical, chemical, and nuclear sources, the atoms of the spacecraft are thermally agitated and emit a stream of infrared electromagnetic radiation into space. The balance between photons of solar origin which are absorbed and those emitted by the craft governs whether the vehicle is several hundred degrees above the boiling point of water or is 200 degrees below the freezing point. Engineers specify the surface coatings of such spacecraft to obtain the temperatures required.

In the interior of the spacecraft, both conduction and radiation play a part. Heat flows from warm to cold regions, transported by conduction and radiation. Convection may also play a role. For example, a high power level electronic device may dissipate 100 watts over an area of a square inch or two. Rather than employ a heavy conduction fin to conduct and then radiate energy away, the engineer may use a heat pipe — a thin-walled, hollow tube in which vapor can evaporate, flow, and condense. Heat and mass are carried by the fluid as it flows from the evaporator end to the condenser end in this convective transfer process. The condenser then radiates away the power dissipated.

1.2.6   Ablation

When a space vehicle approaches a planet, its speed increases sharply due to gravitational forces. Without retro-rockets the vehicle may enter into the planet's atmosphere at over 30,000 ft/sec. Colliding with gas molecules the vehicle slows down giving up some of its kinetic energy to the gas, which becomes very hot. Intense heating of the vehicle surface results.

From the viewpoint of an astronaut returning to earth, air appears to be rushing up to his stationary vehicle, as depicted in Figure 1.3. The oncoming air compresses as it impacts on the forward surface of the vehicle, but pressure disturbances can travel only at the cold gas speed of sound, about 1000 ft/sec, and are thus not able to propagate ahead of the vehicle to decelerate the gases gradually, or accelerate them off to one side. Instead a *shock wave* forms in front of the vehicle, where high velocity, low pressure gas is suddenly and drastically slowed down by collisions with low velocity gas adjacent to the vehicle. As a result, the pressure and temperature of the cold stream rise enormously. Temperatures on the order of 20,000°R are obtained and result in molecular collisions of such violence that molecules dissociate into atoms, some of which are stripped of one or more electrons and ionized. (The sheath of ionized particles causes the familiar *communications blackout* which occurs during entry.) Since dissociation is a highly endothermic process, some internal energy gained by the cold stream is absorbed, serving to lower the peak temperature attainable by the gas. Additional energy leaves by means of electromagnetic radiation which is emitted by gas particles whose vibrational and electronic modes are excited or by ionized particles which are accelerated during collisions. Some of these photon *energy packets* of radiant energy pass through the shock wave and are absorbed by distant gas; the remainder are transmitted toward the surface of the entry vehicle.

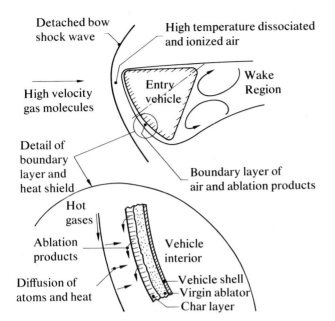

Figure 1.3 Atmospheric entry.

Needless to say, the walls of the vehicle cannot be at the peak gas temperature; if suddenly they were, they would instantly explode. Engineers provide a thermal protection system for the vehicle by blanketing its forward surfaces with a *heat shield* made from a composite mixture of fused fiberglass and phenolic resins. As heat penetrates the mixture, the resins pyrolyze, soaking up a large amount of energy as the solid constituents vaporize and dissociate. The pyrolysis products transpire through the porous, fiberglass reinforced *char* and absorb additional energy as they heat up. On reaching the char surface, these gases are blown into and "swept" along by the outer flow of hot gases, forming a layer of relatively cool gas adjacent to the vehicle. The char surface itself recedes as carbon oxidizes and fiberglass melts and vaporizes, absorbing more energy.

In the cooler *boundary layer* adjacent to the heat shield, a number of transfer processes occur. Atomic species, which have a high concentration at the *outer edge* of the boundary layer, diffuse inward and undergo both recombination at the reduced temperatures as well as reactions with pyrolysis gases and the char itself, all of which release heat. Similarly, the temperature is higher at the boundary layer edge than it is at the solid surface; thus heat also diffuses (that is, is conducted) inward. Collisions between gas molecules within the boundary layer transport momentum so that cool gases adjacent to the vehicle are not swept away as fast as they would be otherwise. Thus the viscous nature of the flow also plays a role in protecting the vehicle. Finally, radiation transport occurs in the boundary layer; photons from the hot gases enter the boundary layer, some being absorbed and less energetic ones reradiated.

The total mass which the heat shield loses is the crucial design parameter of interest to the engineer. The mass loss is found by calculating the rate at which energy is delivered to the pyrolysis zone by diffusive and radiative transfer processes, from which the rate of decomposition is determined. A calculation performed over the complete entry trajectory yields the total mass loss and surface recession.

1.3 OBJECTIVES

The objectives of this text are to introduce the subject of transfer processes to the student and to permit him to exercise mathematical skills and knowledge of elementary physics and chemistry which he has just acquired. In what follows there is first a development of heat conduction. This subject was chosen because the phenomenon is close to everyday experience and therefore readily understood. After establishing the basic ideas of steady-state conduction, transient conduction is considered.

Diffusion of species of matter, mass transfer, is then introduced. For simplicity the presentation is first restricted to one dimension. Simultaneous diffusion of heat or mass in some simple laminar flows is treated. The examples are limited to those tractable to ordinary differential formulation. The concept of a boundary layer and the basic ideas of similarity then complete the first half of the text. An Appendix on the formulation of multidimensional problem is referred to in conjunction with this material. Emphasis is placed on formulating problems and understanding similarity; but no attempt is made to present methods of solution for multidimensional problems. In the first half of Chapter 6, radiation transfer is introduced, without recourse to microscopic concepts, and problem-solving skills are developed.

In the second half of Chapter 6 the student is given some understanding of the *microscopic* underlying mechanisms of radiative transport. The Maxwell-Boltzmann and Planck distribution laws are presented and used for simple cases of free molecule and photon transport. The student is expected to gain an appreciation for how a large number of carriers, each with a different energy, can transmit energy, momentum, and mass at a statistically averaged steady rate. Chapter 7 continues the microscopic treatment by showing how collisions interrupt the transport process and lead to, on a statistically averaged basis, a flux proportional to a potential gradient. Chapter 8 introduces the basic concepts of turbulence, building on the ideas of statistical averages and mean free paths introduced in Chapters 6 and 7.

Transfer from one flow to another is taken up in the last two chapters. Chapter 9 deals with heat transfer between two flows on opposite sides of a wall. Chapter 10 is concerned with mass transfer between two contacting, flowing phases. These last two chapters deal with transfer equipment. As early as Chapter 2 concepts such as *fin effectiveness*, *overall transfer coefficient*, and the use of networks to model complex problems are introduced. These concepts are built upon and extended in the last two chapters to include relations between *number of transfer units* and *exchanger effectiveness*. These concepts lay the basis of heat and mass exchanger design.

CHAPTER 2

ONE-DIMENSIONAL HEAT CONDUCTION IN STATIONARY MEDIA

2.1 INTRODUCTION

In this chapter we introduce transfer processes by considering one-dimensional heat conduction in stationary media. By heat conduction in stationary media we mean that the only mechanism for heat transfer is Fourier's law of heat conduction, which simply is an observation that heat flows in the direction of decreasing temperature. We shall see in Chapter 4 that if a medium being considered is moving, a second transfer mechanism called *convection* contributes, and we rule out consideration of this mechanism for now by considering only stationary materials. By one dimensional we mean that temperature changes occur in only one spatial direction. We present several examples of engineering situations where one spatial dimension permits adequate description: heat transfer through a furnace wall (conduction across a slab which is thin in comparison with its lateral extent), cooling of an electrical resistor (radial conduction in a long cylinder or cylindrical shell), and heat transfer from fins (axial conduction along slender bodies with surface transfer).

Here, as well as throughout the book, we offer no apology for a one-dimensional approach; for it is our conviction that the beginning student is better served by minimizing the mathematical complexity of the subject in favor of stressing engineering content and the physical phenomena. Additionally, it is our objective in this chapter, as in others, to ready the student for the discomfiture which often besets the beginning practitioner of engineering when he finds that it frequently is necessary to sacrifice rigor for expediency. This expediency is particularly evident in Section 2.11 where we present an approximate analysis of a somewhat complex system.

2.2   FOURIER'S LAW OF HEAT CONDUCTION

A universal learning experience is finding that a hot object warms a cold one or vice versa. This chapter is concerned with predicting quantitatively the rate at which heat is conducted within a given medium subject to prescribed thermal conditions at its bounding surfaces. The basic phenomenological principle involved is **Fourier's law of heat conduction**, which states that the local heat flux in a homogeneous medium is proportional to the negative of the local temperature gradient. In terms of Cartesian coordinates x, y, z, Fourier's law may be written in mathematical form as

$$-q_x \propto \frac{\partial T}{\partial x}, \quad -q_y \propto \frac{\partial T}{\partial y}, \quad -q_z \propto \frac{\partial T}{\partial z}$$

where q is the heat flux (rate of heat transfer per unit cross-sectional area) and T the absolute temperature. For an isotropic medium these expressions may be written with a single constant of proportionality,

$$q_x = -k\frac{\partial T}{\partial x}, \quad q_y = -k\frac{\partial T}{\partial y}, \quad q_z = -k\frac{\partial T}{\partial z} \qquad \textbf{(2.1a, b, c)}$$

where the constant of proportionality k is called the thermal conductivity and is observed experimentally to depend on temperature level and other thermophysical properties of a given material. Note that q is positive when the local temperature decreases in a given direction, which is consistent with our observation that heat flows from hot to cold regions. Equations (2.1a, b, c) are components of the vector equation

$$\bullet \qquad \mathbf{q} = -k\nabla T \qquad (2.2)$$

which is the three-dimensional form of Fourier's law. From Eq. (2.2) it is seen that the thermal flux vector $\mathbf{q}$ represents a current of thermal energy which flows in the direction of the steepest temperature gradient.

With regard to units for the thermal quantities defined above, we shall employ for the most part the British Engineering system for which

$$\mathbf{q} \leftrightarrow [\text{Btu/ft}^2\,\text{hr}]$$
$$T \leftrightarrow [^\circ\text{R}]$$
$$x, y, z \leftrightarrow [\text{ft}]$$
$$\nabla \leftrightarrow [\text{ft}^{-1}]$$
$$k \leftrightarrow [\text{Btu/ft hr}\,^\circ\text{R}]$$

Equivalent units in the metric (SI) system are [joules/m² sec], [°K], [m], [m⁻¹], and [joules/m sec °K], respectively. Alternatively, one can write [w/m²] for q and [w/m °K] for k, since a joule is 1 watt-second or 1 watt is 1 joule per second. Selected mathematical and physical constants and conversion factors are presented in Tables B.1 and B.2 of Appendix B.

From a molecular viewpoint, the mechanism by which heat is conducted is that of unseen interactions between molecules. For gases these interactions consist primarily of binary collisions between pairs of molecules; thus the principal mode of energy exchange or transfer in gases is attributed to translational movements and collisions of molecules. However, for nonmetallic liquids and solids, molecules are more closely packed and energy is transferred by means of vibrational interactions. Metallic liquids and solids are good thermal conductors in that the relatively high mobility of free electrons leads to rapid energy transport. In general, heat conduction may be viewed as a process in which energy is *transported* by unseen carriers, that is, electrons, atoms, or molecules, whose energy is modified during particle-particle interactions. Based on simple kinetic theory arguments and verified experimentally, these interactions lead to a net flux of thermal energy in the presence of temperature gradients.

From the preceding discussion, it is not surprising that the magnitude of the thermal conductivity k for a given substance is strongly dependent on its molecular structure. For example, at 80°F, a typical gas (air) has a k of 0.015 Btu/hr ft °R, while a typical liquid (H_2O) has a much higher value, 0.35 Btu/hr ft °R. Single crystals and metals have very high values; copper, silver, and aluminum, for example, have values of k in excess of 100 Btu/hr ft °R. Tabulated values of k for various materials, as well as other thermophysical property data, appear in Tables B.3 through B.7 of Appendix B. On examining the data, it would seem that the thermal conductivities of certain solids (the so-called insulating materials) are anomalously low. Actually, these materials are not homogeneous; instead, they are a porous aggregate of solid particles and voids filled with gas such as air. Thus their thermal conductivities are termed *effective* in that they result from the physical nature of a gas-solid mixture.

We note that heat conduction is a diffusive process in the sense that it is the invisible or unseen behavior of molecules in the presence of temperature gradients which leads to net energy flows in matter. We shall see that energy may be transferred by convection and/or by radiation. Convection is energy transfer by thermal conduction into a moving fluid and will be taken up in Chapter 4. Thermal radiation is an energy transport process arising from the thermal motions of electrically charged particles which emit and absorb electromagnetic radiation. Actually, radiation plays a role in conduction, particularly in a porous or semi-transparent solid; but if the electromagnetic energy packets (photons) travel only a short distance within the solid or fluid, they constitute merely another unseen energy carrier and manifest themselves in indirect ways, for example, in the way the apparent thermal conductivity varies with temperature or with pore size. Chapter 6 introduces thermal radiation for situations in which the photons travel in voids between surfaces, and the latter part of Chapter 7 shows how such energy transfer can contribute to conduction.

2.3   THERMAL CONDITIONS AT SURFACES

Practically speaking, problems in heat conduction concern composite regions for which the thermophysical behavior in a given homogeneous region (phase) may differ markedly from that in an adjacent one. For example, in a nuclear reactor a liquid sodium stream flows over a stainless-steel-clad fuel element. Heat is transferred from the uranium oxide fuel to the inner surface of the cladding by means of conduction, through the cladding wall by conduction, and from the outer surface of the cladding by conduction into the moving fluid (convection). To treat such problems we must be able to specify conditions at solid-fluid and solid-solid interfaces.

In certain instances the convection process at a solid-fluid interface can be described by introducing what is termed **Newton's law of cooling**. This law states that the heat flux at a free surface in contact with a convecting fluid is proportional to a suitably selected temperature difference across a *surface film* of the adjacent fluid phase; that is,

$$q \propto (T_s - T_e)$$

where T_s is the temperature of the free surface and T_e is an appropriate temperature external to the surface film. As in Fourier's law, a constant of proportionality may be introduced such that

$$q = h(T_s - T_e) \tag{2.3}$$

where h is called the heat transfer coefficient and is found to depend on the physical properties of the adjacent fluid and its fluid mechanical behavior and, at times, on the radiation properties of the surface and its environs. In practice, values of h are calculated based upon analytical results or experimental ones generalized by using the principles of dimensional analysis. Obtaining values for h will be the subject of Chapter 5.

At solid-solid interfaces, such as the boundary surface between the stainless steel cladding and uranium fuel element cited above, two thermal conditions are required to solve a given problem. Since the interfacial region has negligible mass, it is incapable of *storing* thermal energy. Thus heat which is transferred to an interface from within a given phase must be transferred away from the interface into an adjacent phase at the same rate; that is,

$$q_i^{\mathrm{I}} = q_i^{\mathrm{II}} \tag{2.4a}$$

or, equivalently, using Fourier's law

$$-k\frac{\partial T}{\partial x}\bigg|_i^{\mathrm{I}} = -k\frac{\partial T}{\partial x}\bigg|_i^{\mathrm{II}} \tag{2.4b}$$

where Roman numerals I and II indicate adjacent phases, the subscript i designates an interfacial condition, and x represents the direction normal to the interface directed from I to II. The second condition is simply that of *local* thermodynamic equilibrium

$$• \qquad T_i^I = T_i^{II} \tag{2.5}$$

Actually, Eq. (2.4a) is a general relationship which holds at any interface whether or not convection occurs in one or both of the adjacent phases. In the case of convection parallel to the surface, one could replace one side of Eq. (2.4a) with Newton's law of cooling, Eq. (2.3). In the case of convection normal to the surface, enthalpy changes across the surface must be taken into account.

For example, when one or the other of phases I and II undergoes a change of phase such as ice melting at a rate per unit area of n, there is flow across the interface. Water flows out and ice flows in. The enthalpy difference of the flowing streams is the latent heat of the phase change. The form of the energy balance replacing Eq. (2.4b) is then

$$-k\frac{\partial T}{\partial x}\bigg|_i^I + n\hat{h}_i^I = -k\frac{\partial T}{\partial x}\bigg|_i^{II} + n\hat{h}_i^{II} \tag{2.6}$$

where the flux n[lb/hr ft^2] is taken to be positive when in the positive x-direction. For example, for the case of ice (phase I) flowing toward an interface at which it melts to become water (phase II), heat must be supplied by conduction in the water in the negative x-direction; that is, $k(\partial T/\partial x)|_i^{II}$ must be positive and larger than $k(\partial T/\partial x)|_i^I$ by an amount $n(\hat{h}_i^{II} - \hat{h}_i^I)$, where $(\hat{h}_i^{II} - \hat{h}_i^I)$ is the latent heat of fusion for water-ice, approximately 144 Btu/lb at one atmosphere pressure. This sort of problem, involving as it does both conduction and convection, is treated in Chapter 4.

2.4 VOLUMETRIC HEAT SOURCES

In certain applications, the thermal behavior of a body is affected by the existence of internally generated sources of thermal energy. Recall, for example, the familiar I^2R loss associated with flow of electrical current I in a resistor R. Other volumetric sources which are of commercial importance are nuclear (fission in a uranium fuel rod) and chemical (combustion of natural gas, metabolic processes in biochemical systems, mixing of sulfuric acid and water). As distinguished from the heat flux q, with units Btu/ft^2 hr, the volumetric heat sources $\dot{Q}_V$ (electrical, nuclear, and chemical) have units Btu/ft^3 hr.

2.5 THE ONE-DIMENSIONAL HEAT CONDUCTION EQUATION

Consider, as depicted in Figure 2.1, an element of volume $A\Delta x$ whose orientation in a homogeneous medium is such that $\partial T/\partial x$ is the only nonvanishing component of ∇T. Suppose now that we consider the state of our "system" (the volume $A\Delta x$) between times t and $t + \Delta t$. The change in sensible energy of the

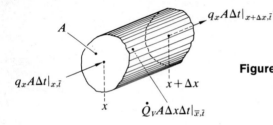

$$q_x A \Delta t|_{x+\Delta x, \bar{t}}$$

A

$$q_x A \Delta t|_{x, \bar{t}}$$

$x + \Delta x$

x

$$\dot{Q}_V A \Delta x \Delta t|_{\bar{x}, \bar{t}}$$

Figure 2.1 Energy balance on a volume element.

system when it is heated at constant pressure is just

$$(\rho c_p A \Delta x)|_{\bar{x}, \bar{t}}[T(\bar{x}, t + \Delta t) - T(\bar{x}, t)]$$

where $x < \bar{x} < x + \Delta x$ and $t < \bar{t} < t + \Delta t$. This change is due to the net energy transferred to or from the system

$$q_x A \Delta t|_{x, \bar{t}} - q_x A \Delta t|_{x+\Delta x, \bar{t}} + \dot{Q}_V A \Delta x \Delta t|_{\bar{x}, \bar{t}}$$

where q_x is presumed to be positive in the positive x-direction and $\dot{Q}_V$ is positive for a source and negative for a sink. From the principle of conservation of energy,

$$(\rho c_p A \Delta x)|_{\bar{x}, \bar{t}}[T(\bar{x}, t + \Delta t) - T(\bar{x}, t)] = -A \Delta t[q_x|_{x+\Delta x, \bar{t}} - q_x|_{x, t}] + \dot{Q}_V A \Delta x \Delta t|_{\bar{x}, t}$$

Dividing by $A \Delta x \Delta t$ and letting Δx and $\Delta t \rightarrow 0$, we obtain

$$\rho c_p \frac{\partial T}{\partial t} = -\frac{\partial q_x}{\partial x} + \dot{Q}_V \qquad (2.7)$$

Substituting for q_x, Eq. (2.1a), we get

$$\rho c_p \frac{\partial T}{\partial t} = \frac{\partial}{\partial x}\left(k \frac{\partial T}{\partial x}\right) + \dot{Q}_V \qquad (2.8)$$

Equation (2.8), or its equivalent (2.7), is the one-dimensional heat conduction equation in rectangular Cartesian coordinates. In this form the thermophysical properties ρ (density, lb/ft³), c_p (heat capacity, Btu/lb °F), and k (thermal conductivity, Btu/ft hr °F) as well as the volumetric heat source $\dot{Q}_V$ (Btu/ft³ hr) may be regarded as functions of both position and time. If k is assumed to be constant, there results

$$\frac{\partial T}{\partial t} = \frac{k}{\rho c_p} \frac{\partial^2 T}{\partial x^2} + \frac{\dot{Q}_V}{\rho c_p} = \alpha \frac{\partial^2 T}{\partial x^2} + \frac{\dot{Q}_V}{\rho c_p} \qquad (2.9a)$$

where $\alpha \equiv k/\rho c_p$ is called the thermal diffusivity (with units ft²/hr) and will shortly be seen to be an important characteristic parameter for unsteady heat conduction problems.

For problems involving radial heat conduction in cylindrical or spherical bodies, Eq. (2.9a) assumes the general form

$$\frac{\partial T}{\partial t} = \frac{\alpha}{r^n} \frac{\partial}{\partial r}\left(r^n \frac{\partial T}{\partial r}\right) + \frac{\dot{Q}_V}{\rho c_p} \qquad (2.9b)$$

where for cylinders $n = 1$ while for spheres $n = 2$. Derivation of Eq. (2.9b) is left to the exercises.

2.6 STEADY-STATE CONDUCTION IN THE SLAB

In various applications, the geometrical form of the conducting body as well as the thermal conditions prescribed over its bounding surface(s) are such that heat flows only in one direction. In addition, the temperature distribution in the body may be invariant with respect to time; that is, the heat flow is steady. Such physical situations give rise to what are termed steady, one-dimensional, heat conduction problems.

Consider the case of a slab of thickness L (Figure 2.2) which extends to infinity in the y- and z-directions. For constant properties and in the absence of sources and sinks, the steady one-dimensional forms of Eqs. (2.7) and (2.8) are

$$-\frac{d}{dx}q_x = k\frac{d^2T}{dx^2} = 0 \qquad\qquad \textbf{(2.10a,b)}$$

Integrating Eq. (2.10b) twice gives the solution

$$T = C_1 x + C_2 \qquad\qquad \textbf{(2.11)}$$

where the constants of integration C_1 and C_2 are to be determined from thermal conditions as prescribed at the bounding surfaces $x = 0$ and $x = L$. We see that for steady heat flow in the slab the heat flux $q_x = -kdT/dx$ is constant provided sources and sinks are absent; hence, the temperature distribution is linear when

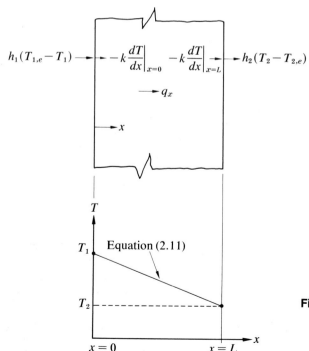

Figure 2.2 Schematic of one-dimensional heat conduction in a slab.

k is a constant. Typically, the form of the thermal boundary condition is

$$B_1 \frac{dT}{dx} + B_2 T = B_3, \qquad x = 0 \quad \text{or} \quad L \tag{2.12}$$

Mathematically speaking, the boundary conditions are termed **Dirichlet** when $B_1 = 0$ (surface temperature prescribed), **Neumann** when $B_2 = 0$ (surface gradient or heat flux prescribed), or **third kind** (Newton's law of cooling). A common exception to Eq. (2.12) exists when thermal radiation is important. In such an event the equation may assume the more general form

$$B_1 \frac{dT}{dx} + B_2 T + B_4 T^4 = B_3 \tag{2.13}$$

(See Chapter 6.)

In the strict sense, Dirichlet boundary conditions apply only in physical situations where the temperature at a surface is known *a priori*, for example, at a vapor-liquid interface during evaporation or condensation. Of course, they do apply, in an indirect sense, for phases in intimate contact when the temperature distribution in one of the phases is known. Occasionally, the Neumann condition is realistic; most often the third kind boundary condition is used such that Newton's law of cooling gives

$$h_1(T_{1,e} - T_1) = -k \frac{dT}{dx}\bigg|_{x=0} \qquad \text{at } x = 0 \tag{2.14}$$

$$-k \frac{dT}{dx}\bigg|_{x=L} = h_2(T_2 - T_{2,e}) \qquad \text{at } x = L \tag{2.15}$$

where h is the convective heat transfer coefficient, k the thermal conductivity of the slab material, T_1 and T_2 the surface temperatures at $x = 0$ and $x = L$, respectively, and $T_{1,e}$ and $T_{2,e}$ the ambient temperatures of the convecting fluid to the left of $x = 0$ and the right of $x = L$, respectively.

2.7   THE CONCEPT OF THERMAL RESISTANCE

2.7.1   Thermal Resistance of a Slab

Suppose the boundary conditions on our slab are simply Dirichlet conditions,

$$T = T_1 \quad \text{at } x = 0$$
$$T = T_2 \quad \text{at } x = L$$

Equation (2.11) yields at $x = 0$ and at $x = L$

$$T_1 = C_2$$
$$T_2 = C_1 L + C_2, \qquad C_1 = (T_2 - T_1)\frac{1}{L}$$

It therefore becomes

$$T = T_1 + (T_2 - T_1)\frac{x}{L} \tag{2.16}$$

Once the solution for a temperature profile has been found, the heat flux can always be determined from Fourier's law. In this case we obtain

$$q_x = -k\frac{dT}{dx} = -k(T_2 - T_1)\frac{1}{L}$$

Multiplying by the slab area A, denoting $q_x A$ by $\dot{Q}$ [Btu/hr], the heat flow per unit time, and rearranging gives

●
$$\dot{Q} = \frac{kA}{L}(T_1 - T_2), \qquad T_1 - T_2 = \dot{Q}\frac{L}{kA} \tag{2.17a, b}$$

Equations (2.17a, b) are seen to be forms of Ohm's law:

[current] = [conductance][potential difference]

[potential difference] = [current][resistance]

The quantity kA/L is thus a thermal conductance. Its reciprocal, L/kA, is said to be the thermal resistance of a slab of thickness L, thermal conductivity k, and area A. The heat transfer rate $\dot{Q}$ is the thermal current, and temperature is thermal potential.

The flow of heat through a slab can thus be represented simply by a conventional representation of a circuit, as shown in Figure 2.3.

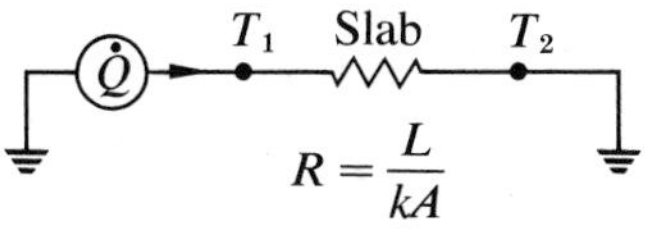

Figure 2.3   Circuit representation of a slab.

Note that the circuit is drawn as if our boundary conditions had been

$$-k\frac{dT}{dx} = \frac{\dot{Q}}{A} \quad \text{at } x = 0 \quad (T_1 \text{ unknown})$$

$$T = T_2 \quad \text{at } x = L$$

But the relationship between $\dot{Q}$, T_1, and T_2 is the same as that indicated in Figure 2.3 whether one regards T_1 and T_2 as imposed voltages and $\dot{Q}$ the resulting current, or $\dot{Q}$ an imposed current and $(T_1 - T_2)$ a resulting potential difference.

2.7.2   Film Resistance

Now suppose the boundary conditions are one Dirichlet condition,

$$T = T_1 \quad \text{at } x = 0$$

and one **film** condition,

$$q_x\big|_{x=L} = -k\frac{dT}{dx}\bigg|_{x=L} = h_2(T_2 - T_{2,e})$$

where h_2 is the Newton law of cooling *film* coefficient and $T_{2,e}$ is the temperature of the fluid some distance from surface 2. The term film is used because h_2 has units of k_f/δ_f where k_f is the thermal conductivity of adjacent convecting fluid and δ_f is a thickness which is imagined to be that of a film of stagnant fluid adhering to the surface. This concept is overly simplistic, as will be discussed in Chapter 5, but it does not affect the mathematical condition embodied in Eq. (2.12). We again apply each boundary condition to Eq. (2.11),

$$T_1 = C_2$$
$$-kC_1 = h_2(C_1 L + C_2 - T_{2,e})$$

solve for the constants of integration,

$$C_2 = T_1, \qquad C_1 = -h_2(T_1 - T_{2,e})/(h_2 L + k)$$

and substitute them into Eq. (2.11) to obtain the temperature profile

$$T = T_1 - [h_2(T_1 - T_{2,e})/(h_2 L + k)]x \qquad (2.18)$$

Equation (2.18) is then inserted into Fourier's law to obtain the heat flux

$$q_x = -k\frac{dT}{dx} = \frac{kh_2}{h_2 L + k}(T_1 - T_{2,e})$$

Multiplying by the area again gives the transfer rate $\dot{Q}$ [Btu/hr]

$$\dot{Q} = \frac{kh_2 A}{h_2 L + k}(T_1 - T_{2,e}) = \frac{1}{\dfrac{L}{kA} + \dfrac{1}{h_2 A}}(T_1 - T_{2,e}) \qquad (2.19)$$

It is instructive to note the form of Eq. (2.19). It is again in the form of Ohm's law. The thermal resistance in this case is the sum of two terms

$$\frac{L}{kA} + \frac{1}{h_2 A}$$

The first term is recognized as the thermal resistance of the slab itself, and the second term is the thermal resistance of the film on boundary 2. The flow of heat (thermal current) may again be represented by a circuit, as shown in Figure 2.4.

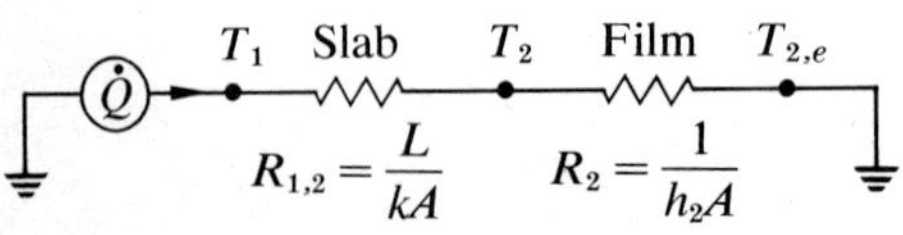

Figure 2.4   Circuit for a slab with a surface film.

2.7.3   Thermal Networks for Composite Walls

Now consider the physical situation illustrated in Figure 2.5. A series of $n-1$ slabs of dissimilar materials in intimate contact are located between two fluids, one at temperature $T_{1,e}$ and the other at $T_{n,e}$. Again assuming no heat sources or sinks, Eq. (2.10) applies, and a first integration yields

$$-q_x = k\frac{dT}{dx} = C_1$$

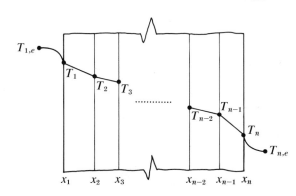

Figure 2.5 Schematic of the composite wall problem.

Thus the heat flux q_x is seen to be a constant at any cross-section (any position x) in Figure 2.5. Moving from left to right in the figure, there exists in each sub-region the equivalent expressions for q_x

$$
\begin{aligned}
q_x &= (h_1)(T_{1,e} - T_1) \\
&= (k_{1,2})(T_1 - T_2)/(x_2 - x_1) \\
&= (k_{2,3})(T_2 - T_3)/(x_3 - x_2) \\
&\qquad\vdots \\
&= (k_{n-1,n})(T_{n-1} - T_n)/(x_n - x_{n-1}) \\
&= (h_n)(T_n - T_{n,e})
\end{aligned}
$$

Multiplying by area, dividing by the coefficients of the temperature differences, and adding yields

$$\dot{Q}\cdot\left(\frac{1}{h_1 A} + \frac{x_2 - x_1}{k_{1,2} A} + \frac{x_3 - x_2}{k_{2,3} A} + \cdots + \frac{x_n - x_{n-1}}{k_{n-1,n} A} + \frac{1}{h_n A}\right) = (T_{1,e} - T_{n,e})$$

We thus find that the heat transfer rate $\dot{Q} = q_x A$ may be written

$$\dot{Q} = \frac{T_{1,e} - T_{n,e}}{\dfrac{1}{h_1 A} + \displaystyle\sum_{i=2}^{n}\dfrac{x_i - x_{i-1}}{k_{i-1,i} A} + \dfrac{1}{h_n A}} \qquad (2.20)$$

If now we define an "overall" heat transfer coefficient U, we can alternatively express $\dot{Q}$ as

$$\dot{Q} = UA(T_{1,e} - T_{n,e}) \tag{2.21}$$

where

$$UA = \cfrac{1}{\cfrac{1}{h_1 A} + \sum_{i=2}^{n} \cfrac{x_i - x_{i-1}}{k_{i-1,i}A} + \cfrac{1}{h_n A}} \tag{2.22}$$

In terms of UA the interfacial temperatures $T_1, T_2 \ldots$ may be written

$$T_1 = T_{1,e} - \frac{UA}{h_1 A}(T_{1,e} - T_{n,e})$$

$$T_2 = T_1 - \frac{(x_2 - x_1)(UA)}{k_{1,2}A}(T_{1,e} - T_{n,e}) \tag{2.23}$$

and so on. The electrical analog of the above *thermal network* is again evident if *resistances* to heat transfer are identified as reciprocals of $h_1 A$, $k_{1,2}A/(x_2 - x_1)$, and so on; overall temperature drop $(T_{1,e} - T_{n,e})$ is viewed as a potential; and heat flow $\dot{Q}$ is regarded as a current. The thermal-electrical analogs are illustrated in Figure 2.6.

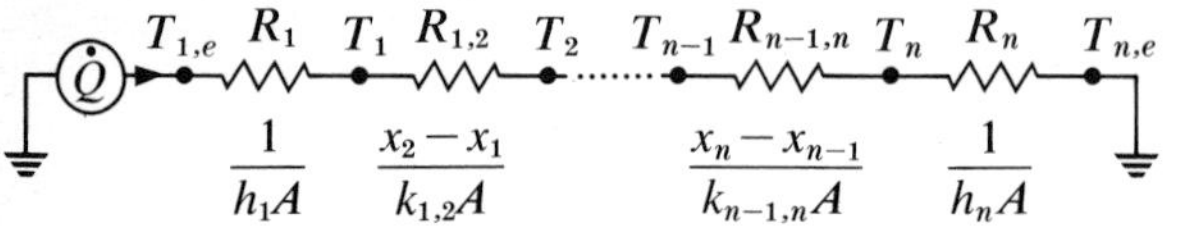

Figure 2.6   The thermal network for a composite slab with surface films.

2.7.4   *EXAMPLE*   Insulation for a Laboratory Oven

A small electrically heated oven, inside dimensions 1 ft × 1 ft × 2 ft, is to operate at temperatures up to 600°F. It is desired to package the oven in a painted steel case 0.060 in. thick which will enclose a layer of fiberglass insulation and the ceramic heater element, which has a total thickness of 0.50 in. and contains the heating elements embedded in the inner half of this thickness. The outside temperature of the case is to remain below 160°F so that severe burns will not be received by inadvertently touching the case. An air conditioning system prevents the laboratory air from rising above 80°F. Assume that both the inner and outer heat transfer coefficients are 2 Btu/hr ft² °F and that the thermal conductivity of the ceramic is 0.8 Btu/hr ft °F. Specify the insulation thickness and the maximum power permitted to be dissipated.

With the aid of a thermal network one may quickly visualize the problem. Assuming no heat loss from the interior, it is seen that the thermal current $\dot{Q}$ flows through a ceramic resistance $R_{1,2}$, a fiberglass resistance $R_{2,3}$, a resistance $R_{3,4}$ of the steel case (undoubtedly negligible), and a resistance R_4 for the surface film. If the areas were constant, the problem would be quite simple, for we would know all resistances save $R_{2,3}$. For example, for a 10 ft² total area, R_4 would be

$$R_4 = \frac{1}{h_4 A_4} = \frac{1}{(2)(10)} = 0.05 \left[\frac{Btu}{hr \,°F}\right]^{-1}$$

The temperature potential across this resistance, $T_4 - T_{4,e}$, is known to be $160 - 80 = 80°F$ under the worst conditions. Then the heat rate $\dot{Q}$ would be

$$\dot{Q} = (T_4 - T_{4,e})/R_4 = 80/0.05 = 1600 \text{ Btu/hr}$$

$$\dot{Q} = 1600[\,Btu/hr\,]/3.41[\,Btu/w\,hr\,] = 470 \text{ watts}$$

It is required that the inside temperature be $600°F$; consequently

$$(R_{1,2} + R_{2,3} + R_{3,4})\dot{Q} = (T_1 - T_4)$$

$$R_{1,2} + R_{2,3} + R_{3,4} = \frac{600 - 160}{1600} = 0.275 \left[\frac{Btu}{hr \,°F}\right]^{-1}$$

If the area were 10 ft²,

$$R_{1,2} = \frac{L_{1,2}}{k_{1,2} A_{1,2}} = \frac{(0.25/12)}{(0.80)(10)} = 0.0026 \left[\frac{Btu}{hr \,°F}\right]^{-1}$$

$$R_{3,4} = \frac{L_{3,4}}{k_{3,4} A_{3,4}} = \frac{(0.060/12)}{(28)(10)} = 0.000018 \left[\frac{Btu}{hr \,°F}\right]^{-1}$$

$$R_{2,3} = 0.275 - (0.0026 + 0.00002) = 0.272 \left[\frac{Btu}{hr \,°F}\right]^{-1}$$

Consequently the thickness of fiberglass required would be

$$L_{2,3} = R_{2,3} k_{2,3} A_{2,3} \text{ from } R_{2,3} = \frac{L_{2,3}}{k_{2,3} A_{2,3}}$$

It would remain to determine $k_{2,3}$. Unfortunately $k_{2,3}$ varies with temperature (see Section 7.6), but for now a value at an intermediate temperature, say $380°F$, will do. By linear extrapolation from Table B.4 one would arrive at a value of 0.065 Btu/hr ft °F for fiberglass at $380°F$ (Section 7.6 will suggest the basis for a rational extrapolation). In this way we find

$$L_{2,3} = (0.272)(0.065)(10) = 0.177 \text{ ft}$$

$$L_{2,3} \doteq 2 \text{ in.}$$

With a rough value of $L_{2,3}$ known, we may allow for the fact that $A_{2,3}$, $A_{3,4}$, and A_4 are somewhat greater than 10 ft². For example,

$$A_{2,3} \doteq \frac{1}{144} [2(15)(15) + 4(15)(27)] = 14.4 \text{ ft}^2$$

$$A_4 \doteq \frac{1}{144} [2(17.1)(17.1) + 4(17.1)(29.1)] = 17.9 \text{ ft}^2$$

The procedure may now be repeated a few times to obtain more correct values of $\dot{Q}$ and $L_{2,3}$.

2.8   STEADY CONDUCTION IN THE CYLINDER

2.8.1   Temperature Profile in a Cylinder with No Heat Sources

As for the slab, consider steady, constant property heat conduction in the absence of heat sources or sinks. Further, assume heat flow in the radial direction only. With these constraints the heat conduction equation in cylindrical coordinates (Eq. 2.9b) reduces to

$$\frac{d}{dr}\left(r\frac{dT}{dr}\right) = 0$$

which may be written

$$\frac{d}{dr}(rq_r) = 0$$

where $q_r = -k\, dT/dr$ is the radial flux in cylindrical coordinates. This equation expresses the fact that, in the absence of sources, the rate at which heat is conducted in the radial direction is constant; that is,

$$\frac{d}{dr}(\dot{Q}_r) = \frac{d}{dr}(2\pi rLq_r) = 0 \tag{2.24}$$

where L is the cylinder length. Integrating once gives

$$2\pi rLq_r = C_1, \qquad -2\pi kLr\frac{dT}{dr} = C_1 \tag{2.25}$$

Dividing by r and integrating a second time yields

$$-2\pi kLT = C_1 \ln r + C_2 \tag{2.26}$$

Thus the temperature profile is logarithmic.

To complete the problem two boundary conditions must be specified and used to solve for C_1 and C_2. Once these are known, the temperature profile can be used in Fourier's law to find q_r. The total heat rate $\dot{Q}_r$ at any radius can then be found by multiplying by $2\pi rL$.

For example, consider an insulated pipeline used to convey hot oil from a tanker to a terminal on shore. The oil is heated to make it easier to pump. Suppose the pipe is insulated by a jacket of insulation of inner radius r_2 and outer radius r_3. The metal pipe is at temperature T_2 and the sea water at T_3. Mathematically we write the boundary conditions as

$$T|_{r=r_2} = T_2, \qquad T|_{r=r_3} = T_3$$

and substitute them into Eq. (2.26),

$$-2\pi kLT_2 = C_1 \ln r_2 + C_2$$

$$-2\pi kLT_3 = C_1 \ln r_3 + C_2$$

Subtracting the first equation from the second gives

$$2\pi kL(T_2 - T_3) = C_1 \ln r_3 - C_1 \ln r_2 = C_1 \ln (r_3/r_2)$$

$$C_1 = \frac{2\pi kL(T_2 - T_3)}{\ln (r_3/r_2)}$$

Either of the first two equations is then used to find C_2.

$$C_2 = -2\pi kLT_2 - \frac{2\pi kL(T_2 - T_3)}{\ln (r_3/r_2)} \ln r_2$$

The power required to balance the heat loss from the pipe is

$$\dot{Q} = -kA \frac{dT}{dr}\bigg|_{r=r_2} = -k2\pi r_2 L \frac{dT}{dr}\bigg|_{r=r_2}$$

From Eq. (2.25)

$$-2\pi r_2 kL \frac{dT}{dr}\bigg|_{r=r_2} = C_1$$

so that the power required is

$$\dot{Q} = \frac{2\pi kL(T_2 - T_3)}{\ln (r_3/r_2)} \tag{2.27}$$

2.8.2   Thermal Resistance of a Cylindrical Shell

Note that Eq. (2.27) is in the form of Ohm's law. The thermal resistance is

$$R_{2,3} = \frac{\ln (r_3/r_2)}{2\pi kL} \tag{2.28}$$

When $r_3 = r_2 + \delta$ where $\delta/r_2 \ll 1$, Eq. (2.28) reduces to the resistance of a slab $\delta/(2\pi r_2 kL)$.

Once Eqs. (2.27) and (2.28) are found, it is possible to treat composite cylinders and/or cylinders with a Newton's law of cooling boundary condition without doing further analytical work. The thermal circuit for an insulated pipe with both internal and external film resistances is

$$R_1 = \frac{1}{h_1 A_1}, \quad R_{1,2} = \frac{\ln (r_2/r_1)}{2\pi k_{1,2}L}, \quad R_{2,3} = \frac{\ln (r_3/r_2)}{2\pi k_{2,3}L}, \quad R_3 = \frac{1}{h_3 A_3}$$

where in this case

$$A_1 = 2\pi r_1 L \quad \text{and} \quad A_3 = 2\pi r_3 L$$

2.8.3   Critical Thickness of Insulation on a Cylinder

Consider adding a cylindrical shell of insulating material over a tube. We may select a material of low conductivity in order to reduce heat losses from (or gains to) the tube. A thermal circuit indicates

$$\dot{Q} = \frac{T_{1,e} - T_{3,e}}{\dfrac{1}{2\pi r_1 L h_1} + \dfrac{\ln r_2/r_1}{2\pi k_{1,2} L} + \dfrac{\ln r_3/r_2}{2\pi k_{2,3} L} + \dfrac{1}{2\pi r_3 L h_3}} \tag{2.29}$$

We wish to determine the effect upon $\dot{Q}$ of varying r_3. Assuming R_1 and $R_{1,2}$ are small and rearranging Eq. (2.29), we find

$$\dot{Q} = \frac{2\pi L h_3 (T_{1,e} - T_{3,e})}{[1/r_3 + (h_3/k_{2,3}) \ln (r_3/r_2)]} = \frac{N}{D(r_3)}$$

Then differentiating gives

$$\frac{d\dot{Q}}{dr_3} = -\frac{N}{D^2} \frac{dD}{dr_3} = -\frac{N}{D^2}\left[\frac{h_3}{k_{2,3} r_3} - \frac{1}{r_3^2}\right] = -\frac{N}{r_3 D^2}\left[\frac{h_3}{k_{2,3}} - \frac{1}{r_3}\right]$$

The derivative of $\dot{Q}$ equals zero when

$$\frac{h_3}{k_{2,3}} - \frac{1}{r_3} = 0; \qquad \text{that is, } r_3 = r_{cr} = \frac{k_{2,3}}{h_3} \tag{2.30}$$

Thus $\dot{Q}$ passes through either a maximum or minimum when $r_3 = r_{cr}$. That the extremum is a maximum for positive $T_{1,e} - T_{3,e}$ (that is, positive N) may be seen from the negative sign of the second derivative

$$\frac{d^2\dot{Q}}{dr_3^2} = -\frac{N}{D^2} \frac{d^2 D}{dr_3^2}\bigg|_{r_3 = r_{cr}} = -\frac{N}{D^2}\left[-\frac{h_3}{k_{2,3} r_{cr}^2} + \frac{2}{r_{cr}^3}\right] = -\frac{N}{r_{cr}^2 D^2}\left[\frac{1}{r_{cr}}\right]$$

The quantity $r_{cr} - r_2$ is called *the critical thickness of insulation* because adding insulation when r_2 is less than r_{cr} actually increases the heat flow. A small tube or wire, particularly when it is bare metal with a low emittance for thermal radiation (discussed in Chapter 6), is best left alone unless a very large layer of insulation is added. The concept of a critical radius is also utilized by a designer bent on increasing heat flow as will be shown in Section 2.8.4.

2.8.4   *EXAMPLE* Thermal Cooling of an Electrical Resistor

Thermal considerations often play a large role in the design of electronic components which must dissipate $I^2 R$ heating. The engineer must often design for minimum temperature rise to avoid damaging the components when they operate at maximum rating. For example, consider a 1-watt, 2-megohm resistor 1.20 in. long as shown in Figure 2.7. Let us calculate the temperature T_2 of the resistor surface with and without an encapsulating micanite layer. The micanite, which is a crushed· mica-phenolic-resin mixture, serves as additional *electrical insulation* and also as a *thermal conductor*; it typically has a thermal

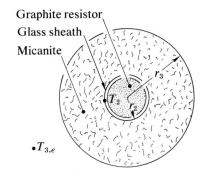

Graphite resistor
Glass sheath
Micanite

$\bullet T_{3,e}$

Figure 2.7 Cross section of a two megohm resistor insulated with micanite.

conductivity of 10^{-3} watt/cm °K. We assume that only 40% of the I^2R heating is dissipated by combined convective and radiative transfer from the outer surface with $h_3 = 3.0$ Btu/hr ft² °R. The remaining 60% of the I^2R heating typically is conducted through the copper leads to a circuit board which acts as a fin. Sections 2.10 and 2.11 show how this lead fin cooling can be estimated. We assume the 60–40% split holds for both the encapsulated and bare case; although in reality, it would be affected by the encapsulating material.

The micanite has maximum cooling effect when its radius is at the critical value which maximizes $\dot{Q}/(T_2 - T_{3,e})$. From Eq. (2.30)

$$r_3 = r_{cr} = \frac{k_{2,3}}{h_3}$$

$$= 10^{-3}\left[\frac{\text{watt}}{\text{cm °K}}\right]3.41\left[\frac{\text{Btu}}{\text{watt hr}}\right]30.5\left[\frac{\text{cm}}{\text{ft}}\right]\frac{1°\text{K}}{1.8°\text{R}}\frac{1}{3.0\left[\frac{\text{Btu}}{\text{hr ft}^2 °\text{R}}\right]}$$

$$= \frac{0.0578}{3.0} = 0.01928 \text{ ft} = 0.23 \text{ in.}$$

The thermal circuit for the micanite and surface film is simply two resistances in series, hence

$$T_2 - T_{3,e} = \dot{Q}\left\{\frac{\ln(r_3/r_2)}{2\pi k_{2,3}L} + \frac{1}{h2\pi r_3 L}\right\}$$

With $r_2 = 0.036$ in. $= 0.0030$ ft, $\dot{Q} = 0.4$ watt $\times 3.41$ Btu/hr watt, $L = 1.2$ in. $= 0.1$ ft, $k_{2,3} = 0.0578$ Btu/hr ft °F, and $T_{3,e} = 100$°F, we obtain

$$T_2 = 100 + 1.364(51.3 + 27.5)$$
$$T_2 = 100 + 107 = 207°\text{F}$$

When no micanite is present

$$T_2 - T_{3,e} = \dot{Q}\left\{\frac{1}{h2\pi r_2 L}\right\}$$

$$T_2 = 100 + 1.364(177)$$
$$T_2 = 100 + 242 = 342°\text{F}$$

2.9   STEADY CONDUCTION IN THE SPHERE

2.9.1   Temperature Profile in a Sphere with No Heat Sources

As in the case of the cylinder, consider steady, constant property, radial heat conduction in the absence of heat sources or sinks. With these constraints Eq. (2.9b) reduces to

$$\frac{d}{dr}\left(r^2\frac{dT}{dr}\right)=0$$

Integrating once gives

$$r^2\frac{dT}{dr}=C_1$$

Dividing by r^2 and integrating a second time

$$T=-\frac{C_1}{r}+C_2$$

As was the case for the cylinder, two boundary conditions must be fixed to determine C_1 and C_2. If a spherical shell with inner radius r_1 and outer radius r_2 has inner temperature T_1 and outer temperature T_2, the boundary conditions give

$$T|_{r=r_1}=T_1=-\frac{C_1}{r_1}+C_2$$

$$T|_{r=r_2}=T_2=-\frac{C_1}{r_2}+C_2$$

Subtracting the second relation from the first yields

$$T_1-T_2=\frac{C_1}{r_2}-\frac{C_1}{r_1},\qquad C_1=-\frac{T_1-T_2}{1/r_1-1/r_2}$$

As before C_2 is found from either of the relations; from the first one

$$C_2=T_1+\frac{C_1}{r_1}=T_1-\frac{T_1-T_2}{1-r_1/r_2}$$

The heat flow at location $r=r_1$ can be found from Fourier's law,

$$\dot{Q}=-4\pi r^2 k\frac{dT}{dr}\bigg|_{r=r_1}$$

$$\dot{Q}=-4\pi r_1{}^2 k\frac{C_1}{r_1{}^2}$$

$$\dot{Q}=\frac{4\pi k(T_1-T_2)}{1/r_1-1/r_2} \tag{2.31}$$

2.9.2 Thermal Resistance of a Spherical Shell

Again we note that Eq. (2.31) is in the form of Ohm's law. The thermal resistance is

$$R = \frac{1/r_1 - 1/r_2}{4\pi k} \tag{2.32}$$

As before this thermal resistance can be used in a thermal circuit.

2.9.3 Lower Bound on the Convective Heat Transfer Coefficient

If the outer radius r_2 goes to infinity, the heat transfer given by Eq. (2.31) is

$$\dot{Q} = \frac{4\pi k(T_1 - T_2)}{\dfrac{1}{r_1}} = 4\pi r_1^2 \frac{k}{r_1}(T_1 - T_2)$$

This relation governs, for example, the cooling rate of a small drop or pellet in still air. Comparing this relation to Newton's law of cooling,

$$\dot{Q} = h_1 A(T_1 - T_2) = h_1(4\pi r_1^2)(T_1 - T_2)$$

shows

$$h_1 = \frac{k}{r_1}$$

It is customary to write this answer in the form

$$\frac{h_1 D}{k} = 2 \tag{2.33}$$

where $D = 2r_1$. The quantity hD/k is called the **Nusselt number** in honor of an early worker in heat transfer.

2.10 FINS

2.10.1 The Fin Concept

A steady heat transfer problem, which is not really one-dimensional but which can be so treated to a good approximation, is that of the **fin**. A fin is a device used to increase heat transfer area. A glance at a car radiator or the condenser commonly placed ahead of car radiators on air-conditioned automobiles shows thin lightweight metal surfaces called fins projecting into air. Heat flows from the base of the fin into the thin metal and out into the air. In this way the effective hA can be made large so that $\dot{Q}$ will be large even though the temperature difference between the solid and the fluid is small.

Fins also appear on the cylinders of air cooled engines such as are used on motorbikes and lawn mowers. Cooling of solid-state diodes used for rectifying A.C. power and of high power transistors is also often promoted by mounting fins on the devices.

Because the fins are thin, the temperature differences in the thin direction are negligible. The approximation is made that the surface heat loss (or gain) can be treated as distributed uniformly over the thickness without introducing significant error. In this way what is a two-dimensional heat conduction problem can be solved in one dimension. Read this paragraph once more, think about it, then continue.

2.10.2   A Pin Fin

Consider a pin fin used to help cool electronic components. Figure 2.8 shows a schematic of such a fin. Analysis is straightforward.

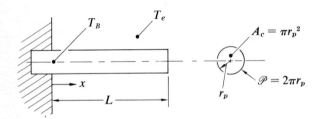

Figure 2.8   Schematic of a pin fin.

The first step is to choose a coordinate system and make a heat balance on a typical element, that between x and $x + \Delta x$.

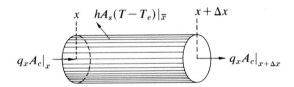

We recognize that energy can enter and leave by conduction through the cross-sectional area of the pin, $A_c = \pi r_p^2$. Energy leaves also from the surface according to Newton's law of cooling. The area in this case is the side area $A_s = \mathscr{P} \, \Delta x = 2\pi r_p \, \Delta x$, where $\mathscr{P} = 2\pi r_p$ is the perimeter exposed to cooling. From a heat balance,

$$q_x A_c \big|_x - q_x A_c \big|_{x+\Delta x} - h \mathscr{P} \, \Delta x (T - T_e) \big|_{\bar{x}} = 0$$

or, dividing by Δx and letting $\Delta x \to 0$,

$$-\frac{d}{dx}(A_c q_x) - h \mathscr{P}(T - T_e) = 0$$

$$q_x = -kA \cdot \frac{dT}{dx}$$

Assume A_c constant, and replace q_x by $-k\, dT/dx$ to obtain

$$\frac{d^2T}{dx^2} - \frac{h\mathscr{P}}{kA_c}T = -\frac{h\mathscr{P}}{kA_c}T_e \tag{2.34}$$

The next step is to set up the boundary conditions. At $x = 0$ the fin is at its base temperature T_B.

$$T|_{x=0} = T_B$$

At $x = L$ the fin loses heat according to Newton's law of cooling. However, the end area A_c is ordinarily negligible compared to the total side area $\mathscr{P}L$. It is therefore not a serious error to take zero heat flow at the end,

$$-A_c k \frac{dT}{dx}\bigg|_{x=L} = 0$$

This condition is somewhat more convenient mathematically.

The next step is to solve the governing differential equation. This equation is a linear ordinary differential equation. The general solution is the sum of a solution to the homogeneous equation and a particular solution. By inspection a particular solution is

$$T = T_e$$

The homogeneous solution is composed of two exponential terms having exponent coefficients which are roots of the characteristic equation

$$\left(D^2 - \frac{h\mathscr{P}}{kA_c}\right) = 0, \qquad D = \pm\sqrt{\frac{h\mathscr{P}}{kA_c}}$$

To simplify notation let

$$\beta^2 = \frac{h\mathscr{P}}{kA_c} \tag{2.35}$$

Then the homogeneous solution can be written in either of two forms,

$$T = C_1 e^{+\beta x} + C_2 e^{-\beta x}$$
$$T = B_1 \sinh \beta x + B_2 \cosh \beta x$$

The second form is more convenient. The general solution is

$$T = B_1 \sinh \beta x + B_2 \cosh \beta x + T_e \tag{2.36}$$

Imposition of the boundary conditions gives two algebraic equations in the unknown constants B_1 and B_2:

$$T_B = B_1 \sinh(0) + B_2 \cosh(0) + T_e, \quad B_2 = T_B - T_e$$

$$\frac{dT}{dx}\bigg|_{x=L} = \beta B_1 \cosh \beta L + \beta B_2 \sinh \beta L = 0, \quad B_1 = -B_2 \tanh \beta L$$

Substituting B_1 and B_2 in Eq. (2.36)

$$T = -(T_B - T_e)[\tanh \beta L \sinh \beta x - \cosh \beta x] + T_e \tag{2.37a}$$

$$T = (T_B - T_e)\frac{\cosh \beta(L-x)}{\cosh \beta L} + T_e \tag{2.37b}$$

The heat dissipated by the fin can be found in two ways. The obvious, but awkward, way is to find

$$\dot{Q} = \int_0^L h\mathscr{P}(T - T_e)\, dx$$

A less obvious, but more convenient, way is to use Fourier's law. From Eq. (2.37a) we obtain

$$\dot{Q} = -kA_c \frac{dT}{dx}\bigg|_{x=0} = kA_c\beta(T_B - T_e)\tanh \beta L \tag{2.38}$$

2.10.3   Fin Effectiveness

If a fin is made of a good thermal conductor and is thick ($kA_c \gg h\mathscr{P}$), it will be nearly isothermal at the base temperature T_B, since β will be very small. Such a fin is said to be 100% effective because all of its area is as hot as the base. If a fin is too long, it will cool to the air temperature and not transfer any more heat than a shorter one. Such a fin is ineffective. Effectiveness is defined as

$$\eta = \frac{\dot{Q}_{\text{actual}}}{\dot{Q}_{\text{isothermal fin}}} = \frac{kA_c\beta(T_B - T_e)\tanh \beta L}{\mathscr{P}Lh(T_B - T_e)}$$

or, since $\beta^2 = h\mathscr{P}/kA_c$,

$$\eta = \frac{1}{\beta L}\tanh \beta L \tag{2.39}$$

When βL is small, η is nearly unity; when βL is larger than 4, η is essentially $1/\beta L$, since then $\tanh \beta L \simeq 1$. Note that a thick fin with an effectiveness of nearly 100% will not usually be an optimum fin from the standpoint of heat transferred per unit weight or unit cost. The concept of fin effectiveness refers only to the ability of a fin to transfer heat per unit exposed area.

2.10.4   Thermal Resistance of a Fin

Note that Eq. (2.38) is in the form of Ohm's law. The thermal resistance of a pin fin is

$$R_{\text{fin}} = \frac{1}{kA_c\beta \tanh \beta L} = \frac{1}{h\mathscr{P}L\eta} \tag{2.40}$$

Note that the fin with its film is a *single* resistance. Table 2.1 summarizes the thermal resistances derived in this chapter.

Table 2.1 Thermal Resistances for Steady Heat Conduction without Volume Heating

Slab	L/kA
Film	$1/hA$
Cylindrical annulus	$(\ln r_2/r_1)/2\pi kL$
Spherical shell	$(1/r_1 - 1/r_2)/4\pi k$
Fin with constant $\mathscr{P}, A_c, h$	$1/kA_c\beta \tanh \beta L$

2.10.5 EXAMPLE Pin Fins to Cool a Transistor

Consider an array of eight pin fins used to cool a transistor. They are black anodized aluminum, $\frac{1}{8}$ in. wide, 0.015 in. thick, and 1.5 in. long. When the bases of the fins are 150°F and the ambient air is 85°F, how much power do they dissipate? Take $h = 1.5$ Btu/hr ft² °F and $k = 110$ Btu/hr ft °F.

A first step is to compute β using Eq. (2.35). The cross-sectional area is $(0.125/12)(0.0150/12) = 1.30 \times 10^{-5}$ ft². The perimeter is $0.28/12 = 0.0233$ ft.

$$\beta^2 = \frac{(1.5)(0.0233)}{(110)(1.30 \times 10^{-5})} = 24.5, \qquad \beta = 4.95$$

From Eq. (2.39), the fin effectiveness is therefore

$$\eta = \frac{1}{(4.95)(1.5/12)} \tanh\left[(4.95)\left(\frac{1.5}{12}\right)\right] = \frac{0.550}{0.618} = 0.89$$

All eight fins have an area A_T exposed to air of $8(1.5/12)(0.0233) = 0.0233$ ft².

From Newton's law of cooling, if the fins were 100% effective, they would lose

$$hA_T(T_B - T_e) = (1.5)(0.0233)(150-85) = 2.27 \text{ Btu/hr}$$

Since they are 89% effective, they lose

$$\dot{Q} = \eta hA_T(T_B - T_e) = (0.89)(2.27) = 2.02 \text{ Btu/hr} = 0.59 \text{ watt}$$

2.10.6 EXAMPLE Cooling Coil Spacing for a Heat Barrier

Consider next a *heat barrier* consisting of a $\frac{1}{16}$ in. thick brass plate to which $\frac{1}{4}$ in. copper tubing is soldered as shown in Figure 2.9. The tubes are spaced 4 in. apart. Cooling water passing through the tubes keeps them at approximately 100°F. The underside of the brass wall is insulated with $\frac{1}{2}$ in. of asbestos which in turn contacts a hot wall at 500°F. From Appendix B (Tables B.3 and B.4) $k = 64$ Btu/hr ft °F for brass and 0.09 for asbestos. Assuming the convective heat transfer from the *cold* side of the plate is negligible, calculate the temperature of the hottest spot on the brass wall.

The sketch reveals that the brass, from its base where it contacts the

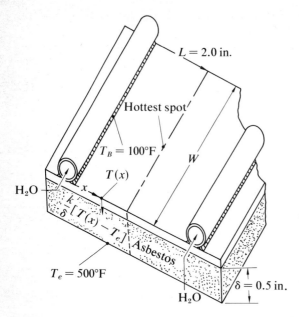

Figure 2.9   A heat barrier.

copper tube to the midpoint between the tubes, is a sort of a fin. It is 2 in. long, W in. wide, and $\frac{1}{16}$ in. thick. The base temperature is 100°F, and the tip temperature midway between the tubes is the hottest spot. The heated perimeter is W. The $\frac{1}{2}$ in. thick asbestos gives a surface heat flux of

$$\frac{k}{\delta}(T(x) - T_e) = h_{eq}(T(x) - T_e)$$

where T_e is 500°F and $h_{eq} = k/\delta$ is an equivalent heat transfer coefficient whose value is $0.09/(0.5/12) = 2.16$ Btu/hr ft² °F. The quantity β is then

$$\beta = \left(\frac{h_{eq}\mathscr{P}}{kA_c}\right)^{1/2} = \left(\frac{h_{eq}W}{kW \cdot \delta_b}\right)^{1/2} = \left[\frac{2.16}{64(0.0625/12)}\right]^{1/2} = 2.54$$

From Eq. (2.37b) the tip temperature at $x = L$ is

$$T_L = T_e - (T_e - T_B)\frac{\cosh(0)}{\cosh \beta L}$$

$$T_L = T_e - (T_e - T_B)\frac{1}{\cosh \beta L}$$

$$T_L = 500 - (500 - 100)\frac{1}{\cosh[2.54(2/12)]}$$

$$T_L = 500 - \frac{400}{1.091} = 133°F$$

To check our assumption of negligible heat transfer from the cold side of the plate to the ambient fluid on that side we could rederive Eq. (2.34) for a flat fin with a different T_e on each side. If a heat transfer coefficient of 2 Btu/hr ft² °F and an ambient T_e of 80°F existed on the cold side, we would find $T_L = 130°F$ instead of 133°F. Before proceeding to the next section the student is advised to carefully study the *engineering approach* used in this example.

2.11 ANALYSIS OF A SYSTEM

2.11.1 A Problem Involving Several Concepts

A final steady-state example is chosen to show that, although we have restricted ourselves to only simple one-dimensional situations, we can still analyze approximately a fairly complex system. We consider the cooling of an integrated circuit element sketched in Figure 2.10.

An integrated circuit is formed by depositing electrically conducting and nonconducting layers of semiconductors on a *chip*. The chip is encapsulated in an electrical insulator from which electrical leads project. The capsule is on the order of $\frac{1}{4}$ in. square by 0.1 in. thick. The power rating for such a device is typically 120 mw. The chip itself is roughly cubical in shape with a size of only 0.040 in. or so. The chip should not operate much above 200°F. The problem is to analyze the heat transfer process from the chip to the surrounding air and to recommend guidelines for the circuit board designer using what little we have learned of heat conduction in this chapter so far.

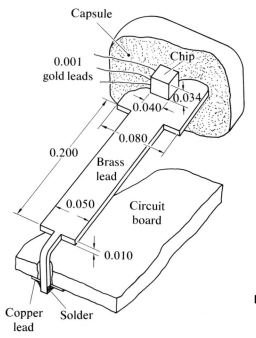

Figure 2.10 A thermal sink lead for an integrated circuit. (All dimensions are in inches.)

2.11.2   Losses from the Capsule Surface

We might first ask ourselves how much heat is dissipated by conduction through the insulating capsule. As a rough idealization, consider the capsule to be a sphere with the same surface area as the actual geometry; $r_2 \doteq 0.134$ in. We idealize the chip as being a sphere with the same volume; $r_1 \doteq 0.0233$ in. Assuming the conductivity of the encapsulating material is 0.058 Btu/ft hr °F (see Section 2.8.4) allows a thermal resistance to be estimated from Eq. (2.32).

$$R_c = \frac{12}{4\pi(0.058)}\left[\frac{1}{0.0233} - \frac{1}{0.134}\right] \doteq 600 \,[\text{Btu/hr} \,°\text{F}]^{-1}$$

To this resistance must be added that of the air film. To estimate h we use Eq. (2.33). This equation gives the minimum possible value

$$h_{\min} = 2\frac{k_{\text{air}}}{2r_2} = \frac{0.016}{0.134/12} = 1.43 \,[\text{Btu/hr ft}^2 \,°\text{F}]$$

We arbitrarily double this value to account roughly for free convection and radiation treated in Chapters 5 and 6. The film then has resistance

$$R_f = \frac{1}{hA_2} = \frac{144}{(2.86)(4\pi)(0.134)^2} = 220[\text{Btu/hr} \,°\text{F}]^{-1}$$

The total resistance is therefore $R = R_c + R_f = 820 \,[\text{Btu/hr} \,°\text{F}]^{-1}$. Assuming an ambient temperature of 100°F, the resulting temperature difference of 100°F would allow the dissipation of

$$\dot{Q} = \frac{100}{820} \,[\text{Btu/hr}]\frac{1}{3.41[\text{Btu/hr watt}]} = 36 \,\text{mw}$$

We check this answer against our rule of thumb that 60% of the power dissipation from small electronic components is through the leads and find that this must be so or the device would be too hot.

2.11.3   Losses through the Leads

To determine the conduction paths for lead losses, we break the capsule open and identify the structure depicted in Figure 2.10.

Integrated circuit chips are often on beryllia (BeO) substrates, because beryllia has an enormous thermal conductivity for a dielectric, approximately that of aluminum. If this is so, the chip is essentially isothermal. The resistance of the chip would be

$$R_1 = \frac{L}{kA} = \frac{0.034/12}{(110)(0.040/12)^2} = 2.32 \,[\text{Btu/hr} \,°\text{F}]^{-1}$$

That of the heat sink lead is

$$R_2 = \frac{L}{kA} = \frac{(0.200/12)}{(70)(0.010/12)(0.050/12)} = 68.6[\text{Btu/hr} \,°\text{F}]^{-1}$$

If copper were used, R_2 would be 70/240 as large. That the fine gold whisker leads do not contribute significantly to lead conduction can be found by comparing the value of R for one such lead with the above value.

2.11.4   Thermal Resistance of a Circuit Board

We come now to the interesting part. The lead is grounded with a well-soldered contact (let us hope) to a circuit board. Boards are commonly made by laminating 0.003 in. copper foil on a phenolic-filled cloth board typically 0.056 in. thick. A photographic exposure of the board coated with a film causes only certain portions of the board to be protected from subsequent washing and acid etching processes. The final result is a board of phenolic having long copper conductors left on the opposite side from which the components are mounted. The leads of the components are inserted through holes in the board, and solder is applied, fusing the leads to the copper. A well-designed board will have wide conductors to join to the thermal sink leads, and the conductors will be reasonably spaced.

At first sight analyzing the heat transfer from the circuit board appears quite difficult. But let us use the **fin concept**. Surely, a long copper conductor on the board is a sort of a fin, because temperature is varying in mainly one direction, along the length. But how does heat leave this fin? It leaves partly by conduction out into the phenolic and thence, by Newton's law of cooling, to the surrounding air, and partly by Newton's law, directly from the copper surface to the fluid. But is not the process of conduction into the board also somewhat like that which occurs in a fin?

Assume that our board has a single 0.10 in. wide conductor strip connected to our integrated circuit, and suppose that the thermal conductor strips are spaced at least 0.5 in. apart. Figure 2.11 shows a sketch. The board acts as a fin on either side of the copper conductor. Let us estimate its effectiveness. A small length dz of copper is assumed for the width of the fin as sketched in Figure 2.12. The effectiveness is, from Eqs. (2.35) and (2.39),

$$\eta = \left(\frac{h\,\mathscr{P}L^2}{kA_c}\right)^{-1/2} \tanh \left(\frac{h\,\mathscr{P}L^2}{kA_c}\right)^{+1/2}$$

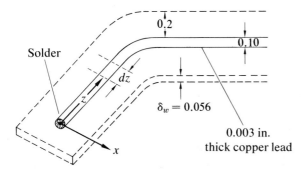

Figure 2.11   A copper conductor on a circuit board. (All dimensions are in inches.)

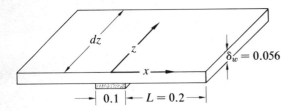

Figure 2.12   An element of circuit board acting as a fin for a conductor. (All dimensions are in inches.)

We take $h = 1$ [Btu/hr ft² °F], $\mathscr{P} = 2\,dz$, $L = 0.20/12$ [ft], $k = 0.10$ [Btu/hr ft °F], $A_c = \delta_w\,dz$, $\delta_w = 0.056/12$ [ft]. We find

$$\eta = 0.73$$

Now we consider the conductor with its adjoining board to be a fin running in the direction of the copper conductor, the z-direction. The length of this fin is as long as the conductor length L_c, and it has an *effective* perimeter equal to (see Figure 2.12)

$$\mathscr{P} = 2(0.10) + 2(0.40)\,(0.73) = 0.784 \text{ in.}$$

The conductivity-area product of the copper dominates conduction in the z-direction

$$kA_c = 240(0.003/12)\,(0.10/12) = 0.5 \times 10^{-3} \text{ Btu ft/hr °F}$$

From Eqs. (2.35) and (2.40) the thermal resistance of the circuit board is

$$R_3 = \frac{1}{(h\mathscr{P}kA_c)^{1/2} \tanh\left(\dfrac{h\mathscr{P}L_c^2}{kA_c}\right)^{1/2}}$$

If L_c is long, so that the tanh term is nearly one, we obtain

$$R_3 = \frac{1}{(h\mathscr{P}kA_c)^{1/2}} = \frac{1}{[(1)\,(0.784/12)\,(0.5 \times 10^{-3})]^{1/2}} = 175 \text{ [Btu/hr °F]}^{-1}$$

For the tanh to be nearly one, $(h\mathscr{P}L_c^2/kA_c)^{1/2}$ should perhaps be 4. Hence,

$$\frac{h\mathscr{P}L_c^2}{kA_c} = 16$$

$$L_c = 4[(0.5 \times 10^{-3})\,(12)/0.784]^{1/2} = 0.35 \text{ ft} = 4.2 \text{ in.}$$

Thus L_c is not unreasonably long.

The total resistance of the leads and circuit board is then approximately the sum $R_1 + R_2 + R_3$,

$$R = 2.3 + 68.6 + 175 = 246 \text{ [Btu/hr °F]}^{-1}$$

A temperature difference of 100°F will then give

$$\dot{Q} = \frac{100(1000)}{(246)\,(3.41)} = 119 \text{ mw}$$

The total from the capsule and the lead is then

$$\dot{Q} = 36 + 119 = 155 \text{ mw versus } 120 \text{ mw rating}$$

We must conclude that the manufacturer's rating is conservative but perhaps not overly so. For our calculation we assumed wide thermal conductors (0.1 in.), vertical spacing of 0.5 in., and horizontal spacing of $2 \times 4 = 8$ in. A designer might well employ a closer spacing. On the other hand, if the thermal conductor runs both ways on the board from the circuit element, there would be two fins of the type shown in Figure 2.11 to dissipate heat from each element. A final observation is that we have assumed $T_e = 100°F$. A blower and a supply of cool air might be required to maintain this temperature between circuit boards, depending upon the density of the elements, the cooling port area provided, and other such considerations.

2.11.5   Summary of the Example

In this example we used the concept of a thermal network. Two parallel paths were imagined. One was through two thermal resistors in series, a spherical shell resistance, and a film resistance. The other path was through three resistances in series. Two of these resistances were slabs, one of beryllia (we assumed) and one of brass. The third resistance was that of the circuit board acting as a two-dimensional fin. We analyzed the two-dimensional fin by dissecting it into two one-dimensional fin problems. The main fin consisted of the copper conductor. Adjacent to it was phenolic board which added to the effective heat transfer area. The effectiveness of this area was estimated by treating the heat conduction from the copper base out into the board as another fin problem. The student is encouraged to reread Section 2.10.6, which is quite straightforward, and think about how that two-dimensional heat conduction problem was simplified into a one-dimensional one.

2.12   TRANSIENT CONDUCTION IN ONE SPATIAL DIMENSION

2.12.1   The Semi-Infinite Solid with Step Change of Surface Temperature

In the preceding sections steady-state conduction in one spatial dimension was considered. In many processes involving heat conduction, the spatial temperature distribution changes with time. Consider, for example, the behavior of a forging which is thrust into a quench bath. Initially the hot metal loses heat to the bath fluid and ultimately reaches thermal equilibrium. Obviously, the forging cools more rapidly near its surface than it does in its interior. Thus temperature gradients are set up in the metal, heat being transferred from the relatively hot inner regions to the cold surface. However, since energy is being continuously removed, the temperature at any given point within the piece decreases with time.

In some applications, it is possible to neglect the thermal behavior at modest distances from the free surface of a body, assuming that, for a reasonably short period of time following changes in the free surface boundary conditions, only the near surface region is disturbed thermally. Consider, for example, a slab of material, initially at uniform temperature T_0, whose left face is suddenly raised to the ambient temperature T_s. From Eq. (2.9a), we have

$$\frac{\partial T}{\partial t} = \alpha \frac{\partial^2 T}{\partial x^2} \tag{2.41}$$

where x is the coordinate direction normal to the face of the slab and is assumed to be positive from left to right. We assume the slab to be infinite in the y and z directions. Boundary conditions for the problem are as follows:

$$T(0, t) = T_s \tag{2.42}$$

$$\lim_{x \to \infty} T(x, t) = T_0 \tag{2.43}$$

The initial condition is

$$T(x, 0) = T_0 \tag{2.44}$$

The solution to Eq. (2.41) may be obtained by attempting a *similarity* solution. It is assumed that heat diffuses into the slab a distance $\delta(t)$ and that the temperature profile is scaled to this distance. In other words, it is assumed that temperature is not a function of x and t acting independently, but only of the ratio of the actual depth in the body x to the depth of penetration δ of a heat wave which has been soaking in for time t,

$$\eta = \frac{x}{\delta(t)} \tag{2.45}$$

If this attempt succeeds at solving the equation, the temperature profiles are said to be *self-similar*.

Before transforming from x and t to η it is convenient to make the boundary conditions simple numbers. For this reason we introduce a dimensionless temperature

$$T^* = \frac{T - T_0}{T_s - T_0} \tag{2.46}$$

Now we try

$$T^*(x, t) = T^*(\eta) \tag{2.47}$$

The transformation is easily accomplished. To obtain the term on the right-hand side of Eq. (2.41) we need the second derivative of T with respect to x. This term will be $(T_s - T_0)$ times the second derivative of T^* with respect to x. Differentiating T^* a first time with respect to x and making use of Eq. (2.45) yields

$$\frac{\partial T^*}{\partial x}\bigg|_t = \frac{dT^*}{d\eta} \frac{\partial \eta}{\partial x}\bigg|_t = \frac{1}{\delta} \frac{dT^*}{d\eta}$$

Differentiating a second time with respect to x, we find

$$\frac{\partial^2 T^*}{\partial x^2}\bigg|_{t,} = \frac{1}{\delta^2}\frac{d^2 T^*}{d\eta^2}$$

To obtain the term on the left-hand side of Eq. (2.41) we note that it will be $(T_s - T_0)$ times the derivative of T^* with respect to t,

$$\frac{\partial T^*}{\partial t}\bigg|_x = \frac{dT^*}{d\eta}\frac{\partial \eta}{\partial t}\bigg|_x = -\frac{x\delta'(t)}{\delta^2}\frac{dT^*}{d\eta}$$

where the prime denotes differentiation with respect to t. Eq. (2.41) becomes

$$\frac{-x\delta'(t)}{\delta^2}\frac{dT^*}{d\eta} = \frac{\alpha}{\delta^2}\frac{d^2 T^*}{d\eta^2}$$

If T^* is a function of η only, the equation which gives T^* must contain no x- or t-dependent terms. Equation (2.45) can be used to eliminate x. Eliminating x and multiplying both sides by δ^2 gives

$$-\delta(t)\delta'(t)\eta \frac{dT^*}{d\eta} = \alpha \frac{d^2 T^*}{d\eta^2}$$

In order for the t-dependency to disappear we find that the product $\delta\delta'$ must be a constant,

$$\delta(t)\delta'(t) = \delta \frac{d\delta}{dt} = C$$

where C is any constant to be chosen at our whim. We will simply choose a convenient C; the value chosen does not affect the final result. We find, upon separating variables, that

$$\delta \, d\delta = C \, dt, \qquad \frac{\delta^2}{2} = Ct$$

The equation for T^* is then

$$-C\eta \frac{dT^*}{d\eta} = \alpha \frac{d^2 T^*}{d\eta^2}$$

To solve this equation, let $p = dT^*/d\eta$. Then

$$-C\eta p = \alpha \frac{dp}{d\eta}, \qquad \frac{dp}{p} = -\frac{C}{\alpha}\eta \, d\eta = -\frac{C}{\alpha}d\left(\frac{\eta^2}{2}\right)$$

Let C be chosen to make $C/\alpha = 2$. Then $dp/p = -d(\eta^2)$, and

$$p = \frac{dT^*}{d\eta} = C_1 e^{-\eta^2}$$

$$T^* = C_2 + C_1 \int_0^\eta e^{-u^2} \, du$$

Imposition of the boundary conditions gives C_1 and C_2. From Eq. (2.42) at $\eta = 0$,

$T^* = 1$; so $C_2 = 1$. From Eq. (2.43) at $\eta = \infty$, $T^* = 0$; therefore C_1 is given by

$$0 = 1 + C_1 \int_0^\infty e^{-u^2} du = 1 + C_1 \frac{\sqrt{\pi}}{2}, \qquad C_1 = -\frac{2}{\sqrt{\pi}}$$

The solution is thus found to be

$$T^* = \frac{T - T_0}{T_s - T_0} = 1 - \frac{2}{\sqrt{\pi}} \int_0^\eta e^{-u^2} du$$

$$T^* = \text{erfc}\,(\eta) \tag{2.48}$$

where from Eq. (2.45) and from $\delta^2 = 2Ct$ found above,

$$\eta = \frac{x}{(2Ct)^{1/2}} = \frac{x}{(4\alpha t)^{1/2}} \tag{2.49}$$

The quantity

$$1 - \frac{2}{\sqrt{\pi}} \int_0^\eta e^{-u^2} du = 1 - \text{erf}\,(\eta) = \text{erfc}\,(\eta)$$

is known as the complementary error function, erfc (η), and is tabulated in Table 2.2.

The heat flux at the surface is found from Fourier's law

$$q_x = -k \frac{\partial T}{\partial x}\bigg|_{x=0} = -k(T_s - T_0) \frac{\partial T^*}{\partial x}\bigg|_{x=0}$$

$$q_x = -k(T_s - T_0) \left(-\frac{2}{\sqrt{\pi}} e^{-\eta^2} \frac{1}{\delta(t)} \right)\bigg|_{\eta=0}$$

$$q_x = \frac{2}{\sqrt{\pi}} \frac{k}{(4\alpha t)^{1/2}} (T_s - T_0) \tag{2.50}$$

2.12.2   *EXAMPLE* **Freezing of a Buried Pipe**

Consider the design problem of burying water pipes underground in a geographical area whose mean winter temperature is about 40°F, but which is subject to sudden drops in air temperature to about 10°F for a maximum duration of 48 hours. Determine the minimum depth of the pipes to prevent freezing. From Eq. (2.48)

$$T^* = \frac{T - T_0}{T_s - T_0} = \frac{32 - 40}{10 - 40} = 0.267 = \text{erfc}\,(\eta)$$

Table 2.2 shows $\eta = 0.79$. Assuming that $\alpha = 0.30/(128)(0.44)$ for dry soil (Table B.4) we obtain

$$x = [4(0.00532)(48)]^{1/2}(0.79) = 0.80 \text{ ft}$$

Table 2.2   The Complementary Error Function

$$\text{erfc}(\eta) = 1 - \frac{2}{\sqrt{\pi}} \int_0^\eta e^{-u^2}\, du$$

η	erfc(η)	η	erfc(η)
0.0	1.0000	1.1	0.11980
0.05	0.9436	1.2	0.08969
0.1	0.8875	1.3	0.06599
0.15	0.8320	1.4	0.04772
0.2	0.7773	1.5	0.03390
0.25	0.7237	1.6	0.02365
0.3	0.6714	1.7	0.01621
0.35	0.6206	1.8	0.01091
0.4	0.5716	1.9	0.00721
0.45	0.5245	2.0	0.00468
0.5	0.4795	2.1	0.00298
0.55	0.4367	2.2	0.00186
0.6	0.3961	2.3	0.001143
0.65	0.3580	2.4	0.000689
0.7	0.3222	2.5	0.000407
0.75	0.2889	2.6	0.000236
0.8	0.2579	2.7	0.000134
0.85	0.2293	2.8	0.000075
0.9	0.2031	2.9	0.000041
0.95	0.1791	3.0	0.000022
1.00	0.1573		

Thus freezing will not occur if the top of the pipe is more than 10 in. below the surface.

2.12.3   The Semi-Infinite Solid with Surface Pulse of Energy

Suppose an amount of energy E per unit area were released instantaneously at the surface of a semi-infinite solid, and none of this energy were lost from the surface. It is stated without proof that the temperature distribution would be

$$T(x, t) = T_0 + \frac{E}{\rho c_p (\pi \alpha t)^{1/2}} e^{-x^2/(4\alpha t)} \tag{2.51}$$

The student may substitute this relation into Eq. (2.41) and verify that it is a solution. That the energy E is constant and contained within the solid is shown by integrating

$$E = \int_0^\infty \rho c_p (T - T_0)\, dx$$

$$E = \frac{E}{(\pi \alpha t)^{1/2}} \int_0^\infty e^{-x^2/(4\alpha t)}\, dx = E \frac{2}{\pi^{1/2}} \int_0^\infty e^{-u^2}\, du$$

$$E = E \operatorname{erf}(\infty) = E[1 - \operatorname{erfc}(\infty)]$$

$$E = E$$

2.12.4   The Finite Solid

We see from Table 2.2 that at $\eta = 1.20$ or so, the temperature is only affected 9%. At the center of a finite slab of thickness L heated or cooled from both sides, this value of η corresponds to a certain value of time

$$\eta = 1.20 = \frac{L/2}{(4\alpha t)^{1/2}}$$

$$t = \frac{L^2}{23\alpha}$$

For time periods shorter than $L^2/23\alpha$, the slab may be regarded as semi-infinite. For longer time periods the slab should be treated as finite.

We will not in this introductory text consider the problem of the finite slab in detail. Let us, however, set up the governing differential equation and boundary conditions. For a slab being cooled by a fluid, the governing equation is Eq. (2.9a),

$$\frac{\partial T}{\partial t} = \alpha \frac{\partial^2 T}{\partial x^2} \tag{2.52}$$

Suppose the boundary conditions are

$$q_x(0, t) = -k \frac{\partial T}{\partial x}\bigg|_{x=0} = -h(T(0, t) - T_e)$$

$$q_x(L, t) = -k \frac{\partial T}{\partial x}\bigg|_{x=L} = +h(T(L, t) - T_e)$$

and the initial condition is

$$T(x, 0) = T_0$$

We wish to use a spatial coordinate that goes from 0 to 1, so we let

$$x^* = \frac{x}{L}, \quad \frac{\partial T}{\partial x} = \frac{\partial T}{\partial x^*} \frac{dx^*}{dx} = \frac{1}{L} \frac{\partial T}{\partial x^*}, \quad \frac{\partial^2 T}{\partial x^2} = \frac{1}{L^2} \frac{\partial^2 T}{\partial x^{*2}}$$

By the same token we wish to have a dependent variable that varies between 0 and 1, so we let

$$T^* = \frac{T - T_e}{T_0 - T_e}$$

To make t dimensionless we use the group $t^* = \alpha t / L^2$ which we found at the beginning of this section. (If $\alpha t / L^2$ is less than $1/23$, we can use the semi-infinite solution.) Equation (2.52) becomes

$$\frac{\partial T^*}{\partial t^*} = \frac{\partial^2 T^*}{\partial x^{*2}} \tag{2.53}$$

The boundary conditions are then

$$\left. \frac{\partial T^*}{\partial x^*} \right|_{x^*=0} = \frac{hL}{k} T^*(0, t^*) = Bi \cdot T^*(0, t^*) \tag{2.54}$$

$$\left. \frac{\partial T^*}{\partial x^*} \right|_{x^*=1} = -\frac{hL}{k} T^*(1, t^*) = -Bi \cdot T^*(1, t^*) \tag{2.55}$$

One sees that the solution, when found, will be only a function of x^*, t^*, and $hL/k = Bi$. In heat conduction, hL/k is called the **Biot number** and t^* is often called the **Fourier number**. The observation that T^* is a function of x^*, t^*, and Bi is the basis of several useful graphs which can be found, for example, in the *Boelter Heat Notes*.

2.13   FINITE DIFFERENCE METHODS

While many simple steady-state and transient problems can be solved analytically, practical calculations for involved geometries are best made numerically. The procedure in such a case is to discretize the spatial and time coordinates and use finite difference approximations for the derivatives appearing in the governing energy conservation equation. In steady-state problems a set of linear algebraic equations is found with as many unknowns as mesh points. Solution may be found by matrix inversion or iteration, for example. In transient state problems the temperatures at the next time step are found using the values obtained from the preceding time step. There follows only one example, a simple explicit solution to transient conduction in the finite solid.

Let a slab be subdivided into N strips with $N + 3$ node points as shown in Figure 2.13. Let $T^*(x_i^*, t_j^*)$ be denoted by $T_{i,j}^*$, where $x_i^* = (i - 2) \Delta x^*$ and $t_j^* = j \Delta t^*$. We will be concerned with *locations* for $i = 2$ to $N + 2$. For an interior node point Eq. (2.53) is approximated by

$$\frac{T_{i,j+1}^* - T_{i,j}^*}{\Delta t^*} \doteq \frac{1}{\Delta x^*} \left[\left(\frac{T_{i+1,j}^* - T_{i,j}^*}{\Delta x^*} \right) - \left(\frac{T_{i,j}^* - T_{i-1,j}^*}{\Delta x^*} \right) \right]$$

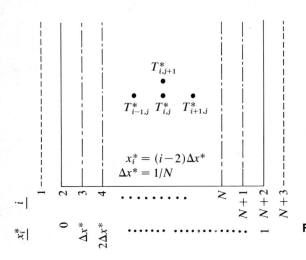

$$T^*_{i,j+1}$$

$$T^*_{i-1,j} \quad T^*_{i,j} \quad T^*_{i+1,j}$$

$$x^*_i = (i-2)\Delta x^*$$
$$\Delta x^* = 1/N$$

Figure 2.13   Discretization of a slab.

or, solving for $T^*_{i,j+1}$,

$$T^*_{i,j+1} = T^*_{i,j} + r[T^*_{i+1,j} - 2T^*_{i,j} + T^*_{i-1,j}] \tag{2.56}$$

where $r \equiv \Delta t^*/(\Delta x^*)^2$ and $i = 2, 3, \ldots, N+2$. To allow for the boundary conditions we must derive special relations for the exterior node points. We approximate the boundary conditions, Eqs. (2.54) and (2.55), by

$$\frac{T^*_{3,j} - T^*_{1,j}}{2\Delta x^*} = Bi \cdot T^*_{2,j}$$

and

$$\frac{T^*_{N+3,j} - T^*_{N+1,j}}{2 \Delta x^*} = -Bi \cdot T^*_{N+2,j}$$

respectively. That is,

$$T^*_{1,j} = T^*_{3,j} - 2Bi \Delta x^* T^*_{2,j} \tag{2.57}$$

$$T^*_{N+3,j} = T^*_{N+1,j} - 2Bi \Delta x^* T^*_{N+2,j} \tag{2.58}$$

Equations (2.56) through (2.58) are solved subject to the initial condition

$$T^*_{i,0} = (T(x_i,0) - T_e)/(T_0 - T_e) = (T_0 - T_e)/(T_0 - T_e) = 1$$
$$i = 2, 3, \ldots, N+2 \tag{2.59}$$

We illustrate the solution procedure with the following example.

Suppose a slab of meat 2 in. thick, with a thermal conductivity of 0.333 Btu/hr ft °F and a density-specific heat product (ρc_p) of 60 Btu/ft³ °F, is removed from a refrigerator at 40°F, wrapped in aluminum foil, and placed in an oven at 400°F. As a first approximation we neglect radiation heat transfer and moisture loss (evaporative cooling). The heat transfer coefficient in the oven is 1 Btu/hr ft² °F.

Let Δx be 0.5 in. ($N = 4$). The Biot number is

$$Bi = \frac{hL}{k} = \frac{(1)(2/12)}{0.333} = 0.5$$

We choose $r = 0.25$ which then fixes the time step.

$$r = \frac{\Delta t^*}{\Delta x^{*2}} = \frac{\alpha\,\Delta t}{L^2(\Delta x/L)^2}$$

$$\Delta t = r\frac{(\Delta x)^2}{\alpha} = 0.25\frac{(1/24)^2}{(0.333/60)} = 0.0781 \text{ hr} = 4.69 \text{ min}$$

For this case $\Delta x^* = \frac{1}{4}(N = 4)$, $r = \frac{1}{4}$, $Bi = 0.5$. Then Eqs. (2.56) through (2.58) become

$$T^*_{1,j} = T^*_{3,j} - \tfrac{1}{4}T^*_{2,j} \qquad\qquad \textbf{(2.60)}$$

$$T^*_{7,j} = T^*_{5,j} - \tfrac{1}{4}T^*_{6,j} \qquad\qquad \textbf{(2.61)}$$

and for $i = 2, 3, \ldots 6$,

$$T^*_{i,j+1} = T^*_{i,j} + (T^*_{i-1,j} - 2T^*_{i,j} + T^*_{i+1,j})/4 \qquad\qquad \textbf{(2.62)}$$

The solution is marched forward by applying in order Eqs. (2.60), (2.61), and (2.62) repeatedly. For example, let $j = 0$. Then Eq. (2.59) gives starting values for $i = 2$ through 6. Equations (2.60) and (2.61) give

$$T^*_{1,0} = 1 - \tfrac{1}{4}(1) = \tfrac{3}{4}, \qquad T^*_{7,0} = 1 - \tfrac{1}{4} = \tfrac{3}{4}$$

Equation (2.62) gives

$$T^*_{2,1} = 1 + (\tfrac{3}{4} - 2 \times 1 + 1)/4 = \frac{15}{16}$$

$$T^*_{3,1} = 1 + (1 - 2 \times 1 + 1)/4 = 1$$

$$T^*_{4,1} = 1 + (1 - 2 \times 1 + 1)/4 = 1$$

$$T^*_{5,1} = 1 + (1 - 2 \times 1 + 1)/4 = 1$$

$$T^*_{6,1} = 1 + (1 - 2 \times 1 + \tfrac{3}{4})/4 = \frac{15}{16}$$

Thus the solution, as is to be expected from the boundary conditions, is symmetric about $i = 4$.

For $j = 1$ we obtain,

$$T^*_{1,1} = 1 - \frac{1}{4}\frac{15}{16} = \frac{49}{64}, \qquad T^*_{7,1} = \frac{49}{64}$$

$$T^*_{2,2} = \frac{15}{16} + \left(\frac{49}{64} - 2 \times \frac{15}{16} + 1\right)\Big/4 = \frac{233}{256}$$

and so on for $T^*_{3,2}$ through $T^*_{6,2}$. The results for the first few steps are summarized below:

$$
\begin{array}{c|ccccccc}
 & & & & & & & \\
j=3\;(14.07) & \dfrac{3647}{4096} & \dfrac{993}{1024} & \dfrac{95}{96} & \dfrac{993}{1024} & \dfrac{3647}{4096} & & \\
j=2\;(9.38) & \dfrac{775}{1024} & \dfrac{233}{256} & \dfrac{63}{64} & 1 & \dfrac{63}{64} & \dfrac{233}{256} & \dfrac{775}{1024} \\
j=1\;(4.69) & \dfrac{49}{64} & \dfrac{15}{16} & 1 & 1 & 1 & \dfrac{15}{16} & \dfrac{49}{64} \\
j=0\;(0) & \dfrac{3}{4} & 1 & 1 & 1 & 1 & 1 & \dfrac{3}{4} \\
\hline
i= & 1 & 2 & 3 & 4 & 5 & 6 & 7 \\
x_i\,(\text{in})= & 0 & \tfrac{1}{2} & 1 & 1\tfrac{1}{2} & 2 & &
\end{array}
$$

(t_j [min]; j)

After three steps (14.07 min), according to our rough numerical solution, the surface temperature is

$$T_s = T_e - (T_e - T_0)(T^*_{2,3}) = 400 - (400 - 40)\left(\frac{3647}{4096}\right) = 79°F$$

while the center temperature is

$$T_c = T_e - (T_e - T_0)(T^*_{4,3}) = 44°F$$

These results are quite approximate, first, because our solution technique permits the interior temperatures to respond to the boundary conditions only after a time step has elapsed and, second, because we have chosen to eliminate radiation and conduction between the meat and a pan from the analysis. More accurate results would be obtained if the node spacing were reduced.

The above procedure is called an *explicit* method, due to the fact that the solution marches forward using only values of the temperature at a previous time step. Unfortunately, this simplicity must be paid for in terms of a limit on the magnitude of the time step which may safely be taken (i.e., on the magnitude of $r = \Delta t^*/\Delta x^{*2}$). If r is too large, the solution procedure is *unstable*. Such is the case for the present problem if $r = \frac{1}{2}$ rather than $\frac{1}{4}$. For the explicit procedure outlined here, r must be chosen to be less than $\frac{1}{2}$. As a final note, we observe that the symmetric nature of the boundary conditions resulted in a symmetric temperature distribution in the slab. Thus we could have solved the problem on the interval $0 \leq x^* \leq \frac{1}{2}$, using as a boundary condition $\partial T^*/\partial x^* = 0$ at $x^* = \frac{1}{2}$. For reasons of simplicity this interval was not used here, but the numerical calculations can be made more efficiently by eliminating the duplication in the larger interval.

2.14   SUMMARY

Fourier's law of conduction was introduced. Several one-dimensional steady-state conduction problems were then solved. To treat such problems an energy balance was made on a differential element to obtain a differential equation governing the temperature distribution in a body. Boundary conditions between two solids were discussed. Newton's law of cooling was described as a boundary condition often occurring.

In the absence of volume heat sources the solutions to the steady-state heat conduction problems were found to be of the form $\dot{Q} = \Delta T/R$. The ideas of thermal resistances and thermal circuits were thus introduced. The resistances were found for plane, cylindrical, and spherical shells and for fins of constant perimeter and cross section, and a sample application was made to the problem of thermal analysis of a circuit board.

Transient conduction in one spatial dimension was treated only for the semi-infinite solid. A step change in surface temperature was seen to evoke a temperature response given by the complementary error function of a ratio of distance to the square root of a time-diffusivity product. The parameters which occur upon making the transient conduction equation dimensionless for a finite body were briefly discussed as was an explicit finite difference method of numerical solution.

EXERCISES

1. The local temperature distribution in an aluminum bar, as measured by a set of embedded thermocouples, results in the temperature gradients

$$\frac{\partial T}{\partial x} = 6.72\frac{°R}{ft}, \quad \frac{\partial T}{\partial y} = 3.18\frac{°R}{ft}, \quad \text{and} \quad \frac{\partial T}{\partial z} = -1.94\frac{°R}{ft}$$

   Calculate q_x, q_y, q_z, $\mathbf{q}$, and ∇T. Sketch in rectangular Cartesian coordinates the current of energy $\mathbf{q}$ and the gradient of temperature ∇T.

2. A facet of a piece of ice in a body of water is found to be melting at a rate of 1.0 gm/min when its surface area is 3.6 cm². If the ice has been melting for some time, calculate the average heat flux q_i^{II}, where II refers to the water phase, on the water side of the ice-water interface. Find the temperature gradient in the water very near the ice.

3. An electrical current, $I = 15$ amp, flows in an 18 gauge copper wire. Calculate the volumetric heat source $\dot{Q}_V$ in Btu/ft³ hr.

4. Taking an element of volume $A \cdot \Delta r$, where A is $2\pi rL$ for cylinders and $4\pi r^2$ for spheres, derive Eq. (2.9b).

5. Estimate the heating load due to conduction through the walls for a single family dwelling in a cold climate when the outside temperature is 0°F. The walls and ceilings of the residence are a composite of 0.5 in. wall board ($k = 0.25$ Btu/ft hr °F), 3.5 in. of vermiculite insulation ($k = 0.019$ Btu/ft hr °F), and 1 in. wood ($k = 0.08$ Btu/ft hr °F). Assume that the inside and outside heat transfer coefficients are 1.5 and 6.0 Btu/ft² hr °F, respectively. The total surface area of walls and ceiling is 3500 ft² and the inside temperature is 72°F. (In actuality, the total heating load will be higher because of transpiration of air through cracks in the structure. Chapter 4 will give some insight into this mechanism of heat loss.)

6. In Exercise 5, suppose that the occupant wishes to replace one shaded wall of his home with soda-lime glass. Compare the heat fluxes through the wall when it is

    (a) $\frac{1}{8}$ in. thick plate glass,
    (b) thermopane, consisting of an 0.25 in. thick stagnant air gap between two $\frac{1}{8}$ in. thick glass plates, and
    (c) the original wall.

    Use the data of Exercise 5 as required.

7. A thermal conductivity cell consists of concentric cylinders. Thermal energy, generated by an electrical heater within the inner tube, flows radially through the test specimen, a fluid contained between the two cylinders. The temperature on the outside surface of the inner cylinder is T_2. The temperature on the inside surface of the outer cylinder is T_3. Temperatures T_2 and T_3 are measured by means of surface thermocouples.

    Given that r_2 and r_3 are 1.0 and 1.25 in., respectively, calculate the thermal conductivity of a test specimen for which T_2 and T_3 are 180 and 176°F, respectively, and the power generated is 8.6 watts per foot of cylinder.

8. Freon-12 at -40°F flows in a copper refrigerant line, 0.210 in. inside diameter by 0.250 in. outside diameter. Determine the effect of polystyrene foam "insulation" with an aluminized outer surface on heat leakage to the refrigerant per foot of line as a function of insulation thickness. Assume the internal and external heat transfer coefficients are 100 and 1 Btu/ft² hr °F, respectively. The line is in a room where the air temperature is 64°F. Calculate the thickness of insulation required to reduce heat leakage to 50% of the bare tube result. Should the insulation be used?

9. Superheated steam at 440°F flows in Schedule 40 steel pipe, nominal size 6 in. Determine the effect of adding asbestos insulation to the pipe as a function of insulation thickness and external heat transfer coefficient. Assume an external temperature of 80°F and an inside film coefficient of 2000 Btu/hr ft² °F. Submit your results in the form of a family of curves on a single graph with $\dot{Q}$ as ordinate, insulation thickness as abscissa, and h_3 as a parameter. (Let $2 \leqslant h_3 \leqslant 20$.)

10. Obtain a solution for the temperature profile inside a long cylindrical nuclear fuel rod with volumetric heat source $\dot{Q}_V$, thermal conductivity k, and radius r_1. At r_1 take the temperature to be known at T_1.

11. Develop the thermal circuit for radial heat conduction in composite spherical shells.

12. In order to find the thermal resistance of an opaque honeycomb aircraft wall material, a 100 watt light bulb was allowed to burn for some time in a hollow spherical shell with inner and outer radii of 12 in. and 13.25 in., respectively. Compute the heat flux q_r at the outer surface of the shell. If the inner surface was measured to be 121°F and the outer one 97°F, what was the effective conductivity of the honeycomb?

13. During the flight of Apollo 12, plutonium oxide ($Pu^{238}O_2^{16}$) was used to generate electrical power for various instruments. Heat was uniformly produced at a rate $\dot{Q}_V$ watts/cm³ by virtue of the loss of kinetic energy of alpha particles emitted from the fuel (Pu^{238}).

     Consider a spherical mass of this material covered with thermo-electric elements for converting heat to electrical energy. The physical properties of these elements (tellurides) and heat rejection considerations dictate that the outside temperature should be 200°C. The ceramic nature of the plutonium oxide fuel ($Pu^{238}O_2^{16}$) is such that it may achieve a temperature of 1750°C. With these constraints calculate the following for a fuel element 1 in. in diameter:

     (a) The maximum allowable volumetric heating rate $\dot{Q}_V$.
     (b) The electrical power generated, assuming a thermal electric efficiency of 4%.

The thermal conductivity of the fuel may be taken as 0.01 cal/cm sec °C.

14. Find the heat transfer per foot of tube under the following conditions: The fluid within the tube is 212°F and the heat transfer coefficient h_1 is 1000 Btu/hr ft² °F. The tube is 1.0 in. inside diameter and has a copper wall 0.060 in. thick. Over the tube is an aluminum sleeve with a 0.060 in. wall and having 240 pin fins per inch of length. The pin fins are 0.067 in. in diameter and project 1.5 in. from the outer wall of the sleeve. The heat transfer coefficient on the fins is 1.5 Btu/hr ft² °F, and the outside fluid temperature is 400°F.

15. A radial fin consisting of 0.010 in. thick annulus of aluminum with an inside radius of 0.125 in. and an outside radius of 0.625 in. is used to cool a transistor. Set up the differential equation and boundary conditions which govern the temperature profile. (*Optional*: Solve the equations in terms of modified Bessel functions.)

16. Find the average rate at which heat flows into a semi-infinite slab. Compare the answer to the instantaneous rate.

17. A piece of veneer 0.07 in. thick is to be bonded to a slab of wood by clamping it in a hot press. The wood is originally 70°F, and the hot press is 200°F. The glue used must be heated to 180°F to activate it. How long a time is required before the glue reaches the desired temperature?

18. Discuss the possibility of using a laser to spot weld a thin piece of metal to a thick one. What burst of energy E should be used?

19. It is desired to make a finite difference calculation of the heat conduction in a sheet of plastic ($k = 0.1$, $\rho = 50$, $c_p = 0.3$). The behavior at a time of 1 second is desired, so it is decided to use $\Delta t = 0.01$ second. Recommend a size for Δx.

20. Re-solve the problem of Section 2.13, using for the boundary condition on the right $\partial T^*/\partial x^* = 0$ at $x^* = \frac{1}{2}$.

21. Re-solve the problem of Section 2.13, using $r = \Delta t^*/(\Delta x^*)^2 = \frac{1}{2}$. (Generate enough steps to show that the solution is unstable.)

REFERENCES

Heat Conduction

The preeminent treatise in the field is:

> Carslaw, H. S., and J. C. Jaeger, *Conduction of Heat in Solids*, 2nd ed. New York: Oxford University Press, 1959.

This text contains a large number of exact solutions for transient and steady heat flow problems.

Works which contain some useful graphs for hand estimates of transient heat conduction are:

> Boelter, L. M. K., V. H. Cherry, H. A. Johnson and R. C. Martinelli, *Heat Transfer Notes*. New York: McGraw-Hill, 1965. (See particularly pp. 199–203.)
> Kreith, F., *Principles of Heat Transfer*, 2nd ed. Scranton, Pa.: International Textbook Co., 1965 (Chs. 2–4.)

Another standard text which contains, in particular, a good treatment of fins is:

> Eckert, E. R. G., and R. M. Drake, *Heat and Mass Transfer*. New York: McGraw-Hill, 1959. (See pp. 55–58.)

These above three references are particularly easy for the beginning student to read.

Recent texts stressing integral transform techniques, such as the Laplace transform, are:

> Arpaci, V. S., *Conduction Heat Transfer*. Reading, Mass.: Addison-Wesley, 1966.
> Luikov, A. V., *Analytical Heat Diffusion Theory*. New York: Academic Press, 1968.
> Ozisik, M. N., *Boundary Value Problems of Heat Conduction*. Scranton, Pa.: International Textbook Co., 1968.

Both Arpaci and Ozisik have chapters on approximate variational solution techniques, for example, the Ritz and Galerkin methods. Numerical techniques are expounded briefly also in both Arpaci and Ozisk.

CHAPTER 3

MASS TRANSFER

3.1 INTRODUCTION

In this chapter we are concerned with the movement of a chemical species through a mixture or solution and across phase boundaries. The detailed physical mechanisms are several; however, it is convenient to begin with the consideration of only the two most important, which are ordinary diffusion and convection. These transfer mechanisms play an important part in many familiar physical phenomena.

An elementary experiment of physics is the placing of a crystal of potassium permanganate in a beaker of stagnant water; as the permanganate dissolves, it may be seen to diffuse through the water. The local concentration of permanganate is indicated by color, the deepest purple being adjacent to the crystal. The diffusion is always in the direction of decreasing concentration. This process of ordinary diffusion occurs whenever there is a concentration gradient in a liquid solution; it occurs also in solid solutions and gas mixtures. According to kinetic theory, molecules are in a state of random motion. If an imaginary plane is placed normal to the concentration gradient of the species in question, it follows that more molecules of that species cross from the side with the higher concentrations of that species than from the other side. Consequently, there is a net transfer of the species across the plane in the direction of decreasing concentration. This transfer is simply due to the difference in concentrations on each side of the plane; there may be no movement of the mixture as a whole. If the water in the beaker is now stirred, the rate of dispersion of the permanganate is greatly increased. This increase is due to the bulk motion of the water bodily transporting the permanganate; this transfer due to bulk motion is what we call convection.

In the treatment which follows we will initially consider transport by diffusion only; later in the chapter simple convective situations will be considered. Mass transfer in more general convective situations will be dealt with in Chapters 4 and 5.

3.2   DEFINITIONS OF CONCENTRATION

3.2.1   Number, Mass, and Mole Fractions

In a multicomponent system the local concentration of a species can be expressed in a number of ways. Figure 3.1 shows an elemental volume dV surrounding the location under consideration. The problem is to describe the composition of the material within the volume. One method would be to determine somehow the number of molecules of each species present and divide by dV to obtain the number of molecules per unit volume; hence a *number density* can be defined.

Number density of species i = number of molecules of i per unit volume

$$\equiv \mathcal{N}_i \tag{3.1}$$

Alternately, if the total number of molecules of all species per unit volume is denoted as $\mathcal{N}$, then we define the *number fraction* of species i as

$$n_i \equiv \frac{\mathcal{N}_i}{\mathcal{N}} \tag{3.2}$$

Equations (3.1) and (3.2) describe *microscopic* concepts; they are essential when the kinetic theory of gases is employed to describe transfer processes.

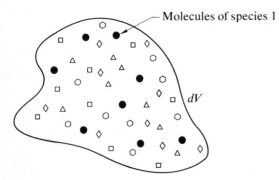

Molecules of species 1

dV

Figure 3.1 Elemental volume used to define concentrations at a point.

It is usually more convenient to treat matter as a continuum; the smallest volume considered is sufficiently large so that macroscopic properties such as pressure and temperature have their usual meaning. Thus we require macroscopic definitions of concentration, first on a mass basis:

Mass concentration of species $i \equiv$ *partial density* of species i

$$\equiv \rho_i \text{ mass/unit volume} \qquad (3.3)$$

The total mass concentration is the total mass per unit volume; that is, the density $\rho = \Sigma \rho_i$, where the summation is over all species in the mixture, $i = 1, 2, \ldots, n$. The *mass fraction* of species i is defined as

$$m_i = \frac{\rho_i}{\rho} \qquad (3.4)$$

Second, on a molar basis:

Molar concentration of species $i \equiv$ number of moles of i per unit volume

$$c_i = \frac{\rho_i}{M_i} \text{ moles/unit volume} \qquad (3.5)$$

where M_i is the molecular weight of species i. The total molar concentration is the molar density $c = \Sigma \, c_i$. Then the *mole fraction* of species i is defined as

$$x_i \equiv \frac{c_i}{c} \qquad (3.6)$$

A number of important relations follow directly from these definitions and are listed below. The mean molecular weight of the mixture is denoted M and may be expressed as

$$M = \frac{\rho}{c} = \Sigma \, x_i M_i \qquad (3.7)$$

Also

$$\frac{1}{M} = \Sigma \frac{m_i}{M_i} \qquad (3.8)$$

By definition the following summation rules hold true:

$$\Sigma \, m_i = 1 \qquad (3.9)$$

$$\Sigma \, x_i = 1 \qquad (3.10)$$

It is often necessary to have the mass fraction of species i expressed explicitly in terms of mole fractions and molecular weights; this relation may be derived to be

$$m_i = \frac{x_i M_i}{\Sigma \, x_j M_j} \qquad (3.11)$$

The corresponding expression for the mole fraction is

$$x_i = \frac{m_i/M_i}{\sum m_j/M_j} \qquad (3.12)$$

Note: The foregoing definitions of concentration are not the only ones in common usage.

3.2.2   *EXAMPLE* Specifying the Composition of Air

The composition of air is commonly described by stating that the partial pressures of O_2 and N_2 are in the ratio 0.21 to 0.79. Assuming an ideal gas mixture, determine the mass fractions of O_2 and N_2.

Dalton's law of partial pressures states that

$$P = \sum P_i, \quad P_i = \rho_i R_i T$$

Dividing by total pressure and substituting $R_i = \mathscr{R}/M_i$ gives

$$\frac{P_i}{P} = \frac{\rho_i}{M_i} \frac{\mathscr{R}T}{P} = c_i \frac{\mathscr{R}T}{P} = x_i \frac{c\mathscr{R}T}{P}$$

But

$$P = \rho \frac{\mathscr{R}}{M} T = c\mathscr{R}T$$

Therefore

$$\frac{P_i}{P} = x_i$$

That is, for an ideal gas mixture, the mole fraction and partial pressure are equivalent measures of concentration (as is the number fraction). Thus for air

$$x_{O_2} = \frac{0.21}{0.21 + 0.79} = \frac{0.21}{1} = 0.21$$

$$x_{N_2} = \frac{0.79}{0.21 + 0.79} = \frac{0.79}{1} = 0.79$$

Using Eq. (3.11) we find

$$m_{O_2} = \frac{(0.21 \times 32)}{(0.21 \times 32) + (0.79 \times 28)} = 0.232$$

$$m_{N_2} = \frac{(0.79 \times 28)}{(0.21 \times 32) + (0.79 \times 28)} = 0.768$$

3.3   FICK'S LAW: STEADY DIFFUSION

It is convenient to introduce Fick's law as a phenomenological relation; in Chapter 7 the physical basis of the law will be examined in greater detail. In 1855 Adolph Fick rediscovered Berthollet's analogy between heat conduction and mass diffusion and proposed what became known as his **first law**, a linear relation between the rate of diffusion and the local concentration gradient. The concept of a linear relation between a flux and the corresponding driving force had already been introduced by Newton in his law of viscosity, by Fourier in his law of heat conduction, and by Ohm in his law of electrical conduction. We will first consider a stationary medium, for example, a solid or stagnant gas, consisting of species 1 and 2, and propose the following form of Fick's first law:

$$\bullet \qquad \mathbf{j}_1 = -\rho \mathscr{D}_{12} \nabla m_1 \qquad (3.13)$$

where the terms are:

> $\mathbf{j}_1$,   diffusive mass flux of species 1, with units of mass/area-time
> ρ,   local density of mixture, with units mass/volume
> m_1,   mass fraction of species 1, dimensionless
> $\mathscr{D}_{12}$,   the constant of proportionality, referred to as the binary diffusion coefficient or mass diffusivity, with units of $(\text{length})^2/\text{time}$

The corresponding law written for species 2 is

$$\mathbf{j}_2 = -\rho \mathscr{D}_{21} \nabla m_2 \qquad (3.14)$$

As is the case with Fourier's law of heat conduction, Eq. (2.2), Fick's law is a vector equation, and the vector $\mathbf{j}_1$ is in the same direction as $-\nabla m_1$. For a one-dimensional situation with a gradient of concentration in the z-direction only, Eq. (3.13) simplifies to

$$j_1 = -\rho \mathscr{D}_{12} \frac{dm_1}{dz} \qquad (3.15)$$

A valid question at this point is the suitability of Eq. (3.13) as a statement of Fick's law. After all we have introduced six measures of the composition of a mixture (number density, number fraction, partial density, mass fraction, molar concentration, mole fraction), so why should mass fraction be the appropriate driving potential for diffusion? And why should diffusion be expressed as a mass flux; would not a molar flux be more suitable? We can, simply by introducing the pertinent relations, derive alternative forms of Eq. (3.13). After we have carefully examined what we mean by a stationary medium, we may, for example, derive

$$\bullet \qquad \mathbf{J}_1 = -c \mathscr{D}_{12} \nabla x_1 \qquad (3.16)$$

where $\mathbf{J}_1$ is the diffusion molar flux of species 1, with units of moles/area-time. On the other hand if we were simply to write down the equation

$$\mathbf{j}_1 = -\mathscr{D}_{12} \nabla \rho_1 \qquad (3.17)$$

we would find that the diffusion coefficient defined by Eq. (3.17) is not the same as the one appearing in Eq. (3.13). It cannot be demonstrated in a simple manner that Eq. (3.13) is indeed the most appropriate mathematical statement of Fick's law. There is no single physical mechanism of diffusion; in particular there are radical differences between the mechanisms in gases, liquids, and solids. The problem must be examined in the light of physical theory, the results of irreversible thermodynamics, and experimental data. At the present stage it will be stated without demonstration that the kinetic theory of gases shows that Eq. (3.13) is appropriate for gas mixtures at low pressures; furthermore, experimental data shows that it is appropriate for dilute liquid solutions. Since the majority of engineering problems fall within these two categories, it is convenient to accept this expression as a fundamental statement of Fick's first law of diffusion.

Diffusion coefficients of gases at low pressures are almost composition independent, increase with temperature, and vary inversely with pressure. Liquid and solid diffusion coefficients are markedly concentration dependent and generally increase with temperature. In Chapter 7 we will derive an expression for $\mathscr{D}_{12}$ from the kinetic theory of gases; selected data for $\mathscr{D}_{12}$ are tabulated in Appendix B, Table B.12.

3.4   THE EQUATION OF CONSERVATION OF SPECIES IN A STATIONARY MEDIUM

In order to simplify the required calculus as much as possible, we shall consider diffusion in one spatial dimension only. To be specific let us suppose that within our medium is a binary system of species 1 and 2, and there exist concentration gradients in the z-direction only. Figure 3.2 depicts the situation. An element Δz in length is embedded in the medium which may be considered to be of unit width in the x- and y-directions. The verbal statement of the principle of conservation of species 1 within the element of volume $1 \cdot 1 \cdot \Delta z$ is:

$$\text{accumulation} = \text{net inflow} + \text{production}$$

Considering each term in the above equation for the elemental time interval t, $t + \Delta t$ we have

$$\text{accumulation} = (\rho_1|_{\bar{z}, t+\Delta t} - \rho_1|_{\bar{z}, t}) \cdot (1 \cdot 1 \cdot \Delta z)$$

$$\text{net inflow} = \text{flow across area } 1 \cdot 1 \text{ at } z - \text{flow across area } 1 \cdot 1 \text{ at } z + \Delta z$$

$$= (j_1|_{z, \bar{t}} - j_1|_{z+\Delta z, \bar{t}}) \cdot (1 \cdot 1 \cdot \Delta t)$$

$$= \left(-\rho \mathscr{D}_{12} \frac{\partial m_1}{\partial z}\bigg|_{z, \bar{t}} + \rho \mathscr{D}_{12} \frac{\partial m_1}{\partial z}\bigg|_{z+\Delta z, \bar{t}} \right) \cdot (1 \cdot 1 \cdot \Delta t)$$

$$\text{production} = \dot{r}_1 \cdot (1 \cdot 1 \cdot \Delta z \cdot \Delta t)$$

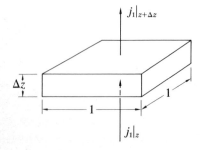

Figure 3.2   Control volume for species conservation in a stationary medium.

where j_1 has been eliminated by means of Fick's law, Eq. (3.15), and $\dot{r}_1$ is the rate of production of species 1 due to chemical reactions, with dimensions of mass/volume-time.

By substituting into the verbal statement of the conservation principle and dividing through by $1 \cdot 1 \cdot \Delta z \cdot \Delta t$ there results

$$\left.\frac{\rho_1|_{t+\Delta t} - \rho_1|_t}{\Delta t}\right|_{\bar{z}} = \frac{-\rho\mathscr{D}_{12}\left.\frac{\partial m_1}{\partial z}\right|_z + \rho\mathscr{D}_{12}\left.\frac{\partial m_1}{\partial z}\right|_{z+\Delta z}}{\Delta z}\Bigg|_{\bar{t}} + \dot{r}_1|_{\bar{z},\bar{t}}$$

which becomes in the limit as $\Delta z \to 0$ and $\Delta t \to 0$

$$\frac{\partial \rho_1}{\partial t} = \frac{\partial}{\partial z}\left(\rho\mathscr{D}_{12}\frac{\partial m_1}{\partial z}\right) + \dot{r}_1 \qquad (3.18)$$

The corresponding equation in molar terms is

$$\frac{\partial c_1}{\partial t} = \frac{\partial}{\partial z}\left(c\mathscr{D}_{12}\frac{\partial x_1}{\partial z}\right) + \dot{R}_1 \qquad (3.19)$$

3.5 TRANSIENT DIFFUSION

3.5.1 The Governing Equation

Fick's so-called *second law* of diffusion is simply a restricted form of Eq. (3.18) which is obtained if homogeneous chemical reactions are absent and the density of the medium and diffusivity may be assumed to be constant. (The word *homogeneous* denotes that the reactions occur within the medium.) Then, since $\rho_1 = m_1\rho$, Eq. (3.18) reduces to

$$\frac{\partial m_1}{\partial t} = \mathscr{D}_{12}\frac{\partial^2 m_1}{\partial z^2} \qquad (3.20)$$

or on a molar basis,

$$\frac{\partial x_1}{\partial t} = \mathscr{D}_{12}\frac{\partial^2 x_1}{\partial z^2} \qquad (3.21)$$

Equations (3.20) and (3.21) are similar in form to the heat conduction equation, thereby establishing an analogy between heat conduction and diffusion for this situation. We have, say,

$$\frac{\partial m_1}{\partial t} = \mathscr{D}_{12}\frac{\partial^2 m_1}{\partial z^2}$$

and

$$\frac{\partial T}{\partial t} = \alpha\frac{\partial^2 T}{\partial z^2}$$

and we see that the mass diffusivity $\mathscr{D}_{12}$ and the thermal diffusivity α have the same dimensions, (length)^2/time. It follows that, when the various assumptions made hold true, solutions for diffusion problems may be obtained directly from the corresponding heat conduction problem. In the mathematics literature both the heat conduction equation and Fick's second law of diffusion are usually simply referred to as the *diffusion equation*, written

$$\frac{\partial \phi}{\partial t} = \alpha \frac{\partial^2 \phi}{\partial z^2} \tag{3.22}$$

3.5.2   Transient Diffusion in a Semi-Infinite Solid

Figure 3.3 depicts a semi-infinite solid with species 1 in solution in species 2. For example, we might consider the solution of carbon in iron. The initial concentration of species m_1 is $m_{1,0}$. At time $t = 0$, the concentration of m_1 at the surface $z = 0$ is suddenly raised to $m_{1,s}$. The problem is to describe the concentration profiles of species 1 as a function of time and position.

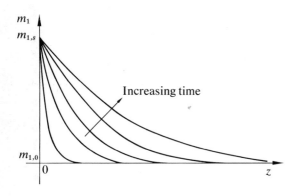

Figure 3.3   Transient diffusion in a semi-infinite solid.

The equation to be solved is

$$\frac{\partial m_1}{\partial t} = \mathscr{D}_{12} \frac{\partial^2 m_1}{\partial z^2}$$

subject to the initial condition

$$m_1 = m_{1,0} \quad \text{at } t = 0$$

and boundary conditions

$$m_1 = m_{1,s} \quad \text{at } z = 0$$
$$m_1 \to m_{1,0} \quad \text{as } z \to \infty$$

The analogous heat conduction problem was solved in Chapter 2; thus we may write down the solution immediately. See Eqs. (2.48) and (2.49).

$$\frac{m_1 - m_{1,0}}{m_{1,s} - m_{1,0}} = \text{erfc} \frac{z}{\sqrt{4\mathscr{D}_{12}t}} \tag{3.23}$$

Since most solid diffusion coefficients are very low, Eq. (3.23) is particularly useful for estimating the *penetration depth* of species 1 for most geometries encountered in practice.

3.5.3 *EXAMPLE* Case Hardening of Steel

A preheated mild steel rod is to be case hardened. Its initial concentration of carbon is 0.2% by weight, and it is packed in a carbonaceous material. From thermodynamics the equilibrium surface concentration of carbon at the rod surface is 1.5%. Calculate the time required for the location 0.025 in. below the surface to have a concentration of 0.8% carbon. The diffusion coefficient of carbon in steel at the process temperature may be taken to be 5.8×10^{-6} cm²/sec. ·
From Eq. (3.23) we have

$$\frac{m_1 - m_{1,0}}{m_{1,s} - m_{1,0}} = \mathrm{erfc} \frac{z}{\sqrt{4\mathcal{D}_{12}t}}$$

$$\frac{0.008 - 0.002}{0.015 - 0.002} = \mathrm{erfc} \left(\frac{(0.025/12)}{2\sqrt{(5.8 \times 10^{-6})(3600/30.5^2)\,t}} \right) \quad \text{where } t \text{ is in hours}$$

$$\frac{0.006}{0.013} = \mathrm{erfc} \left(\frac{0.22}{\sqrt{t}} \right)$$

$$0.461 = \mathrm{erfc} \left(\frac{0.22}{\sqrt{t}} \right)$$

From Table 2.2

$$\frac{0.22}{\sqrt{t}} = 0.52$$

$$t = \left(\frac{0.22}{0.52} \right)^2 \text{ hr}$$

$$= 0.18 \text{ hr}$$

3.6 ONE-DIMENSIONAL DIFFUSION WITH A HETEROGENEOUS REACTION

3.6.1 The Process

The phenomenon of heterogeneous catalysis is often encountered in engineering problems; it is the process whereby a chemical reaction is promoted by contact of the reactants with an exposed surface. (The adjective *heterogeneous* means the reaction takes place *on the surface*; it is in contradistinction to the adjective *homogeneous*, which, as noted earlier, means the reaction takes place *within the fluid*.) (For example, the reaction $2NO + 2CO \rightarrow N_2 + 2CO_2$ can, in the absence of oxygen, be promoted in the exhaust gases of an automobile by passing the gases through a catalytic reactor. Alternatively, the relatively cool metal of a reentry nose cone acts as catalyst for the recombination of oxygen and nitrogen, the atomic species having resulted from dissociation in the high temperature region aft of the bow shock.) The essential characteristics of heterogeneous catalysis may be demonstrated by analyzing the simple one-dimensional model depicted in Figure 3.4.

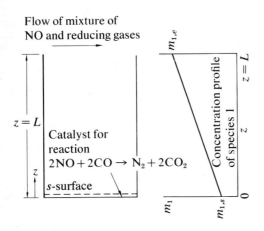

Figure 3.4   Diffusion with heterogeneous catalysis. The reaction $2NO + 2CO \rightarrow N_2 + 2CO_2$ on the catalytic base of a Stefan tube in the presence of an inert carrier gas.

A tube has height L, and over its mouth flows a mixture of gases including CO and a relatively small concentration of NO. The surface at the base of the tube is a catalyst for the reaction

$$2NO + 2CO \rightarrow N_2 + 2CO_2 \tag{3.24}$$

We label the nitric oxide NO as species 1. Now Fick's first law, as we have defined it, is strictly valid only for binary mixtures; hence the appearance of a binary diffusion coefficient $\mathscr{D}_{12}$. However, when species 1 is in small concentration, we may safely define an *effective binary diffusion coefficient* for species 1 diffusing through the mixture, $\mathscr{D}_{1m}$. This diffusion coefficient is an average value for the various species in the mixture. The rate at which the reaction proceeds at the catalyst surface is, for a first order reaction,

$$k'' m_{1,s} \rho$$

where $m_{1,s}$ is the concentration of NO in the gas phase adjacent to the surface, and k'' is the rate constant, a function of temperature and pressure; the precise value for given conditions is obtained from chemical data. The flow across the mouth is assumed to maintain the concentration there at $m_{1,e}$ and to cause no spurious convection currents. For a timewise steady state and for no reactions in the gas phase (no homogeneous reactions), Eq. (3.18) which governs the diffusion of NO in the mixture, becomes

$$\frac{d}{dz}\left(\rho \mathscr{D}_{1m} \frac{dm_1}{dz}\right) = 0 \tag{3.25}$$

We now assume the system to be approximately isothermal; then $\mathscr{D}_{1m} \cong$ a constant, and the mixture density may also be taken to be constant since the range of species molecular weights is not large. Equation (3.25) becomes

$$\frac{d^2 m_1}{dz^2} = 0 \tag{3.26}$$

This differential equation is to be solved subject to the boundary conditions:

$$z = 0, \quad j_1 = -\rho \mathcal{D}_{1m} \frac{dm_1}{dz}\bigg|_{z=0} = -k'' \rho m_{1,s}$$

$$z = L, \quad m_1 = m_{1,e}$$

Integrating Eq. (3.26) yields

$$m_1 = C_1 z + C_2$$

Evaluating the constants from the boundary conditions yields the concentration distribution

$$\frac{m_1}{m_{1,e}} = \frac{1 + \dfrac{k'' z}{\mathcal{D}_{1m}}}{1 + (k'' L / \mathcal{D}_{1m})} \tag{3.27}$$

The rate of reaction may be calculated most easily from the first boundary condition:

$$j_1 = -k'' \rho m_{1,s}$$

$$= -\frac{k'' \rho m_{1,e}}{1 + \dfrac{k'' L}{\mathcal{D}_{1m}}}$$

Dividing by $\rho k''$ and dropping the minus sign from here on, we obtain the *removal* rate of species 1,

$$j_1 = \frac{m_{1,e}}{\dfrac{1}{\rho k''} + \dfrac{L}{\rho \mathcal{D}_{1m}}} \tag{3.28}$$

Note that Eq. (3.28) is in the form of Ohm's law for the passage of current j_1 through two resistances in series, the reaction rate resistance $1/\rho k''$ and the diffusion resistance $L/\rho \mathcal{D}_{1m}$. The potential driving the transfer is $m_{1,e}$, since we have assumed that the reaction can go to completion at the surface, if $\rho k'' \gg \rho \mathcal{D}_{1m}/L$. We can therefore use a circuit representation for this situation much the same as that in Figure 2.4 in Chapter 2.

Two special cases can be distinguished, one in which the total resistance is dominated by the reaction rate resistance and the other in which the total resistance is dominated by the diffusion rate resistance.

Special Case (a)   If the reaction rate is slow, that is, $k'' \ll \mathcal{D}_{1m}/L$, the reaction is said to be *rate controlled*. From Eq. (3.28), $j_1 = k'' \rho m_{1,e}$. Note that $m_{1,e} \cong m_{1,s}$ for this situation.

Special Case (b)   If the reaction rate is fast, that is, $k'' \gg \mathscr{D}_{1m}/L$, the reaction is said to be *diffusion controlled*. The rate at which the reaction proceeds is governed by the rate at which NO can diffuse to the surface. From Eq. (3.28)

$$j_1 = \frac{\rho \mathscr{D}_{1m}}{L} m_{1,e}$$

Note that $m_{1,s} \doteq 0$ for this situation.

3.6.2   *EXAMPLE*   Removal of NO from Automobile Engine Exhaust

A simple analysis of the reduction of NO in an automobile catalytic reactor may be performed if the real flow of gases over the catalytic surface is modeled by visualizing the diffusion process to occur across a thin stagnant film of thickness δ adjacent to the surface. The appropriate equivalent value of δ depends upon the flow conditions in the reactor, the most relevant of which is the flow velocity. One can often take δ to be proportional to velocity raised to the minus eight-tenths power. Chapters 5 and 8 will explain this feature. Consider a situation where the exhaust gases are at 1000°F and 17.1 psia, have a mean molecular weight of 28.0, and contain 0.19% NO by weight. At this temperature an *effective* rate constant k'' may be 750 ft/hr (see Exercise 14), and the diffusion coefficient $\mathscr{D}_{1m}$ is 3.9 ft²/hr. Determine the maximum value of δ allowable if a NO reduction rate of 0.024 lb/ft² hr is to be achieved.

From Eq. (3.28) we have

$$j_1 = \frac{\rho m_{1,e}}{\dfrac{1}{k''} + \dfrac{\delta}{\mathscr{D}_{1m}}}$$

where we have replaced L by δ. Solving for δ there is obtained

$$\delta = \mathscr{D}_{1m} \left(\frac{\rho m_{1,e}}{j_1} - \frac{1}{k''} \right), \quad \rho = \frac{P}{\dfrac{\mathscr{R}T}{M}}$$

$$= (3.9 \text{ ft}^2/\text{hr}) \left[\frac{\dfrac{(17.1 \text{ lb}_f/\text{in}^2)(144 \text{ in}^2/\text{ft}^2)}{(1545/28 \text{ ft lb}_f/\text{lb }°\text{R})(1460 °\text{R})}(0.0019)}{0.024 \text{ lb/ft}^2 \text{ hr}} - \frac{1}{750 \text{ ft/hr}} \right]$$

$$= 3.9 \left(\frac{1}{413} - \frac{1}{750} \right)$$

$$= 0.0042 \text{ ft}$$

$$\delta = 0.051 \text{ in.}$$

This value, though quite small, could (as will be shown in Chapter 5) be achieved in practice.

3.7 SIMULTANEOUS DIFFUSION AND CONVECTION

3.7.1 Convective Transfer of Mass Species

So far we have considered the transport of a mass species by diffusion only, that is, we restricted our attention to a stationary medium. Actually we did not even bother to define precisely what we meant by stationary. Since we have seen that in a mixture the various species can move relative to each other, a suitable definition of stationary is not immediately obvious. What we had in mind was an imprecise idea based on a physical experience. For example, diffusion of a trace species in a solid would qualify as diffusive transport. On the other hand the dispersion of smoke issuing from the stack of a power plant is clearly dominated by convective motions due to wind currents and thermal buoyancy forces; molecular diffusion plays but a secondary role. Unfortunately, the concepts which arise in the analysis of simultaneous diffusion and convection require some thought before they can be grasped. It was for this reason that the special case of pure diffusive transport was treated first, as a stepping stone to the complete picture, glossing over such details as a precise definition of a stationary medium.

3.7.2 Definitions of Velocities

In a multicomponent system the various species may move at different velocities. Let $\mathbf{v}_i$ denote the absolute velocity of species i; that is, the velocity relative to stationary coordinate axes. In this sense, the velocity is not that of an individual molecule of species i. Rather, it is the local average of the species; that is, the sum of the velocities of species i within an elemental volume divided by the number of such molecules. Then the local *mass-average velocity*, $\mathbf{v}$, is defined as

$$\mathbf{v} = \frac{\sum \rho_i \mathbf{v}_i}{\sum \rho_i} = \frac{\sum \rho_i \mathbf{v}_i}{\rho} = \sum m_i \mathbf{v}_i \qquad (3.29)$$

The quantity $\rho\mathbf{v}$ is the local mass flux; that is, the rate at which mass passes through a unit cross section placed normal to the velocity vector $\mathbf{v}$. From Eq. (3.29), $\rho\mathbf{v} = \sum \rho_i \mathbf{v}_i$; that is, the local mass flux is the sum of the local species mass fluxes. The velocity $\mathbf{v}$ is the velocity which would be measured by a Pitot tube and corresponds to the velocity $\mathbf{v}$ used when considering pure fluids. Of particular importance is that this velocity $\mathbf{v}$ is the velocity field described by the Navier-Stokes equations and thereby has its origin in Newton's second law of motion.

The local *molar-average velocity*, $\mathbf{v}^*$, is defined in an analogous manner:

$$\mathbf{v}^* = \frac{\sum c_i \mathbf{v}_i}{\sum c_i} = \sum x_i \mathbf{v}_i \qquad (3.30)$$

The quantity $c\mathbf{v}^*$ is the local molar flux; that is, the rate at which moles pass through a unit area placed normal to the velocity $\mathbf{v}^*$.

3.7.3   *EXAMPLE* Molar and Mass Average Velocities

A gas mixture contains 50% He and 50% O_2 by weight. The absolute velocities of each species are $v_{He} = -9i$ ft/sec; $v_{O_2} = +9i$ ft/sec, where i denotes the unit vector in the x-direction. Determine the mass and molar average velocities.

From Eq. (3.29) we find v as follows:

$$v = \sum_{i=1}^{2} m_i v_i$$

$$= \tfrac{1}{2}(-9i) + \tfrac{1}{2}(+9i)$$

$$= 0$$

From Eq. (3.30)

$$v^* = \sum_{i=1}^{2} x_i v_i$$

The mole fractions are found from Eq. (3.12).

$$x_{He} = \frac{\tfrac{1}{2} \cdot \tfrac{1}{4}}{\tfrac{1}{2} \cdot \tfrac{1}{4} + \tfrac{1}{2} \cdot \tfrac{1}{32}} = \frac{8}{9}, \qquad x_{O_2} = \frac{\tfrac{1}{2} \cdot \tfrac{1}{32}}{\tfrac{1}{2} \cdot \tfrac{1}{4} + \tfrac{1}{2} \cdot \tfrac{1}{32}} = \frac{1}{9}$$

and hence

$$v^* = \tfrac{8}{9}(-9i) + \tfrac{1}{9}(+9i)$$

$$= -7i \text{ ft/sec}$$

It can be seen that although the mass average velocity is zero, the molar average velocity is indeed large. Is the mixture stationary? We will soon be able to resolve this question.

3.7.4   Diffusion Velocities

The velocity of a particular species relative to the mass or molar-average velocity is termed a *diffusion velocity*. We define two such velocities:

$$v_i - v \equiv \text{diffusion velocity of species } i \text{ relative to v} \qquad \textbf{(3.31)}$$

$$v_i - v^* \equiv \text{diffusion velocity of species } i \text{ relative to v*} \qquad \textbf{(3.32)}$$

We shall see that a species can have a velocity relative to v or v^* only if gradients exist in mass fraction or mole fraction; from Fick's law we know that diffusion must then be taking place.

3.7.5   Definitions of Fluxes

The mass (or molar) flux of species i is a vector quantity given by the mass (or moles) of species i that passes per unit time through a unit area normal to the vector. Such fluxes may be defined relative to stationary coordinate axes or to either of the two local average velocities. We define the absolute mass and molar fluxes, that is, relative to stationary coordinate axes, as

$$\text{mass flux:} \quad \mathbf{n}_i \equiv \rho_i \mathbf{v}_i \tag{3.33}$$

$$\text{molar flux:} \quad \mathbf{N}_i \equiv c_i \mathbf{v}_i \tag{3.34}$$

The mass flux relative to the mass-average velocity $\mathbf{v}$ is

$$\mathbf{j}_1 \equiv \rho_i (\mathbf{v}_i - \mathbf{v}) \tag{3.35}$$

The molar flux relative to the mole-average velocity $\mathbf{v}^*$ is

$$\mathbf{J}_i^* \equiv c_i (\mathbf{v}_i - \mathbf{v}^*) \tag{3.36}$$

From a mathematical viewpoint any one of these flux definitions is adequate for all diffusion situations; however, in a given situation there is usually one definition which, when employed, leads to minimum algebraic complexity. The most important such situation is when the convective transport present requires a solution of the conservation of momentum equation; the solution yields the mass-average velocity field, and it is then most convenient to use the mass flux relative to the mass-average velocity, that is, $\mathbf{j}_i$. Conditions of constant pressure and temperature, often encountered by chemical engineers, have usually led to the choice of the absolute molar flux $\mathbf{N}_i$ to take advantage of the constant molar density c which results in such situations.

The definitions of fluxes lead directly to a number of important relations as follows:

$$\sum \mathbf{n}_i = \rho \mathbf{v} \tag{3.37}$$

$$\sum \mathbf{N}_i = c \mathbf{v}^* \tag{3.38}$$

$$\sum \mathbf{j}_i = \sum \mathbf{J}_i^* = 0 \tag{3.39}$$

$$\mathbf{N}_i = \frac{n_i}{M_i} \tag{3.40}$$

$$\mathbf{n}_i = \rho_i \mathbf{v} + \mathbf{j}_i = m_i \sum \mathbf{n}_i + \mathbf{j}_i \tag{3.41}$$

$$\mathbf{N}_i = c_i \mathbf{v}^* + \mathbf{J}_i^* = x_i \sum \mathbf{N}_i + \mathbf{J}_i^* \tag{3.42}$$

3.7.6 More Precise Statement of Fick's First Law

We have noted how a mass species may be transported by convection and diffusion. Convection is of its nature a *bulk* motion and thus transports the mixture as a whole. A given species can be transported relative to this bulk motion only if concentration gradients exist so that diffusion occurs. Thus it is clear that a precise definition of Fick's first law must describe diffusion relative to an average velocity of the mixture. Our definitions of velocities and fluxes, Eqs. (3.29) through (3.36), were of a mathematical nature and depend on no physical laws. However, they were chosen so that the introduction of physics via Fick's first law is straightforward. We now propose that the law should be written

$$\mathbf{j}_1 = -\rho \mathcal{D}_{12} \nabla m_1 \tag{3.43}$$

that is, exactly as Eq. (3.13), but now $\mathbf{j}_1$ has been precisely defined as the mass flux of species 1 *relative to the mass average velocity*. The corresponding law written for species 2 is again

$$\mathbf{j}_2 = -\rho \mathcal{D}_{21} \nabla m_2 \tag{3.44}$$

Since $\nabla m_1 = -\nabla m_2$ in a binary system and, from Eq. (3.39), $\mathbf{j}_1 + \mathbf{j}_2 = 0$, it then follows immediately that

$$\mathcal{D}_{12} = \mathcal{D}_{21} \tag{3.45}$$

The molar form of Fick's first law equivalent to Eq. (3.43) is

$$\mathbf{J}_1{}^* = -c \mathcal{D}_{12} \nabla x_1 \tag{3.46}$$

A simple algebraic manipulation will show that Eqs. (3.46) and (3.43) are mathematically identical. However, we see that the diffusive molar flux $\mathbf{J}_1{}^*$ is the molar flux *relative to the molar average velocity*.

It is now possible to give a correct interpretation of what we mean by a stationary medium. If we are working in mass units, we require that the mass average velocity be zero; if we are working in molar units, we require that the molar average velocity be zero. Since a Pitot tube or anemometer measures the mass average velocity, the interpretation of stationary as zero mass average velocity is more in accord with our physical intuition. The student is left to convince himself that a zero molar average velocity is really a microscopic concept and corresponds to a zero *molecule average velocity*.

Finally, the substitution of Fick's first law into Eqs. (3.41) and (3.42), written for a binary system, yields two important relations:

- $\quad \mathbf{n}_1 = \rho_1 \mathbf{v} - \rho \mathcal{D}_{12} \nabla m_1 = m_1 (\mathbf{n}_1 + \mathbf{n}_2) - \rho \mathcal{D}_{12} \nabla m_1 \tag{3.47}$

- $\quad \mathbf{N}_1 = c_1 \mathbf{v}^* - c \mathcal{D}_{12} \nabla x_1 = x_1 (\mathbf{N}_1 + \mathbf{N}_2) - c \mathcal{D}_{12} \nabla x_1 \tag{3.48}$

We see that the absolute flux of a species can always be conveniently expressed as the sum of two components, one due to convection and the other due to diffusion.

3.8   THE ONE-DIMENSIONAL EQUATION OF CONSERVATION OF SPECIES

We now derive an equation of conservation of species which applies both to a moving fluid and our previously considered stationary medium. To avoid algebraic complexity, or the introduction of vector calculus, the derivation will again consider transport in one spatial direction only. We will assume that there is a velocity component and concentration gradients in the z-direction only. Appendix A contains a derivation for the general three-dimensional situation. Figure 3.5 depicts an element Δz in length, stationary in space, and of unit width in the x

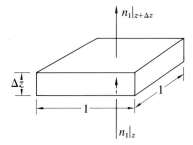

Figure 3.5   Control volume for conservation of species.

and y directions. The verbal statement of the principle of conservation of species i within the element of volume $1 \cdot 1 \cdot \Delta z$ is as before,

$$\text{accumulation} = \text{net inflow} + \text{production}$$

The only term in our previous derivation which needs to be modified is the expression for the net inflow. We now have the possibility of mass transport due to both diffusion and convection; thus the diffusive flux j_i in our previous derivation must be replaced by the absolute flux $n_i = \rho_i v_z + j_i$. As before, substituting into the verbal statement of the conservation principle and dividing through by $1 \cdot 1 \cdot \Delta z \cdot \Delta t$ yields

$$\left.\frac{\rho_i|_{t+\Delta t} - \rho_i|_t}{\Delta t}\right|_z = \left.\frac{n_i|_z - n_i|_{z+\Delta z}}{\Delta z}\right|_t + \dot{r}_i|_{z,\bar{t}}$$

which becomes in the limit as $\Delta z \to 0$ and $\Delta t \to 0$

$$\frac{\partial \rho_i}{\partial t} = -\frac{\partial}{\partial z}(n_i) + \dot{r}_i$$

or rearranging,

$$\frac{\partial \rho_i}{\partial t} + \frac{\partial}{\partial z}(n_i) = \dot{r}_i, \qquad i = 1, 2, \ldots n \tag{3.49}$$

which is identical to Eq. (3.18) except that the absolute flux n_i has replaced the diffusive flux j_i. Equation (3.49) may be written for each species, $i = 1, 2, 3, \ldots, n$. Summing shows that

$$\frac{\partial}{\partial t}\left(\sum \rho_i\right) + \frac{\partial}{\partial z}\left(\sum n_i\right) = \sum \dot{r}_i$$

But

$$\sum \rho_i = \rho$$

$$\sum n_i = n = \rho v$$

$$\sum \dot{r}_i = 0 \text{ (since mass is conserved)}$$

Therefore

$$\frac{\partial \rho}{\partial t} + \frac{\partial}{\partial z}(\rho v) = 0 \tag{3.50}$$

This is the equation of conservation of mass, called the continuity equation in fluid mechanics texts.

The preceding development could equally well have been done on a molar basis; the equations corresponding to Eqs. (3.49) and (3.50) are, respectively,

$$\text{species conservation:} \quad \frac{\partial c_i}{\partial t} + \frac{\partial}{\partial z}(N_i) = \dot{R}_i, \quad i = 1, 2, \dots, n \qquad \textbf{(3.51)}$$

$$\text{molar conservation:} \quad \frac{\partial c}{\partial t} + \frac{\partial}{\partial z}(\rho v^*) = \sum \dot{R}_i \qquad \textbf{(3.52)}$$

where $\dot{R}_i$ is the rate of production of moles of i per unit volume. In general moles are not conserved in a chemical reaction, and thus $\sum \dot{R}_i \neq 0$.

3.9   DIFFUSION WITH ONE COMPONENT STATIONARY

3.9.1   The Gas-Controlled Heat Pipe

A **heat pipe** is the name given to an evaporator-condenser system where the liquid is returned from the condenser to the evaporator by capillary action. In its simplest form it is a hollow tube with a few layers of wire screen along the wall to serve as a wick. The screen is filled with a wetting liquid such as sodium or potassium for high temperature applications, or with water, ammonia, or methyl alcohol for moderate temperature applications. If one end of the pipe is heated and the other end cooled, the liquid evaporates at the hot end and condenses at the cold end. As the liquid is depleted in the evaporator section, cavities form in the surface due to the liquid clinging to the wires of the screen. In the condenser section, meanwhile, the screen becomes flooded. The surface tension acting on the concave liquid-vapor interfaces in the evaporator causes the pressure in the vapor to be higher than that in the liquid. This pressure is transmitted by the vapor to the flooded condenser section so that liquid is driven from that section through the screen wick to the evaporator. Such devices are particularly attractive for space vehicle applications where gravitational forces are very weak. In a gravity field the evaporator may be placed below the condenser to assist the liquid flow. In such a case the wick may not be needed at all, and the device is more properly called a *reflux boiler* rather than a heat pipe.

The working fluids chosen for heat pipes have high latent heats of vaporization. Thus the motion of a small mass of vapor down the tube can transport a large amount of thermal energy. The total temperature difference necessary is that required to conduct the heat through the evaporator wall and wick, through the condenser wick and wall, plus an amount necessary to provide the vapor pressure difference which drives the vapor from the evaporator to the condenser. This temperature difference is quite small compared to that which would be necessary to transfer an equal amount of heat through the length of the pipe by conduction, even if the pipe were solid copper.

It is sometimes necessary to determine the effect of a small amount of noncondensable gas added to the vapor in a heat pipe. The gas might be for control purposes, that is, to reduce the heat transfer capability of a given heat pipe; corrosion may have generated gas within the pipe; or a construction defect might have allowed gas to leak into the pipe. A simple one-dimensional analysis gives insight into this phenomenon. We will consider the simple case of a heat pipe with the evaporator and condenser located at the *ends* of the pipe only. Figure 3.6 depicts an idealized model of such a heat pipe.

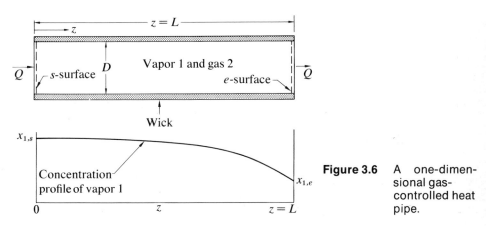

Figure 3.6 A one-dimensional gas-controlled heat pipe.

The vapor is designated as species 1 and the gas as species 2. Heat is applied at the surface $z = 0$, and heat is removed at the surface $z = L$. The sides of the heat pipe are assumed insulated. The temperatures at $z = 0$ and $z = L$ are T_s and T_e, respectively. The corresponding mole fractions of vapor adjacent to the liquid surfaces are $x_{1,s}$ and $x_{1,e}$. These values are the equilibrium values obtained from the saturation vapor pressures corresponding to the liquid surface temperatures, T_s and T_e, respectively. The solubility of gas in the liquid is assumed zero, and thus species 1 is the only substance transferred across the interfaces. At steady state there is net transfer of vapor along the pipe, while conservation of species requires that the gas 2 remains stationary. The variations of pressure and absolute temperature along the pipe are small; thus if an ideal gas mixture is assumed, the molar concentration c is virtually independent of position z. Advantage of this fact may be taken by performing the analysis on a molar basis. For this problem Eq. (3.51) reduces to

$$\frac{d}{dz}(N_1) = 0 \qquad (3.53)$$

Likewise Eq. (3.51) written for species 2 reduces to

$$\frac{d}{dz}(N_2) = 0 \qquad (3.54)$$

Integration yields $N_2 = $ constant. Since the gas is insoluble in the liquid, $N_2|_{z=0} = 0$, and it follows from Eq. (3.54) that N_2 is identically zero; that is, the gas 2 is stationary.

The molar flux N_1 can be expressed in the usual manner as the sum of a convective and diffusive component; from Eq. (3.48)

$$N_1 = x_1(N_1 + N_2) - c\mathcal{D}_{12}\frac{dx_1}{dz} \tag{3.55}$$

But $N_2 = 0$; thus solving for N_1 yields

$$N_1 = -\frac{c\mathcal{D}_{12}}{1 - x_1}\frac{dx_1}{dz} \tag{3.56}$$

Substitution of this expression in the differential equation (3.53) then yields

$$\frac{d}{dz}\left(-\frac{c\mathcal{D}_{12}}{1 - x_1}\frac{dx_1}{dz}\right) = 0$$

The molar concentration c and the binary diffusion coefficient are assumed constant ($\mathcal{D}_{12}$ is independent of composition). Therefore, the differential equation governing the concentration distribution may be written as

$$\frac{d}{dz}\left(\frac{1}{1 - x_1}\frac{dx_1}{dz}\right) = 0 \tag{3.57}$$

which is to be solved subject to the boundary conditions:

$$\text{boundary condition 1:} \quad z = 0, \quad\quad x_1 = x_{1,s}$$
$$\text{boundary condition 2:} \quad z = L, \quad\quad x_1 = x_{1,e}$$

Integrating twice and evaluating the constants of integration from the boundary conditions yields the concentration distribution

$$\left(\frac{1 - x_1}{1 - x_{1,s}}\right) = \left(\frac{1 - x_{1,e}}{1 - x_{1,s}}\right)^{z/L} \tag{3.58}$$

Alternatively, the concentration distribution of the gas 2 is given by

$$\left(\frac{x_2}{x_{2,s}}\right) = \left(\frac{x_{2,e}}{x_{2,s}}\right)^{z/L} \tag{3.59}$$

The concentration distributions are illustrated in Figure 3.6. The rate of vapor transfer through the pipe is the rate at which the liquid evaporates and may be evaluated as follows

$$N_1|_{z=0} = N_{1,s} = -\frac{c\mathcal{D}_{12}}{1 - x_{1,s}}\frac{dx_1}{dz}\bigg|_{z=0}$$

Differentiation of Eq. (3.58) and evaluation at $z = 0$ gives

$$N_1 \equiv N_{1,s} = \frac{c\mathcal{D}_{12}}{L}\ln\frac{1 - x_{1,e}}{1 - x_{1,s}} \tag{3.60}$$

The rate of heat transfer along the pipe is simply $M_1 N_1 \hat{h}_{fg}(\pi/4)D^2$, where $\hat{h}_{fg}$ is the latent heat of vaporization:

$$\frac{\dot{Q}}{(\pi/4)D^2} = \frac{M_1 c \mathcal{D}_{12} \hat{h}_{fg}}{L} \ln\frac{1-x_{1,e}}{1-x_{1,s}} = \frac{M_1 c \mathcal{D}_{12} \hat{h}_{fg}}{L} \ln\frac{x_{2,e}}{x_{2,s}} \tag{3.61}$$

An alternative form of Eq. (3.60) may be obtained by introducing the **logarithmic mean value** of the mole fraction x_2, defined as

$$(x_2)_{l.m.} = \frac{x_{2,e} - x_{2,s}}{\ln(x_{2,e}/x_{2,s})} \tag{3.62}$$

Algebraic manipulation of Eq. (3.60) and substitution of Eq. (3.62) then yields

$$\bullet \qquad N_1 = \frac{c\mathcal{D}_{12}}{L(x_2)_{l.m.}}(x_{1,s} - x_{1,e}) \tag{3.63}$$

     In addition to heat pipes there are other applications where one species is transferred and the second is stationary. For example, evaporation from a reservoir may be viewed as transfer across a stagnant air film, of thickness δ, adjacent to the water surface. Equation (3.63) applies to such cases if the length L is replaced by film thickness δ. For low transfer rates ($N_1 \to 0$), the value of $(x_2)_{l.m.}$ approaches its limiting value of unity, and a linear Ohm's law type relation holds between N_1 and $(x_{1,s} - x_{1,e})$. As the concentration difference Δx_1 and transfer rate increase, $(x_2)_{l.m.}$ decreases, and N_1 is no longer a linear function of Δx_1; this nonlinearity at high mass transfer rates is an important characteristic of mass transfer.

     Even though we have said that one component is stationary, there is of course diffusion of both species. The concentration gradients of components 1 and 2 are equal and opposite, and hence gas 2 diffuses towards the evaporating surface. Although the gas is insoluble, its concentration does not build up at the liquid surface. There is a convective motion in the positive z-direction such that at steady state, the net flux N_2, which is the sum of convective component $x_2 N$ and diffusive component $-c\mathcal{D}_{12}(dx_2/dz)$, is exactly zero. In Chapter 4 we shall see that a small total pressure gradient is required to overcome wall friction in driving this convective flow along the pipe. Convection is, by its nature, a bulk movement of the mixture and hence vapor 1 is also convected in the positive z-direction; this was, of course, accounted for in Eq. (3.55) by writing the total flux of vapor 1, N_1, as a sum of convective and diffusive components. This convective flow which augments the diffusive flux of vapor 1 is often referred to in the literature as the **Stefan flow**.

3.9.2 *EXAMPLE* Effect of Air on Heat Pipe Performance

     A water heat pipe, 30 in. long and 0.75 in. inside diameter, is taken from a storehouse and its performance checked prior to use. For a heat load of 100 watts and an evaporator temperature of 212°F, the condenser temperature is measured to be 201°F. An air leak or gas generation is suspected; calculate the amount of air which might have leaked into the pipe.

First we determine the expected temperature drop if no air were present. The vapor velocity V is calculated from the heat load

$$\rho V\left(\frac{\pi}{4}\right)D^2\hat{h}_{fg} = \dot{Q} = \frac{(100\text{ watts})\,(3.41\text{ Btu/hr watt})}{3600\text{ sec/hr}} = 0.0947\text{ Btu/sec}$$

$$V = \frac{\dot{Q}}{\rho(\pi/4)D^2\hat{h}_{fg}} = \frac{0.0947\text{ Btu/sec}}{(1/26.8\text{ lb/ft}^3)\,(\pi/4)\,(0.75/12\text{ ft})^2(970.3\text{ Btu/lb})} = 0.85\text{ ft/sec}$$

The Reynolds number is

$$Re_D = \frac{VD}{\nu} = \frac{(0.85\text{ ft/sec})\,(0.75/12\text{ ft})}{(23.4\times 10^{-5}\text{ ft}^2/\text{sec})} = 227$$

and, as we shall see in Chapter 4, the friction factor is

$$f = \frac{64}{Re_D} = \frac{64}{227} = 0.282$$

The pressure drop is, from Eq. (4.26),

$$\Delta P = f\frac{L}{D}(\tfrac{1}{2}\rho V^2) = (0.282)\left(\frac{30\text{ in.}}{0.75\text{ in.}}\right)\left(\frac{(1/2)\,(1/26.8\text{ lb/ft}^3)\,(0.85\text{ ft/sec})^2}{(32.2\text{ lb ft/lb}_f\text{ sec}^2)}\right)$$

$$= 0.00473\text{ lb}_f/\text{ft}^2$$

or

$$\Delta P = 3.28\times 10^{-5}\text{ lb}_f/\text{in.}^2$$

Steam tables show that the corresponding temperature drop is quite negligible.

Now consider the situation when air is present. We shall assume (and check later) that the air concentration at the evaporator end is very small; then since the evaporator is at 212°F, the total pressure there is 14.7 psia and varies negligibly along the pipe. From steam tables the saturation pressure corresponding to 201°F is 11.8 psia, and the mole fraction of air there is

$$x_{2,e} = \frac{P_{2,e}}{P_{\text{total}}} = \frac{14.7 - 11.8}{14.7} = 0.198$$

Equation (3.61) may be rearranged to read

$$\ln\frac{x_{2,e}}{x_{2,s}} = \frac{\dot{Q}L}{M_1 c \mathscr{D}_{12}\hat{h}_{fg}(\pi/4)D^2}$$

For water vapor-air mixtures a useful formula for the binary diffusion coefficient is

$$\mathscr{D}_{12} = 0.765\frac{P_0}{P}\left(\frac{T}{T_0}\right)^{1.685}\text{ ft}^2/\text{hr}, \quad P_0 = 1\text{ atm}, \quad T_0 = 460°\text{R}$$

$$\mathscr{D}_{12} = 0.765\left(\frac{672}{460}\right)^{1.685} = 1.45\text{ ft}^2/\text{hr}$$

The total molar concentration c is

$$c = \frac{P}{\mathscr{R}T} = \frac{(14.7\text{ lb}_f/\text{in.}^2)\,(144\text{ in.}^2/\text{ft}^2)}{(1545\text{ ft lb}_f/\text{lb-mole}°\text{R})\,(672°\text{R})} = 2.03\times 10^{-3}\text{ lb-mole/ft}^3$$

where, for convenience, both $\mathscr{D}_{12}$ and c have been evaluated at the evaporator temperature. Let the ratio $x_{2,e}/x_{2,s}$ be denoted by r; then

$$\ln r = \frac{(100 \text{ watts})(3.41 \text{ Btu/hr watt})(2.5 \text{ ft})}{(18 \text{ lb/lb-mole})(2.03 \times 10^{-3} \text{ lb-mole/ft}^3)(1.45 \text{ ft}^2/\text{hr})(970 \text{ Btu/lb})(\pi/4)(0.75/12 \text{ ft})^2}$$

$$= 5.41 \times 10^3$$

which is very large; our assumption of a negligibly small value of $x_{2,s}$ was thus fully justified.

The total amount of air in the pipe is the volume integral of the air concentration,

$$\mathscr{W} = (\pi/4)D^2 \int_0^L x_2 c \, dz, \quad \text{lb-moles}$$

Using Eq. (3.59) we obtain

$$\mathscr{W} = (\pi/4)D^2 c x_{2,s} L \int_0^1 r^{z/L} d(z/L)$$

$$= (\pi/4)D^2 c x_{2,s} L \frac{r-1}{\ln r}$$

$$= (\pi/4)D^2 c x_{2,e} \frac{L}{r} \frac{r-1}{\ln r}$$

and since r is large,

$$= \frac{(\pi/4)D^2 c x_{2,e} L}{\ln r}$$

$$= \frac{(\pi/4)(0.75/12 \text{ ft})^2 (2.03 \times 10^{-3} \text{ lb-mole/ft}^3)(0.198)(2.5 \text{ ft})}{5.41 \times 10^3}$$

$$\mathscr{W} = 5.70 \times 10^{-10} \text{ lb-moles}$$

The mass of air in the pipe, w, is equal to $\mathscr{W} M_1$; thus,

$$w = (5.70 \times 10^{-10} \text{ lb-moles})(29 \text{ lb/lb-mole}) = 16.5 \times 10^{-9} \text{ lb} = 7.50 \times 10^{-6} \text{ gm}$$

Note: That such a small amount of air as 7.5 micrograms would cause a 11°F temperature drop is due to the geometrical arrangement considered. Location of the condenser on the end of the pipe results in direct confrontation between the gas and the vapor attempting to condense. Since a minute amount of gas can completely shut down the system, precise control would be almost impossible. A superior design would be one where the condenser is located along the side wall of the pipe; the working fluid can then simply push the noncondensable aside out of the main stream to an unused portion of the condenser. In this case a large amount of gas is often desirable: on one hand, the gas would prevent the evaporator from becoming too cold when the heat load $\dot{Q}$ drops; on the other hand, it would prevent the system from becoming overheated when $\dot{Q}$ is large. When $\dot{Q}$ decreases, the temperature and, hence, the pressure decrease, allowing the noncondensable to expand up the pipe and block a larger fraction of the condenser; the reverse occurs when $\dot{Q}$ increases, and a smaller fraction of the condenser is blocked.

3.10   THE MASS TRANSFER COEFFICIENT

3.10.1   Definition

In Chapter 2 Newton's law of cooling introduced a heat transfer coefficient to describe the transport of energy from a surface to a fluid. Analogously we define a mass transfer coefficient to describe transport of a mass species from a surface (or phase interface) into a fluid. The results we have just obtained for diffusion with one component stationary serve to introduce the mass transfer coefficient in a convenient manner. Transport of vapor away from the liquid surface is described by Eq. (3.63),

$$N_1 = \frac{c\mathcal{D}_{12}}{L(x_2)_{l.m.}}(x_{1,s} - x_{1,e})$$

or, had our analysis been performed on a mass basis under the (poorer) assumption of $\rho\mathcal{D}_{12}$ constant,

$$n_1 = \frac{\rho\mathcal{D}_{12}}{L(m_2)_{l.m.}}(m_{1,s} - m_{1,e}) \tag{3.64}$$

We now restrict attention to situations where the concentration of the species being transferred, x_1 or m_1, is everywhere much less than unity. Actually, this condition is very often met in engineering practice; some examples are power plant cooling towers, gas scrubbers and air conditioners. By definition

$$(x_2)_{l.m.} \equiv \frac{x_{2,e} - x_{2,s}}{\ln\left(\frac{x_{2,e}}{x_{2,s}}\right)}$$

$$= \frac{x_{2,e} - x_{2,s}}{\ln\left(1 + \frac{x_{2,e} - x_{2,s}}{x_{2,s}}\right)} \tag{3.65}$$

For x_1 small, $x_2 \simeq 1$, so that the denominator may be expanded as $\ln(1+u) = u - \frac{1}{2}u^2 + \frac{1}{3}u^3 + \cdots$ and only the first two terms retained to obtain

$$(x_2)_{l.m.} \simeq (x_{2,s} + x_{2,e})/2$$

$$\simeq 1$$

Substituting into Eq. (3.63) there is obtained

$$N_1 \simeq \frac{c\mathcal{D}_{12}}{L}(x_{1,s} - x_{1,e}) \tag{3.66}$$

But in Section 3.9.1 it was noted that, as $(x_2)_{l.m.} \to 1$, the convective flow which augments the diffusive flux of species 1 becomes negligible; that is, $N_1 \simeq J_1^*$. Thus we may write

$$N_1 \simeq J_1^* \simeq \frac{c\mathcal{D}_{12}}{L}(x_{1,s} - x_{1,e}) \tag{3.67}$$

We now define the mass transfer coefficient K by an equation which resembles

Newton's law of cooling, namely

$$J_1^* = K(x_{1,s} - x_{1,e}) \tag{3.68}$$

thereby identifying $(x_{1,s} - x_{1,e})$ as a driving force for mass transfer. Comparison of Eqs. (3.67) and 3.68) yields

$$K \simeq \frac{c\mathcal{D}_{12}}{L} \tag{3.69}$$

that is, as one might expect, the mass transfer coefficient increases with the molar concentration c and the diffusion coefficient $\mathcal{D}_{12}$, but is inversely proportional to the thickness L of the *slab* of gas through which transport occurs.

The expression for the mass transfer coefficient derived here is for the *special situation* where species 1 is diffusing through an essentially stagnant mixture of width L. In engineering practice, it has often proven useful to view more complex problem situations in terms of diffusion through a *stagnant film* of equivalent width δ. Such an approach was demonstrated in Section 3.6.2. Determination of appropriate values of δ then becomes the major task.

Note

1. We have been careful to define K in terms of J_1^* via Eq. (3.68); this practice is in accord with more advanced mass transfer theory. In general, the approximation $N_1 \simeq J_1^*$ does not hold true, and it would then be incorrect to define K in terms of N_1. However, for the situations encountered in engineering practice where N_1 is approximately equal to J_1^*, we find it convenient to use, as a working equation

   - $$N_1 \doteq K(x_{1,s} - x_{1,e}) \tag{3.70}$$

2. Equation (3.68) defines the mass transfer coefficient K in terms of molar units; the analogous definition in mass units is

   - $$j_1 = K(m_{1,s} - m_{1,e}) \tag{3.71}$$

   where K has dimensions mass/area-time. Since problem solving should always be done wholly in molar units or wholly in mass units, we shall not bother to introduce notation for distinguishing the K defined by Eq. (3.68) from that defined by Eq. (3.71).

3. Combining Eqs. (3.64) and (3.71) via Eq. (3.47) yields, after some manipulation,

   $$n_1 = K\frac{m_{1,e} - m_{1,s}}{m_{1,s} - 1}$$

   that is, the total mass transfer rate of species 1 is given by the mass transfer coefficient multiplied by a driving force

   $$\frac{m_{1,e} - m_{1,s}}{m_{1,s} - 1}$$

   This result is central to more advanced mass transfer theory.

4. In Chapter 5, we will give expressions for K which apply to a variety of flow situations encountered in engineering practice.

3.10.2 *EXAMPLE* Determination of the Mass Transfer Coefficient for a Catalytic Reactor

Consider the mass transfer coefficient prevailing in the catalytic reactor described in Section 3.6.2, assuming it is operating in the diffusion controlled regime.

For the diffusion controlled regime, Eq. (3.28) may be written as

$$j_{1,s} = \frac{\rho \mathcal{D}_{1m}}{\delta} m_{1,e}$$

In the diffusion controlled limit, $m_{1,s} = 0$; thus the above equation may be written

$$j_{1,s} = \frac{\rho \mathcal{D}_{1m}}{\delta} (m_{1,e} - m_{1,s})$$

Hence, by definition, $K = \rho \mathcal{D}_{1m}/\delta$.

Using the numerical data of Section 3.6.2, we have

$$K = \frac{\dfrac{(17.1)(144)}{(1545/28)(1460)}(3.9)}{0.0042} = 28 \text{ lb/ft}^2 \text{ hr}$$

3.11   MASS TRANSFER IN POROUS CATALYSTS

3.11.1   Diffusion in Porous Solids

Engineering applications of catalysis, such as hydrodesulfurization in petroleum refining or oxidation of carbon monoxide (CO) in automobile exhausts, require a large area of catalytic surface per unit volume of reactor. In common use are fixed bed reactors where the catalyst is in the form of porous pellets of size ranging from $\frac{1}{16}$ to $\frac{1}{2}$ in. in diameter. For example, for CO oxidation a copper oxide on alumina pellet may be used; the alumina provides a suitable support for the catalyst as it is relatively inert and structurally stable to high temperatures. In one procedure the pellet is prepared by impregnating alumina powder with copper nitrate solution. When the powder is heated to 800°F the nitrate decomposes, evolving gases and leaving cupric oxide. Finally, the pellet is formed by molding at high pressure. The resulting pellet has a density in the range 0.6 to 1.2 gm/cm³ and a surface area ranging from 100–400 m²/gm. A pore diameter of 1μ is typical.

Mass transport of gaseous reactants within the pores of a catalyst pellet occurs by two mechanisms: **ordinary diffusion** and **Knudsen diffusion**. Ordinary diffusion, as described by Fick's law, dominates when the pores are large and the gas relatively dense. However, when the pores are small or the gas density low, the molecules collide with the pore walls more frequently than with each other, and diffusion of molecules along the pore walls is described by the equations for free molecule or Knudsen flow. At intermediate pressures and pore sizes both types of collisions are important. Mass transport within the pellet is further complicated by the presence of gradients of total pressure and temperature. Notwithstanding such complexity, relative success has been achieved in the analysis of mass transport within catalyst pellets by assuming the transport to be governed by a linear relation

$$N_1 = -cD_{1,\text{eff}}\nabla x_1 \tag{3.72}$$

where the phenomena of ordinary diffusion and Knudsen diffusion are accounted for in an approximate manner, by simply assuming additive resistances, that is,

$$\frac{1}{D_{1,\text{eff}}} = \frac{1}{\mathscr{D}_{12,\text{eff}}} + \frac{1}{D_{K1,\text{eff}}} \tag{3.73}$$

The first of the two effective diffusivities is taken to be

$$\mathscr{D}_{12,\text{eff}} = \frac{\epsilon_v}{\tau}\mathscr{D}_{12} \tag{3.74}$$

where $\mathscr{D}_{12}$ is the binary diffusion coefficient of Fick's law; ϵ_v is the porosity, or volume void fraction, of the pellet and accounts for the reduction in area posed by solid material; τ is the tortuousity factor and accounts for the increased diffusion length due to the tortuous paths of real pores, and for the effects of constrictions and dead end pores. The second is

$$D_{K1,\text{eff}} = \frac{\epsilon_v}{\tau}D_{K1} \tag{3.75}$$

where D_{K1} is the Knudsen diffusion coefficient for species 1,

$$D_{K1} = \tfrac{2}{3}r_e\bar{v}_1 \tag{3.76}$$

where r_e is the effective pore radius and $\bar{v}_1$ is the average molecular speed of species 1,

$$\bar{v}_1 = \left(\frac{8\mathscr{R}T}{\pi M_1}\right)^{1/2}$$

Substituting in a value for the gas constant gives a dimensional equation

$$D_{K1} = 9700r_e\left(\frac{T}{M_1}\right)^{1/2} \text{cm}^2/\text{sec} \tag{3.77}$$

for r_e in centimeters and T in °K.

3.11.2 The Pellet Itself: Effectiveness

When a chemical reaction takes place within a porous pellet, a concentration gradient is set up, and surfaces on pores deep within the pellet are exposed to lower reactant concentrations than surfaces near the pore openings.

Since the rate at which a chemical reaction proceeds is dependent upon reactant concentration, the average reaction rate throughout a catalyst pellet will be less than if all the available catalyst surface were exposed to reactant at the concentration prevailing at the exterior of the pellet. Figure 3.7 depicts a spherical catalyst pellet of a fixed bed reactor where it is immersed in a gas flow containing species 1 in dilute concentration. The catalyst promotes a reaction which consumes species 1, for example, the catalyst might be cupric oxide or platinum, which promotes the reaction

$$2CO + O_2 \rightarrow 2CO_2$$

in oxygen-rich automobile exhaust gases; CO would be species 1. The pellet has a radius R and a catalytic surface area a per unit volume of pellet. The reaction rate is $k''cx_1$ moles of 1/unit area-unit time. The gas flow maintains the concentration of species 1 at the exterior of the pellet at the value $x_{1,s}$.

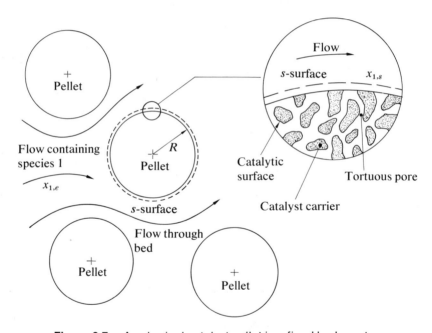

Figure 3.7   A spherical catalyst pellet in a fixed bed reactor.

In order to set up the differential equation governing the concentration distribution of species 1, we apply the principle of conservation of species to an elemental control volume located between spherical surfaces at radii r and $r + \Delta r$. At steady state

[net inflow of species 1] + [rate of production of species 1] = 0

$$[(4\pi r^2 N_1)_r - (4\pi r^2 N_1)_{r+\Delta r}] - [k''acx_1 4\pi r^2 \Delta r]_r = 0$$

Dividing by $4\pi\Delta r$ and letting $\Delta r \to 0$ gives

$$-\frac{d}{dr}(r^2 N_1) - r^2(k''acx_1) = 0$$

Substituting for N_1 from Eq. (3.72) and rearranging yields

$$\frac{1}{r^2}\frac{d}{dr}\left(r^2 cD_{1,\text{eff}}\frac{dx_1}{dr}\right) - k''acx_1 = 0$$

If the pellet is nearly isothermal and is at nearly constant total pressure, $cD_{1,\text{eff}}$ may be assumed constant; then

$$\frac{1}{r^2}\frac{d}{dr}\left(r^2\frac{dx_1}{dr}\right) - \frac{k''a}{D_{1,\text{eff}}}x_1 = 0 \qquad (3.78)$$

This differential equation governs the concentration distribution of species 1; it must be solved subject to the boundary conditions:

$$r = 0: \quad x_1 \text{ bounded (not infinite)}$$

$$r = R: \quad x_1 = x_{1,s}$$

In order to simplify notation we define $b = (k''a/D_{1,\text{eff}})^{1/2}$, and then introduce a new dependent variable $\theta = rx_1$. Equation (3.78) becomes

$$\frac{d^2\theta}{dr^2} - b^2\theta = 0 \qquad (3.79)$$

We recognize that this homogeneous, linear, second order, ordinary differential equation has a general solution of

$$\theta = C_1 \sinh br + C_2 \cosh br \qquad (3.80)$$

or, since $\theta = rx_1$,

$$x_1 = \frac{C_1}{r}\sinh br + \frac{C_2}{r}\cosh br$$

Since $\cosh 0 = 1$, from the first boundary condition $C_2 = 0$. The constant C_1 is evaluated from the second boundary condition, and the resulting concentration distribution is

$$\frac{x_1}{x_{1,s}} = \frac{R\sinh br}{r\sinh bR}, \quad b = \left(\frac{k''a}{D_{1,\text{eff}}}\right)^{1/2} \qquad (3.81)$$

The rate at which species 1 is *consumed* within the catalyst pellet is given by $-4\pi r^2 N_1|_{r=R}$ where

$$4\pi r^2 N_1|_{r=R} = -4\pi R^2 cD_{1,\text{eff}}\frac{dx_1}{dr}\bigg|_{r=R}$$

$$= 4\pi RcD_{1,\text{eff}}x_{1,s}\left[1 - \frac{bR}{\tanh bR}\right]$$

But if the complete internal surface were exposed to reactant at concentration $x_{1,s}$, the rate of consumption would be

$$(\tfrac{4}{3}\pi R^3)k''acx_{1,s}$$

The effectiveness $\eta (0 < \eta \leq 1)$ of the catalyst pellet is the ratio of the actual reaction rate divided by that for a pellet with infinite diffusion coefficient.

$$\eta = \frac{3}{(bR)}\left[\frac{1}{\tanh (bR)} - \frac{1}{(bR)}\right] \tag{3.82}$$

Notice that this effectiveness is used in the same sense as the fin effectiveness introduced in Chapter 2, which was the ratio of actual heat transfer to that for a fin of infinite thermal conductivity. Following chemical engineering practice we recast this result in final form by introducing the Thiele modulus Λ,

$$\Lambda = \frac{V_p}{S_p}\left(\frac{k''a}{D_{1,\text{eff}}}\right)^{1/2} = \frac{V_p}{S_p}b \tag{3.83}$$

where V_p is the pellet volume and S_p is the area of the pellet exterior. For a sphere $V_p/S_p = \tfrac{4}{3}\pi R^3/4\pi R^2 = R/3$. Substituting in Eq. (3.82)

$$\eta = \frac{1}{\Lambda}\left[\frac{1}{\tanh 3\Lambda} - \frac{1}{3\Lambda}\right] \tag{3.84}$$

We know from the definition of effectiveness that when the Thiele modulus is small compared to unity, η approaches unity. When the Thiele modulus is large compared to unity, Eq. (3.84) reduces to

$$\eta \simeq \frac{1}{\Lambda} \tag{3.85}$$

For nonspherical pellets Eq. (3.84) may be applied approximately by using the appropriate value for V_p/S_p in the Thiele modulus. For example, a cylindrical pellet of radius R and length L will have

$$\frac{V_p}{S_p} = \frac{\pi R^2 L}{2\pi RL + 2\pi R^2} = \frac{RL}{2(L+R)}$$

3.11.3   *EXAMPLE*   Effectiveness of a CuO-on-Alumina Pellet

The catalyst bed of an automobile exhaust reactor is packed with $\tfrac{1}{4}$ in. diameter spherical pellets of CuO on alumina. The pellets have a volume void fraction $\epsilon_v = 0.80$, tortuousity factor $\tau = 4.0$, average pore radius $r_e = 1\ \mu$, and catalytic surface per unit volume ($a = 7.12 \times 10^5$ cm^2/cm^3). Calculate the effectiveness of the pellet when promoting CO oxidation at 1000°F and 1 atm.

We first calculate the effective diffusion coefficient $D_{1,\text{eff}}$. From Eq. (3.74)

$$\mathscr{D}_{12,\text{eff}} = \frac{\epsilon_v}{\tau}\mathscr{D}_{12} = \frac{(0.8)(4.22\ \text{ft}^2/\text{hr})}{(4.0)(3600\ \text{sec/hr})} = 2.35 \times 10^{-4}\ \text{ft}^2/\text{sec}$$

where $\mathscr{D}_{12}$ is taken from Appendix B, Table B.12b, as the value for a CO-air mixture. And from Eqs. (3.75) and (3.76),

$$D_{K1,\text{eff}} = \frac{2}{3} \frac{\epsilon_v}{\tau} r_e \bar{v}_1$$

$$r_e = 1 \; \mu = 10^{-4} \text{ cm} = 3.28 \times 10^{-6} \text{ ft}$$

$$\bar{v}_1 = \left(\frac{8 \mathscr{R} T}{\pi M_1}\right)^{1/2} = \left\{\frac{(8)(1545 \text{ ft lb}_f/\text{lb-mole °R})(32.2 \text{ ft lb/lb}_f \text{ sec}^2)(1460 \text{ °R})}{(\pi)(28 \text{ lb/lb-mole})}\right\}^{1/2}$$

$$= 2.57 \times 10^3 \text{ ft/sec}$$

$$D_{K1,\text{eff}} = (2/3)(0.8/4.0)(3.28 \times 10^{-6} \text{ ft})(2.57 \times 10^3 \text{ ft/sec})$$

$$= 1.11 \times 10^{-3} \text{ ft}^2/\text{sec}$$

Substituting into Eq. (3.73) we get

$$\frac{1}{D_{1,\text{eff}}} = \left(\frac{1}{2.35} + \frac{1}{11.1}\right) 10^4$$

$$D_{1,\text{eff}} = 1.933 \times 10^{-4} \text{ ft}^2/\text{sec}$$

For CO oxidation by CuO the rate constant is obtained from chemical kinetics data as an Arrhenius relation,

$$k'' = A e^{-E_a/\mathscr{R}T} \text{ cm/sec}$$

where $A = 5.06 \times 10^3$, and the activation energy $E_a = 18.6$ kcal/gm-mole.

$$k'' = 5.06 \times 10^3 \exp\left(-\frac{(18.6 \times 10^3 \text{ cal/gm-mole})(1.8°\text{R}/°\text{K})}{(1.987 \text{ cal/gm-mole°K})(1460°\text{R})}\right)$$

$$= 5.06 \times 10^3 \, e^{-11.54}$$

$$= 4.94 \times 10^{-2} \text{ cm/sec}$$

and thus

$$ak'' = (7.12 \times 10^5 \text{ cm}^2/\text{cm}^3)(4.94 \times 10^{-2} \text{ cm/sec}) = 3.51 \times 10^4 \text{ sec}^{-1}$$

The Thiele modulus Λ may now be calculated; from Eq. (3.83)

$$\Lambda = \frac{R}{3}\left(\frac{k''a}{D_{1,\text{eff}}}\right)^{1/2} = \frac{(1/96 \text{ ft})}{3}\left(\frac{(3.51 \times 10^4 \text{ sec}^{-1})}{(1.933 \times 10^{-4} \text{ ft}^2/\text{sec})}\right)^{1/2} = 46.7$$

Since Λ is large we may use Eq. (3.85) to obtain the effectiveness

$$\eta = \frac{1}{\Lambda} = \frac{1}{46.7} = 2.14\%$$

3.11.4 Mass Transfer in a Catalyst Pellet Bed

So far in Section 3.11 we have been concerned only with diffusion inside the catalyst pellet. We now consider the coupling between the transport of reactant to the exterior surface of the pellet on the one hand and transport within the pellet itself on the other. We shall not be concerned with the overall perform-ance of the reactor bed; such matters belong more properly in Chapter 10, Mass Exchangers. Each pellet may be viewed as being surrounded by a gas flow con-taining a free-stream reactant concentration $x_{1,e}$. Mass transfer to the exterior surface of the pellet is described by Eq. (3.70), viz., $N_1 = K(x_{1,e} - x_{1,s})$, where K is the mass transfer coefficient and $x_{1,s}$ is the concentration of reactant at the s-surface depicted in Figure 3.7. As before the pellet volume is V_p and exterior surface is S_p; the catalytic area per unit volume is a. For a reaction characterized by rate constant k'', the effectiveness at the specified temperature and pressure is η.

We have two diffusive resistances in series: $R = R_o + R_i$.

1. The outside resistance R_o is simply $1/K$ (based on pellet exterior surface area S_p).
2. The inside resistance must also be based on the area S_p; the resist-ance may be expressed via the relation

$$V_p a k'' c x_{1,s} \eta = \frac{S_p}{R_i} x_{1,s}$$

where we have taken the driving potential to be $(x_{1,s} - 0)$. Solving for the resist-ance R_i we get

$$R_i = \frac{1}{(V_p a / S_p) \eta k'' c} \qquad (3.86)$$

The rate of reaction is therefore

$$N_{1,s} = \frac{(x_{1,e} - 0)}{R_o + R_i}$$

$$N_{1,s} = \frac{x_{1,e}}{\dfrac{1}{(V_p a / S_p) \eta k'' c} + \dfrac{1}{K}} \qquad (3.87)$$

The essential equivalence of Eq. (3.87) to Eq. (3.28) is made clear if K is replaced by $c \mathcal{D}_{12} / \delta$, where δ is the thickness of an equivalent stagnant film. Then

$$N_{1,s} = \frac{x_{1,e}}{\dfrac{1}{(V_p a / S_p) \eta k'' c} + \dfrac{\delta}{c \mathcal{D}_{12}}} \qquad (3.88)$$

The quantity $(V_p a / S_p)$ is the area of catalyst per unit area of pellet exterior, that is, of surface, and η is the fraction of this area that is effective in promoting the reaction.

3.11.5   EXAMPLE   CO Oxidation Rate in an Automobile Catalytic Reactor

A catalytic reactor attached to the exhaust system of an automobile is packed with $\frac{1}{4}$ in. diameter spherical CuO-on-alumina catalyst pellets. At a specific location in the bed the mass transfer coefficient K is 21.2 lb-moles/ft^2 hr, the pellet temperature is 1000°F, and the free-stream concentration of CO is 5.0% by mass. Calculate the CO oxidation rate per unit volume if the bed void fraction is 30%.

From Eq. (3.87) we have for the oxidation rate of CO per unit pellet exterior area

$$N_{CO,s} = \frac{x_{CO,e}}{\dfrac{1}{(V_p a/S_p)\eta k''c} + \dfrac{1}{K}}$$

For a spherical pellet

$$\frac{V_p}{S_p} = \frac{R}{3} = \frac{0.125/12}{3} = 3.48 \times 10^{-3} \text{ ft}$$

From Example 3.11.3

$$\eta = 0.0214 \quad \text{and} \quad ak'' = 3.51 \times 10^4 \text{ sec}^{-1}$$
$$= 1.26 \times 10^8 \text{ hr}^{-1}$$

At 1000°F and 1 atm

$$c = \frac{(14.7)(144)}{(1545)(1460)} = 9.4 \times 10^{-4} \text{ lb-moles/ft}^3$$

Since N_2 predominates we assume a mean molecular weight of 28 for the exhaust gases; $M_{CO} = 28$, so $x_{CO,e} = m_{CO,e} = 0.05$. Making use of these values we obtain

$$\left(\frac{V_p}{S_p}\right)\eta ak''c = (3.48 \times 10^{-3} \text{ ft})(0.0214)(1.26 \times 10^8 \text{ hr}^{-1})(9.4 \times 10^{-4} \text{ lb-moles/ft}^3)$$

$$= 8.8 \text{ lb-moles/ft}^2 \text{ hr}$$

$$N_{CO,s} = \frac{0.05}{\dfrac{1}{8.8} + \dfrac{1}{21.2}} = 0.31 \text{ lb-moles/ft}^2 \text{ hr}$$

In order to find the total rate at which CO is oxidized per unit volume of the reactor, we multiply by the area-to-volume ratio A_s/V_{bed},

$$\frac{A_s}{V_{bed}} = \left(\frac{NV_p}{V_{bed}}\right)\left(\frac{S_p}{V_p}\right) = (1 - \epsilon_{v,bed})\left(\frac{3}{R}\right)$$

$$= (0.70)/(3.48 \times 10^{-3}) = 201 \text{ ft}^2/\text{ft}^3$$

where N here denotes the number of pellets in the bed. The CO consumed per unit volume is then

$$\frac{\dot{M}_{CO}}{V_{bed}} = \frac{N_{CO,s}A_s}{V_{bed}}$$

$$= (0.31)(201) = 62.4 \text{ lb-moles/ft}^3 \text{ hr}$$

3.12 SUMMARY

This chapter has introduced mass diffusion. There was an explanation of how compositions and velocities are described in mixtures. Fick's law of diffusion, which can be written for mass diffusion relative to the mass average velocity or for molar diffusion relative to the molar average velocity, was presented. Steady and transient diffusion problems completely analogous to heat conduction problems were treated. Simultaneous convection and diffusion was introduced for a simple one-dimensional flow situation. The mass transfer coefficient was introduced and interpreted. Finally, catalyst pellet effectiveness was determined and the catalytic conversion process described in terms of mass transfer resistances in series.

EXERCISES

1. Derive Eqs. (3.11) and (3.12), namely

$$m_i = \frac{x_i M_i}{\sum x_j M_j}, \qquad x_i = \frac{m_i/M_i}{\sum m_j/M_j}$$

2. (a) A mixture of noble gases contains equal mole fractions of helium, argon, and xenon; what is the composition in terms of mass fractions?
    (b) If the mixture contains equal mass fractions of He, A, and Xe, what are the corresponding mole fractions?
3. Methane is burned with 20% excess air. At 2250°R the equilibrium composition of the product is:

Species i	x_i
CO_2	0.0803
H_2O	0.160
O_2	0.0325
N_2	0.727
NO	0.000118

Determine the mean molecular weight M and gas constant R of the mixture and the mass fraction of the pollutant nitric oxide in parts per million.
4. A closed cylindrical vessel stands with its axis vertical. Each end is maintained at a uniform temperature with the base colder than the upper end; the cylindrical wall is insulated. The vessel contains stagnant air. Neglecting the variation of hydrostatic pressure, ascertain whether there are vertical gradients of (a) partial density of oxygen, (b) partial pressure of oxygen, and (c) mass fraction of oxygen.
5. A spherical steel tank of 1 liter capacity and wall thickness of 1 mm is used

to store hydrogen at 400°C. The internal pressure is 132 psia and there is a vacuum outside the tank. The diffusion coefficient for hydrogen in steel is about $1.65 \times 10^{-2} e^{-E_0/\mathscr{R}T} cm^2/sec$, where $E_0 = 9200$ cal/gm-mole. At each surface the hydrogen in solution is in equilibrium with the adjoining gas phase; the mass fraction of H_2 in the steel-H_2 solution is $m_{1,s} = K\sqrt{P_{1,s}}$ where $P_{1,s}$ is the pressure in the adjoining gas phase in atmospheres and $K = \exp(-3950/T - 8.48)$ where T is °K. Calculate the rate at which the pressure drops inside the tank.

6. A worm gear cut from 0.1% C steel is to be carburized at 930°C until the carbon content is raised to 0.45% C at a depth of 0.05 cm. The carburizing gas holds the surface concentration at 1% C by weight. If $\mathscr{D} \approx 1.4 \times 10^{-7}$ cm²/sec independent of concentration,

   (a) Calculate the time required at the carburizing temperature.
   (b) What time is required at the same temperature to double this depth of penetration?
   (c) If $\mathscr{D} = 0.25 \exp(-17,400/T°K)$ cm²/sec, what temperature is required to get 0.45% C at 0.10 cm in the same time it was attained at 0.05 cm and 930°C?

7. Exhaust gases at 1000°F and 17 psia flow through an automobile catalytic reactor. The gases have a mean molecular weight of 24.0 and contain 0.13% NO by weight. If the effective rate constant k'' for the reduction of NO is 750 ft/hr at 1000°F and the mass transfer coefficient for diffusion controlled transfer is $K = 37.6$ lb/ft² hr, determine the rate of NO removal.

8. The Chapman-Enskog kinetic theory of gases shows that for ordinary diffusion in a binary mixture, the mass flux relative to the mass average velocity is

$$\mathbf{j}_1 = -\frac{\mathscr{N}^2 m_1 m_2}{\rho} \mathscr{D}_{12} \nabla \mathscr{N}_1$$

where m_1 and m_2 are the masses of the molecules, $\mathscr{N}$ is the total number density and $\mathscr{N}_1$ is the number fraction of species 1. Derive from this relation Fick's first law in the form of Eqs. (3.43) and (3.46), namely

$$\mathbf{j}_1 = -\rho \mathscr{D}_{12} \nabla m_1, \qquad \mathbf{J}_1^* = -c \mathscr{D}_{12} \nabla x_1$$

9. A gas mixture at 1 atm pressure and 300°F contains 20% H_2, 40% O_2, and 40% H_2O by weight. The absolute velocities of each species are -10 ft/sec, -2 ft/sec, and 12 ft/sec, respectively, all in the direction of the z-axis. Calculate $\mathbf{v}$ and $\mathbf{v}^*$ for the mixture. For each species calculate $\mathbf{n}_i, \mathbf{j}_i, \mathbf{N}_i$, and $\mathbf{J}_i^*$.

10. A water heat pipe, 48 in. long, 1 in. inside diameter, is intended to carry a heat load of 100 watts when the evaporator temperature is 212°F. If 30 micrograms of air leak into the pipe and the condenser cannot operate below 202°F, determine the new power transfer rating of the heat pipe.

11. Reread Section 1.2.4 and formulate the problem of finding the backward bias as a function of the backward current for a p-n junction. Use the following

notation and conventions (see Figure 1.2):

Magnitude of charge on electron: e
Positive potential: V
Location of p-material: $-L_p \leqslant z < 0$
Concentration of doped atoms per unit volume, p-material: $c_{a,p}$
Fraction of doped atoms possessing holes, p-material: x_h
Electrical conductivity in p-material: $x_h\sigma_p$
Diffusivity of holes in p-material: $\mathcal{D}_h$
Flux of holes in p-material:

$$N_h = (x_h\sigma_p/e)(-dV/dz) - (c_{a,p}\mathcal{D}_h dx_h/dz)$$

Electrical flux in p-material: $I/A = +eN_h$
Location of n-material: $0 < z \leqslant L_n$
Concentration of doped atoms per unit volume, n-material: $c_{a,n}$
Fraction of doped atoms possessing electrons, n-material: x_e
Electrical conductivity in n-material: $x_e\sigma_n$
Diffusivity of electrons in n-material: $\mathcal{D}_e$
Flux of electrons in n-material:

$$N_e = (x_e\sigma_n/e)(+dV/dz) - (c_{a,n}\mathcal{D}_e dx_e/dz)$$

Electrical flux in n-material: $I/A = -eN_e$
Boundary conditions on V: $V = 0$ and I/A given at $z = -L_p$
Required quantity: $V = V_B$ at $z = +L_n$

12. The Stefan tube is a simple device used to measure diffusion coefficients of vapor-gas mixtures in the laboratory. A liquid is contained at the bottom of a small diameter tube and evaporates into the gas. There is a flow of the gas over the tube mouth sufficient to maintain a zero concentration of the evaporating species there. The mathematical problem is therefore identical to that of Section 3.9; it is diffusion with one component stationary. Suppose a Stefan tube is to be used to investigate the effect of surface monolayers (for example, cetyl alcohol) in retarding evaporation from a water reservoir. The tube is 6 in. long and helium flows over the tube mouth. The total pressure is 3 psia and the system is maintained at 70°F by a thermostat. For calibration purposes evaporation measurements are made for a clean water surface. What is the expected evaporation rate in lb/ft² hr? Also calculate the absolute and diffusion velocities of water vapor at the s-surface.

13. The Wheeler model of a porous solid visualizes the pores as cylinders of one fixed diameter which intersect a plane at an average angle of 45°. Show that the tortuosity factor τ is 2 for this model.

14. Porous catalysts are sometimes manufactured in the form of thin plates. Show that the effectiveness of such a pellet for a first order reaction when only one side of the pellet is exposed to the reactant flow is $\eta = (1/\Lambda) \tanh \Lambda$ where Λ is the Thiele modulus. Assume an isothermal pellet and equimolar counter-diffusion of reactant and products.

15. Ammonia can be catalytically decomposed on a tungsten surface. When the ammonia is at a sufficiently high partial pressure, the catalyst surface is largely covered by adsorbed ammonia and the reaction is consequently of zero order, that is, the reaction rate is independent of ammonia concentration and is given by $Ae^{-Ea/\mathcal{R}T}$ where A has units (moles reactant/unit area-unit time]. Determine the concentration distribution in a spherical catalyst pellet promoting this reaction; assume an isothermal pellet and equimolar counter-diffusion. Hence show that if

$$\frac{V_p}{S_p}\left(\frac{k''a}{D_{1,\text{eff}}c_{1,s}}\right)^{1/2}$$

is less than $\sqrt{2/3}$, the pellet effectiveness is equal to 1.0. (*Warning*: Of course, the reaction does not occur past the radius at which all the reactant has been consumed.)

16. The catalyst bed of an automobile exhaust reactor is packed with CuO-on-alumina cylindrical pellets $\frac{1}{8}$ in. diameter and $\frac{1}{4}$ in. long. The pellets have a volume void fraction of 0.60, tortuosity factor 3.5, average pore radius $0.7\,\mu$, and catalytic surface area per unit volume $6.82 \times 10^5\,\text{cm}^2/\text{cm}^3$. Calculate the effectiveness of the pellet when promoting CO oxidation at 1 atm pressure and (a) 800°F (idle) and (b) 1400°F (full power). The activation energy of the reaction is 18.6 kcal/gm-mole and the preexponential factor in the Arrhenius relation is $5.06 \times 10^3\,\text{cm/sec}$.

REFERENCES

Diffusive Mass Transfer

Most modern treatments derive from:

> Bird, R. B., "Theory of Diffusion," *Advances in Chemical Engineering*, Vol. 1. New York: Academic, 1956. (See pp. 170 et seq.)

Essentially the same material, together with worked examples, is given by:

> Bird, R. B., W. E. Stewart, and E. N. Lightfoot, *Transport Phenomena*. New York: John Wiley, 1960. (See Chs. 16, 17, and 19.)

Convective Mass Transfer

The principles of a unified treatment are presented in:

> Spalding, D. B., "A Standard Formulation of the Steady Convective Mass Transfer Problem," *Int. J. Heat Mass Transfer*, Vol. 1, pp. 192–207, 1960.

A wide range of engineering problems are treated using the simple Reynolds flux model in:

> Spalding, D. B., *Convective Mass Transfer*. New York: McGraw-Hill, 1963.

CHAPTER 4

TRANSFER OF HEAT, MOMENTUM, AND MASS IN SIMPLE FLUID FLOWS

4.1 INTRODUCTION

In the preceding two chapters we considered heat conduction in stationary media and mass diffusion in a stationary medium or in one moving in the direction of the mass transfer. In this latter case the movement of the medium carried along the mass. Such movement can also carry heat and momentum. In this chapter we will show how a simple case of heat conduction and fluid motion in the same direction can be treated. The first topic to be taken up is the simple case of transpiration cooling, flow of heat by conduction and convection in the same direction.

Since fluids stick to the surfaces of solids, the flow of fluid relative to a body gives rise to velocity gradients. These velocity gradients in turn give rise to viscous shear stresses which act to slow down the fluid and whose equal and opposite reaction upon the body exerts a force upon it. The exerting of forces by a moving fluid can be looked upon as momentum transfer. Such momentum transfer itself is an important engineering consideration affecting the size of pumps or propulsion units and imposing structural requirements upon the solids which must sustain the forces. Furthermore, the momentum transport affects the nature of the flow field and consequently affects heat and mass transfer as well. It is thus necessary to account for momentum transfer. The second topic taken up in this chapter is Newton's law of viscosity and a simple viscous flow, Couette flow.

Convection of heat, mass, or momentum is not restricted to the directions in which the heat, mass, or momentum is conducted or diffused. Near a wall or near the edges of jets, plumes, and wakes, the convection acts more nearly at right angles to the diffusive process. Two simple illustrations of this situation are shown to be heat transfer into a laminar flow in a pipe and heat and mass transfer to a film of liquid falling down a vertical wall.

Only simple, laminar fluid flows are treated in this introduction to convective heat, mass, and momentum transfer. Chapter 5 describes more complex flows and shows how engineering calculations can be made for many commonly encountered problems. Turbulent as well as laminar flows are dealt with there, and Chapter 8 extends the treatment of turbulent flow.

4.2 TRANSPIRATION COOLING

4.2.1 The Process

Physically speaking, transpiration cooling is a process in which a fluid is injected through a porous wall whose outer surface is subjected to a deleterious thermal environment. As relatively cold fluid passes through the wall, its temperature increases as it absorbs some of the thermal energy which is being transferred to the outer surface. Depending on the temperature and mass flow rate of the incoming fluid, it is possible to effect substantial reductions in the wall temperature, that is, to "protect" the material of the wall from a severe external thermal environment. An engineering application of transpiration cooling is the design of a thermal protection system for the nose region of an entry vehicle. Entering a planetary atmosphere at high velocities, an entry vehicle is subjected to extremely high temperatures as the vehicle slows down and mechanical energy is continuously converted to thermal energy in the fluid adjacent to the vehicle surface.

4.2.2 Analysis of the Process

Consider the physical situation depicted in Figure 4.1, where the outer surface of a porous wall is subjected to a steady, uniform heat load of q_2 Btu/ft² hr. Cooling is provided by fluid injected through the pores of the wall at a rate n lb/ft² hr. At some distance from the inlet face of the wall, the fluid is at temperature $T_{1,e}$. Since heat flows by conduction from right to left in the figure, $T_2 > T_1 > T_{1,e}$. For the above situation, we ask the following questions:

1. As q_2 becomes large, is it feasible to keep T_2 below the melting point of the wall material or to keep T_1 from exceeding a pre-assigned design temperature?
2. What would be the effect on T_2, T_1, and n if additional cooling (q_1) were provided, say, by passing a refrigerant through coils brazed to the inlet face of the wall?
3. What are the practical implications of such a system?

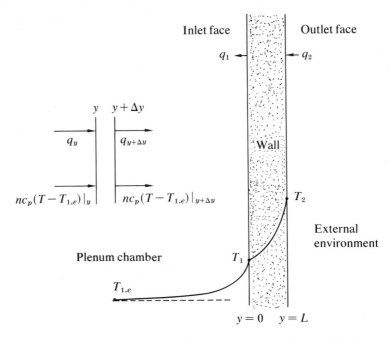

Figure 4.1   One-dimensional transpiration cooling under pre-scribed heat load at the surface of a porous wall.

We begin by making a heat balance over a volume element in the plenum chamber between planes at y and $y + \Delta y$. For an area A normal to the flow we have the following contributions to the rate at which heat is transferred across the surfaces of the element:

1. In at y by convection: $nAc_p(T - T_{1,e})|_y$
2. In at y by conduction: $q_y A$
3. Out at $y + \Delta y$ by convection: $nAc_p(T - T_{1,e})|_{y+\Delta y}$
4. Out at $y + \Delta y$ by conduction: $(q_{y+\Delta y})A$

where $c_p(T - T_{1,e})$ is the specific enthalpy, $\hat{h}$ (Btu/lb), of the fluid relative to the reference temperature $T_{1,e}$. Physically, $\hat{h}$ represents the amount of heat which is transferred to 1 lb of fluid in raising its temperature from $T_{1,e}$ to T at constant pressure. Thus $nAc_p(T_{y+\Delta y} - T_y)$ represents the rate at which heat is absorbed by the flowing fluid as it moves from y to $y + \Delta y$. This rate is due to the difference between the rates at which heat is conducted through the surfaces at y and $y + \Delta y$. At steady state there is no accumulation of thermal energy within the element, and

$$nAc_p(T_{y+\Delta y} - T_y) = -(q_{y+\Delta y} - q_y)A$$

Dividing by $A \, \Delta y$ and letting $\Delta y \to 0$ yields

$$nc_p \frac{dT}{dy} = -\frac{dq}{dy}$$

From Fourier's law, $q = -k\, dT/dy$, and

$$\frac{d^2T}{dy^2} - \frac{nc_p}{k}\frac{dT}{dy} = 0, \qquad -\infty < y < 0 \tag{4.1}$$

where we assume constant properties.

Equation (4.1) applies as well in the porous wall provided we assume that the flow is thermally in equilibrium with the solid matrix at each location y and, in addition, we replace k by k_m where k_m is the *effective* thermal conductivity for the fluid-solid mixture. Thus

$$\frac{d^2T_m}{dy^2} - \frac{nc_p}{k_m}\frac{dT_m}{dy} = 0, \qquad 0 < y < L \tag{4.2}$$

Equations (4.1) and (4.2) are subject to the boundary equations

$$T \to T_{1,e} \text{ as } y \to -\infty \tag{4.3}$$

$$T(0-) = T_m(0+) = T_1 \tag{4.4}$$

$$-k\frac{dT}{dy}\bigg|_{y=0-} = -k_m\frac{dT_m}{dy}\bigg|_{y=0+} + q_1 \tag{4.5}$$

$$T_m(L) = T_2 \tag{4.6}$$

where T_2 can be extracted from an overall enthalpy balance

$$nc_p(T_2 - T_{1,e}) = q_2 - q_1$$

that is,

$$(T_2 - T_{1,e}) = \frac{(q_2 - q_1)}{nc_p} \tag{4.7}$$

Thus we see that T_2 is reduced if heat is removed at the rate q_1 at the inlet face.

The general solution of Eq. (4.1) is seen to be

$$T = A + Bc^{nc_p y/k}, \qquad y < 0 \tag{4.8}$$

while that of Eq. (4.2) is

$$T_m = C + Dc^{nc_p y/k_m}, \qquad 0 < y < L \tag{4.9}$$

The integration constants A through D are obtained by substituting Eqs. (4.8) and (4.9) in the boundary conditions, Eqs. (4.3) through (4.6). Substituting (4.8) in (4.3) gives

$$A = T_{1,e}$$

Substituting Eqs. (4.8) and (4.9) in Eq. (4.5) gives

$$-Bnc_p = -Dnc_p + q_1$$

that is,

$$B - D = \frac{-q_1}{nc_p}$$

Equating (4.8) and (4.9) at $y = 0$ yields

$$C + D = A + B$$

that is,

$$C = A + (B - D) = T_{1,e} - \frac{q_1}{nc_p}$$

Finally, from substituting (4.9) in (4.6), there results

$$C + D e^{ncpL/km} = T_2$$

$$D = (T_2 - C)e^{-ncpL/km} = \left(T_2 - T_{1,e} + \frac{q_1}{nc_p}\right)e^{-ncpL/km}$$

which becomes on replacing $T_2 - T_{1,e}$ as given by Eq. (4.7)

$$D = \left(\frac{q_2}{nc_p}\right)e^{-ncpL/km}$$

The temperature distributions are therefore

$$T = A + Be^{ncpy/k}$$

$$= T_{1,e} + (D - q_1/nc_p)e^{ncpy/k}$$

$$= T_{1,e} + \left(\frac{q_2}{nc_p}\right)e^{ncp(y/k - L/km)} - \left(\frac{q_1}{nc_p}\right)e^{ncpy/k} \qquad (4.10)$$

for $y \leqslant 0$, and

$$T_m = C + De^{ncpy/km}$$

$$= T_{1,e} - \left(\frac{q_1}{nc_p}\right) + \left(\frac{q_2}{nc_p}\right)e^{ncp(y/km - L/km)} \qquad (4.11)$$

for $0 \leqslant y \leqslant L$.

As stated earlier, we are interested in the effects of q_1, q_2, and n on T_1 and T_2. From Eq. (4.11), applied at $y = L$,

$$T_2 = T_{1,e} + \frac{(q_2 - q_1)}{nc_p} \qquad (4.12)$$

From Eq. (4.10), applied at $y = 0$,

$$T_1 = T_{1,e} + \left(\frac{q_2}{nc_p}\right)e^{-ncpL/km} - \left(\frac{q_1}{nc_p}\right) \qquad (4.13)$$

Consider the following special cases:

1. $q_1 = 0$

$$T_1 = T_{1,e} + \left(\frac{q_2}{nc_p}\right)e^{-npL/km} \qquad (4.14a)$$

2. $T_1 = T_{1,e}$

$$q_1 = q_2 e^{-ncpL/km} \qquad (4.14b)$$

We note from Eq. (4.14b) that a substantial reduction in the heat load at $y = 0$ can be effected with transpiration; for, when $nc_p L/k_m = 2.3$, $q_1/q_2 = 1/e^{2.3} = 0.1$, and transpiration has reduced tenfold the heat load q_1 which must be provided at the inlet face if T_1 is to be maintained at $T_{1,e}$. (*Note*: If $n = 0$ and $T_1 = T_{1,e}$, $q_1 = q_2$.) Additionally, we observe from Eq. (4.10) that the plenum chamber temperature distribution decays exponentially (see Figure 4.1). Such a decay is associated with what is called a **thermal boundary layer** in the sense that significant gradients in temperature are confined to a relatively thin layer next to the solid surface. As a final note, we see that Eq. (4.7) may be rearranged to give the dimensionless ratio

$$\frac{q_2 - q_1}{nc_p(T_2 - T_{1,e})} = \frac{q_2 - q_1}{\rho v_y c_p(T_2 - T_{1,e})} = 1$$

which, in the present application, is equal to unity. On examining the ratio, we see that the numerator is a heat flux while the denominator is the product of a mass flow rate ($n = \rho v_y$) times a change in enthalpy ($c_p \Delta T$). Such a dimensionless quantity is called a **Stanton number**, St, and has the general definition

$$St \equiv \frac{q}{\rho V c_p \Delta T}$$

It is an important parameter in the subject matter of convective heat transfer. Although it has appeared here in terms of a flow normal to a conducting wall, the Stanton number usually is associated with a tangential flow over the surface of a body. As will be seen later, the Stanton number in such cases is very much less than unity, the value of unity observed here being atypical.

4.2.3 *EXAMPLE* A Fire Protection Suit

Using the principles discussed above, consider the feasibility of designing a thermal protection system for a fireman who may need to work in a hazardous thermal environment such as would exist during an oil refinery fire.

As a first cut, let us suppose that compressed air at 70°F can be continuously delivered to some sort of porous suit through a flexible asbestos hose which may itself be protected by transpiration. It is assumed that no other cooling is provided, that is, $q_1 = 0$. Further, it is supposed that the inner surface temperature, T_1, of the suit be maintained at 90°F to shield the fireman's body from thermal radiation. Thus from Eq. (4.14a)

$$(T_1 - T_{1,e}) = \left(\frac{q_2}{nc_p}\right) e^{-nc_p L/k_m}$$

where $T_1 - T_{1,e} = 20°F$, q_2 is the heat load at the outer surface of the suit, n is the mass flux of air through the suit, c_p is the heat capacity of air, L is the thickness of the porous suit wall, and k_m its effective thermal conductivity.

For an oil refinery fire, the heat load will be essentially radiation:

$$q_2 = \sigma T^4 = 1.712 \times 10^{-9} T^4 \text{ Btu/ft}^2 \text{ hr} \quad (T \text{ in } °R)$$

Estimating the flame temperature at $1300°F = 1760°R$ (red heat), we have

$$q_2 \simeq 1.65 \times 10^4 \text{ Btu/ft}^2 \text{ hr}$$

For physical properties, we have $c_{p,\text{air}} = 0.24$ Btu/lb °F and, assuming the suit to be an asbestos cloth, $k_m = 0.125$ Btu/ft hr °F. In order to not impair mobility, we choose $L = 0.125$ in. With these assumptions, we can extract n from the above equation:

$$(T_1 - T_{1,e}) = \left(\frac{q_2}{nc_p}\right) e^{-(Lq_2/k_m)/(q_2/nc_p)}$$

$$= \beta e^{-(Lq_2/k_m)/\beta}$$

where

$$\beta \equiv \frac{q_2}{nc_p}$$

Now

$$\frac{Lq_2}{k_m} = \frac{0.125 \text{ in.}}{12 \text{ in./ft}} \frac{1.65 \times 10^4 \text{ Btu/ft}^2 \text{ hr}}{0.125 \text{ Btu/ft hr °F}}$$

$$= 1375°F$$

Thus

$$(T_1 - T_{1,e}) = 20°F = \beta e^{-1375/\beta}$$

To solve by trial and error we construct a table as follows:

β, °F	$e^{-1375/\beta}$	$\beta e^{-1375/\beta}$
400	0.0322	12.9
450	0.0472	21.2
440	0.0440	19.4

Thus $\beta \simeq 443$ and

$$\rho v_y = n = \frac{q_2}{c_p \beta} = \frac{1.65 \times 10^4}{(0.24)(443)} = 155 \text{ lb/ft}^2 \text{ hr}$$

Since $\rho = 0.075$ lb/ft^3 at 70°F,

$$v_y = \frac{155 \text{ lb/ft}^2 \text{ hr}}{0.075 \text{ lb/ft}^3} \frac{\text{hr}}{3600 \text{ sec}} = 0.57 \text{ ft/sec}$$

We observe that the velocity at the inner surface is reasonably small and would not require excessive pressurization of the suit to sustain the flow. (The pressure necessary would have to be computed knowing the permeability of the cloth and is not calculated here.) But what about the outer surface temperature T_2? Applying Eq. (4.12) we obtain

$$(T_2 - T_{1,e}) = \beta = 443°F$$

$$T_2 = 443 + 70 = 513°F$$

Thus T_2 is not excessive from a materials standpoint. In conclusion, we see that such a design is feasible; however, a more sophisticated analysis would be required in practice (1) to account for the additional heat load due to metabolic heating and to account for the cooling due to perspiration, (2) to prescribe a more realistic thermal boundary condition at the outer boundary, and (3) to develop the detailed porous structure of the suit and to provide backup protection such as a suit material which would gasify for a limited time should the air supply fail.

4.3   NEWTON'S LAW OF VISCOSITY

In the previous section, we were able to incorporate an effect of fluid flow on heat transfer without concern for the fluid mechanical details of the problem. This neglect of the fluid mechanics was possible for two reasons. First, the geometrical situation was such that we could ignore the effects of confining walls on the essentially one-dimensional flow for $y < 0$. Second, under the assumption of thermal equilibrium at each $0 < y < L$, we were not interested in the detailed behavior of the flow within the porous wall. As will be seen shortly, this fortuitous state of affairs is not generally applicable. Instead, we will be faced with the task of developing, from the principles of fluid mechanics, the details of a flow and its influence on energy and mass transfer.

Consider the physical situation depicted in Figure 4.2, where a fluid — either a gas or a liquid — is confined between two parallel plates of area A. Imagine that the upper plate is impulsively set in motion with uniform velocity V in such manner as to preserve the uniform spacing L. (This might be done, for example, by considering the upper plate to be a conveyer belt or by having rotating cylinders, whose diameters are large compared to the gap between them of width L.) Now,

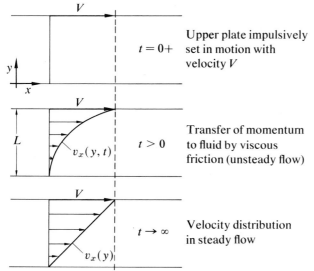

$t = 0+$ Upper plate impulsively set in motion with velocity V

$t > 0$ Transfer of momentum to fluid by viscous friction (unsteady flow)

$t \to \infty$ Velocity distribution in steady flow

Figure 4.2   The impulsively started plate.

fluid which adheres to the upper plate instantaneously assumes the velocity V; and, as time goes on, this layer of fluid as a consequence of viscous friction accelerates (drags along) fluid in layers beneath it. Eventually, a steady-state situation is achieved in which a linear velocity profile exists, the fluid adhering to the lower plate being at rest. Experimentally, it is found that a steady force F must be applied to the area A to sustain the flow, where

$$\frac{F}{A} = \frac{\mu V}{L}$$

that is, the force per unit area is proportional to the velocity increase across the gap L. The constant of proportionality μ is called the *viscosity* and is found to be dependent on the physicochemical nature of the fluid. The quantity $\nu = \mu/\rho$ is called the *kinematic viscosity*.

In the above situation, we note that momentum is transferred from a region of higher to one of lower velocity, much as heat is transferred from warmer to colder regions by means of conduction. Analogous to Fourier's law of heat conduction, we choose to rewrite the expression for F/A as follows:

$$\tau_{yx} = -\frac{F}{A} = -\mu \frac{V - 0}{L - 0} = -\mu \frac{dv_x}{dy} \qquad (4.15)$$

where τ_{yx} is called the momentum flux and dv_x/dy is the velocity gradient. Note that the momentum flux makes itself felt as a shear stress with units of force per unit area. The subscripts and sign convention are explained more fully in Appendix A, Section A.3 and in Chapter 7. (The reader is cautioned that the sign convention followed here is not universally followed.) Equation (4.15) is sometimes referred to as *Newton's Law of Viscosity*. It is applicable under circumstances for which the flow is laminar; that is, when the flow paths of a tracer material (such as smoke particles) are simple, stable geometric curves. With regard to units for the above quantities, we have

SI		Quantity		British
N/m^2	$\Longleftrightarrow$	τ_{yx}	$\Longleftrightarrow$	lb_f/ft^2
m/sec	$\Longleftrightarrow$	v_x	$\Longleftrightarrow$	ft/sec
$kg/m\ sec$	$\Longleftrightarrow$	μ	$\Longleftrightarrow$	$lb_f sec/ft^2$
m^2/sec	$\Longleftrightarrow$	ν	$\Longleftrightarrow$	ft^2/sec

Values of μ and ν for some common engineering fluids are presented in Appendix B.
Although Eq. (4.15) has been introduced under circumstances[1] for which the velocity distribution is linear, it can be generalized. (See Appendix A.)

4.4   HEAT TRANSFER IN A COUETTE FLOW

4.4.1   Temperature Profile

Consider the steady flow situation of Section 4.3 where now we allow the plates to be at dissimilar temperatures T_e and T_s as shown in Figure 4.3.

---

[1]   Flows which result from the relative motion of parallel solid surfaces are sometimes referred to as Couette flows.

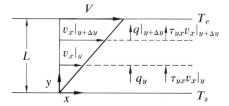

Figure 4.3   Heat transfer in a Couette flow.

As discussed above, the relative motion of the upper plate maintains the flow of the contained fluid as a consequence of fluid "friction." Our experience tells us that there ought to be a heating effect associated with this friction; or, stated another way, if work is being done on the fluid to sustain its motion, what happens to this work?

From the first law of thermodynamics, as applied to the fluid, we have in time Δt

$$\Delta \hat{u} = \hat{Q} + \hat{W} = 0$$

where for viscous heating we prefer to take work done *on* the system as being positive. Since at steady state the temperature and pressure at each point in the fluid is unchanged with respect to time, $\Delta \hat{u} = 0$ and hence $\hat{Q} + \hat{W} = 0$. For a system consisting of fluid contained between planes of area A at positions y and $y + \Delta y$ we identify the following heat and work flow rates:

1. Rate at which heat is transferred to the system by conduction at y: $q_y A$.
2. Rate at which work is done on the system by friction forces at y: $\tau_{yx} A \cdot v_x|_y$.
3. Rate at which heat is transferred from the system by conduction at $y + \Delta y$: $q_{y+\Delta y} A$.
4. Rate at which work is done by the system by friction forces at $y + \Delta y$: $\tau_{yx} A \cdot v_x|_{y+\Delta y}$.

From $\hat{Q} + W = 0$,

$$(q_y - q_{y+\Delta y}) \Delta t + (\tau_{yx} v_x|_y - \tau_{yx} v_x|_{y+\Delta y}) \Delta t = 0$$

Dividing by $\Delta t \, \Delta y$ and taking limits as $\Delta y \to 0$, we obtain

$$-\frac{dq_y}{dy} - \tau_{yx} \frac{dv_x}{dy} = 0$$

since the shear stress τ_{yx} is a constant. But $q_y = -k \, dT/dy$ and $\tau_{yx} = -\mu \, dv_x/dy$ from Fourier's law and Newton's law of viscosity. Thus the governing energy balance equation becomes

$$k \frac{d^2 T}{dy^2} + \mu \left(\frac{dv_x}{dy} \right)^2 = 0 \qquad \textbf{(4.16)}$$

Equation (4.16) is just the steady one-dimensional heat conduction equation with a heat source $\mu\Phi = \mu (dv_x/dy)^2$. The heat source $\mu\Phi$ is called **viscous dissipation** and is a mathematical statement of our conclusion that there is a heating effect associated with the frictional forces caused by the viscous nature of the fluid. Note that $\mu\Phi$ is positive and that mechanical work is being continuously converted to heat. Although the derivation here required that τ_{yx} be a constant, it can be shown that the same form applies when τ_{yx} varies with y.

It is instructive to solve Eq. (4.16) in dimensionless form. Let $V^* = v_x/V$, $y^* = y/L$, and $T^* = (T - T_s)/(T_e - T_s)$. Then

$$\frac{k}{L^2}(T_e - T_s)\frac{d^2T^*}{dy^{*2}} + \frac{\mu V^2}{L^2}\left(\frac{dV^*}{dy^*}\right)^2 = 0$$

that is,

$$\frac{k(T_e - T_s)}{\mu V^2}\frac{c_p}{c_p}\frac{d^2T^*}{dy^{*2}} + \left(\frac{dV^*}{dy^*}\right)^2 = 0$$

Since $dV^*/dy^* = 1$, the above relation becomes

$$\frac{d^2T^*}{dy^{*2}} = 2\frac{\mu c_p}{k}\frac{V^2/2}{c_p(T_s - T_e)}$$

Furthermore, d^2T^*/dy^{*2} is dimensionless and, hence, the right-hand side must also be dimensionless. Since the group

$$\frac{\mu c_p}{k} = \frac{\rho\nu c_p}{k} = \frac{\nu}{\alpha}\frac{[\text{ft}^2/\text{sec}]}{[\text{ft}^2/\text{sec}]}$$

is seen to be dimensionless, $V^2/[2c_p(T_s - T_e)]$ must also be dimensionless (when consistent units are employed). These dimensionless groups are important characteristic parameters in convective heat transfer. They are called the **Prandtl number**,

$$Pr \equiv \frac{\mu c_p}{k} = \frac{\nu}{\alpha}$$

and the **Eckert number**,

$$Ec \equiv \frac{V^2/2}{c_p(T_s - T_e)}$$

after two pioneering investigators in the field of heat transfer.

The effect of viscous dissipation on heat transfer in a Couette flow may now be determined. Since

$$\frac{d^2T^*}{dy^{*2}} = 2\,Pr\,Ec$$

we obtain, assuming constant Pr independent of temperature,

$$T^* = Pr\,Ec\,y^{*2} + C_1y^* + C_2$$

At $y^* = 1$, $T^* = 1$ and at $y^* = 0$, $T^* = 0$. Hence

$$T^* = (1 - Pr\,Ec)y^* + (Pr\,Ec)y^{*2} \tag{4.17}$$

4.4.2   Heat Flux

Once the temperature profile (dimensionless or not) has been determined, the heat flux can always be found from Fourier's law

$$q_s = -k \frac{dT}{dy}\bigg|_{y=0} = -k \frac{dT}{dy^*}\bigg|_{y^*=0} \frac{dy^*}{dy} = -\frac{k}{L} \frac{dT}{dy^*}\bigg|_{y^*=0}$$

Since

$$T^* = \frac{T - T_s}{T_e - T_s}$$

and

$$T = (T_e - T_s)T^* + T_s$$

there results

$$\frac{dT}{dy^*} = (T_e - T_s)\frac{dT^*}{dy^*}$$

Substituting for dT^*/dy^* and recalling that $Ec = (V^2/2)/c_p(T_s - T_e)$, we find

$$q_s = -k \frac{T_e - T_s}{L}(1 - Pr\,Ec + 2\,Pr\,Ec\,y^*)_{y^*=0}$$

$$q_s = -k \frac{T_e - T_s}{L} - \frac{k}{L}Pr\frac{V^2}{2c_p}$$

$$q_s = -\frac{k}{L}\left[\left(T_e + Pr\frac{V^2}{2c_p}\right) - T_s\right] \tag{4.18}$$

4.4.3   The Recovery Temperature Concept

We find that our previous solution for conduction through a slab of thickness L has been modified by replacing the upper surface temperature with a new one

$$T_r = T_e + Pr\frac{V^2}{2c_p} \tag{4.19}$$

This temperature is called the *recovery temperature*, since it is the temperature corresponding to the enthalpy $c_p T_r$ *recovered* from the total enthalpy of a unit mass of fluid near the moving plate,

$$c_p T_e + \tfrac{1}{2}V^2$$

when the stationary plate has no heating or cooling. In other words when $q_s = 0$, the temperature *recovered* on the surface at $y = 0$ is $T_s = T_r$.

For other flows, Eq. (4.19) is written in terms of a recovery factor $r(Pr)$

● $$T_r = T_e + r(Pr)\frac{V^2}{2c_p} \tag{4.20}$$

We have found for our special case of Couette flow that $r = Pr$. For laminar flow over a flat plate $r \doteq Pr^{1/2}$, and for turbulent flow $r \doteq Pr^{1/3}$.

4.4.4   *EXAMPLE* Aerodynamic Heating

Our Couette flow solution gives some insight into the phenomenon of *aerodynamic heating* which results from high speed flow over an aircraft. The skin of the aircraft is heated or cooled not according to the simple form of Newton's law of cooling, but a modified form

•
$$q = h(T_s - T_r) \tag{4.21}$$

At low speeds $T_r = T_e$, but at high speeds T_r is much hotter. Consider air at 100 mph. Note that we must use c_p in units of ft lb$_f$/slug °F, where the slug is 32.2 lb. If Pr were unity, we would have

$$T_r - T_e = \frac{V^2}{2c_p} = \frac{[(100)(88/60)\,\text{ft/sec}]^2}{2(0.24\,\text{Btu/lb°F})(32.2\,\text{ft lb/sec}^2\,\text{lb}_f)(778\,\text{ft lb}_f/\text{Btu})}$$

$$T_r - T_e = 1.79°\text{F} \doteq 1.0\,\text{C}°$$

Thus, in a fluid of $Pr = 1$, it is evident that $T_r - T_e$ is 1.0°C when $V = 100$ mph, 4°C when $V = 200$ mph flow, and 324°C in an 1800 mph flow. Such flows are, for normal air density, turbulent, and $r = (Pr)^{1/3} \doteq 0.9$, since $Pr \doteq 0.7$ for air, so the preceding values are multiplied by 0.9 to obtain the temperature rise which a wing experiences when it is neither heated nor cooled by conduction or radiation.

4.5   FULLY DEVELOPED LAMINAR FLOW IN A PIPE WITH CONSTANT WALL HEAT FLUX

4.5.1   The Velocity Profile and Momentum Transfer

Many applications of convective heat transfer involve flows in ducts, often circular tubes. The purpose of this section is to analyze the effect of a constant wall heat flux on fully developed laminar flow of an incompressible fluid in a pipe. Consider the physical situation depicted in Figure 4.4, where an incompressible fluid is flowing in a horizontal tube at a distance from the inlet which is large as compared with the diameter of the tube. (Under such circumstances, it may be shown that the flow is fully developed; that is, the axial velocity profile is unchanging with respect to z, and the radial velocity is zero.) We assume constant physical properties and neglect viscous dissipation. From the principle of conservation of momentum as applied to the element of volume contained between cylindrical surfaces with inner and outer radii of r and $r + \Delta r$, respectively, we have

[rate of momentum in] − [rate of momentum out] + [sum of forces acting] =

[rate of accumulation of momentum]

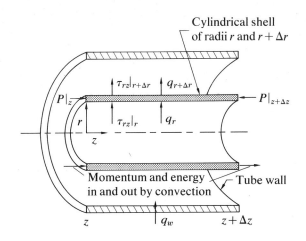

Figure 4.4   Heat and momentum balances for laminar flow in a pipe

We can regard the viscous shears as either forces or momentum fluxes. Listing the individual contributions to z-momentum, we have

rate of momentum in at r (viscous force): $(\tau_{rz}2\pi r\,\Delta z)|_r$
rate of momentum out at $r+\Delta r$ (viscous force): $(\tau_{rz}2\pi r\,\Delta z)|_{r+\Delta r}$
rate of momentum in at z (convection): $(\rho v_z 2\pi r\Delta r)(v_z)|_z$
rate of momentum out at $z+\Delta z$ (convection): $(\rho v_z 2\pi r\,\Delta r)(v_z)|_{z+\Delta z}$
pressure force acting at z: $+(2\pi r\,\Delta r)(P)|_z$
pressure force acting at $z+\Delta z$: $-(2\pi r\,\Delta r)(P)|_{z+\Delta z}$

(Note that we again adopt the sign convention that momentum and forces are positive in the positive directions of the axes.) Since the fluid is assumed to be incompressible and the flow fully developed, v_z does not vary with z, and terms three and four cancel identically. Summing the other terms (recognizing that for steady flow there is no accumulation of momentum):

$$\lim_{\Delta r\to 0}\frac{r\tau_{rz}|_r-r\tau_{rz}|_{r+\Delta r}}{\Delta r}+\lim_{\Delta z\to 0}r\frac{P|_z-P|_{z+\Delta z}}{\Delta z}=0$$

that is,

$$-\frac{d}{dr}(r\tau_{rz})-r\frac{dP}{dz}=0$$

or

$$\frac{1}{r}\frac{d}{dr}(r\tau_{rz})=-\frac{dP}{dz} \tag{4.22}$$

where we have additionally assumed that the variation of the hydrostatic pressure P with position (r,θ) due to gravity is negligible. Now, for steady flow and neglecting entrance and exit effects, the pressure gradient dP/dz is a constant, say $(P_L-P_0)/L$, where L is a representative length of pipe in the fully developed section. Integrating Eq. (4.22), we find

$$\tau_{rz}=\frac{P_0-P_L}{2L}r+\frac{C_1}{r}$$

Since we do not anticipate infinite shear at $r = 0$, $C_1 = 0$. Also, for this case, Newton's law of viscosity is

$$\tau_{rz} = -\mu \frac{dv_z}{dr}$$

and a second integration yields

$$v_z = -\frac{P_0 - P_L}{4\mu L} r^2 + C_2$$

With no slip at the tube wall (fluid sticks to the wall), $v_z = 0$ at $r = R$, and

$$C_2 = \frac{(P_0 - P_L) R^2}{4\mu L}$$

Finally,

$$v_z = \frac{(P_0 - P_L) R^2}{4\mu L} \left[1 - \left(\frac{r}{R} \right)^2 \right] \tag{4.23}$$

which shows that the velocity distribution in the tube is parabolic, with its maximum value occurring at $r = 0$:

$$v_{z,\text{max}} = \frac{(P_0 - P_L) R^2}{4\mu L} \tag{4.24}$$

The reader may show that the area-weighted average value is just

$$v_{z,\text{avg}} = \tfrac{1}{2} v_{z,\text{max}} = \frac{(P_0 - P_L) R^2}{8\mu L} \tag{4.25}$$

We now consider an overall force balance on the fluid between $z = 0$ and $z = L$. Clearly, the net applied force,

$$F \equiv (P_0 - P_L) \pi R^2$$

just balances the viscous drag at the wall, that is,

$$F = (2\pi R L) \left(-\mu \frac{dv_z}{dr} \right) \Big|_{r=R}$$

$$= (2\pi R L) \left[\frac{(P_0 - P_L) R^2}{4L} \right] \left[\frac{2}{R} \right]$$

$$= (P_0 - P_L) \cdot \pi R^2$$

Anticipating the need for treating more general flows for which analytical solutions are not available, we now introduce the idea of a **friction factor** f defined by the relation

$$\bullet \qquad\qquad P_0 - P_L = f \frac{L}{D} \left(\tfrac{1}{2} \rho v_{z,\text{avg}}^2 \right) \tag{4.26}$$

where D is the diameter of the duct. Substituting $v_{z,\text{avg}}$ from Eq. (4.25) into Eq.

(4.26) gives

$$f = \frac{64}{(\rho(2R)v_{z,\text{avg}}/\mu)} = \frac{64}{(\rho D v_{z,\text{avg}}/\mu)}$$

Examining the denominator, we see that it is dimensionless,

$$\frac{\rho D v_{z,\text{avg}}}{\mu} = \frac{v_{z,\text{avg}} D}{(\mu/\rho)} \frac{[\text{ft/sec}][\text{ft}]}{[\text{ft}^2/\text{sec}]}$$

Thus another characteristic dimensionless group appears. It is called the **Reynolds number**, after Osborne Reynolds. Its role here is one of scaling frictional resistance to flow:

- $$Re = \frac{VD}{(\mu/\rho)} = \frac{VD}{\nu} \tag{4.27}$$

- $$f = \frac{64}{Re} \tag{4.28}$$

Often a skin friction coefficient c_f is used instead of or in addition to f

$$\tau_s = -\tau_{rz} = c_f \tfrac{1}{2}\rho v_{z,\text{avg}}^2$$

For pipe flow a force balance equating the pressure difference cross-sectional area product to the shear stress side area product shows $c_f = f/4$. This relation is true whether the flow is laminar or turbulent.[2]

4.5.2   The Temperature Profile and Heat Transfer

We turn now to the heat transfer problem of finding the wall temperature T_w when a uniform heat flux q_w is applied at the tube wall. Consider an element of volume between radii r and $r + \Delta r$. Contributions to a heat balance over the surfaces of the annular element are:

rate of heat transfer in at r (conduction): $(2\pi r \, \Delta z)(q_r)|_r$
rate of heat transfer out at $r + \Delta r$ (conduction): $(2\pi r \, \Delta z)(q_r)|_{r+\Delta r}$
rate of heat transfer in at z (convection): $(\rho v_z 2\pi r \, \Delta r)[c_p(T - T_i)]|_z$
rate of heat transfer out at $z + \Delta z$ (convection):
$$(\rho v_z 2\pi r \, \Delta r)[c_p(T - T_i)]|_{z+\Delta z}$$

where T_i is an arbitrary reference temperature (say the inlet temperature). From conservation of energy

$$\lim_{\Delta z \to 0} r v_z \frac{\rho c_p(T|_{z+\Delta z} - T|_z)}{\Delta z} = -\lim_{\Delta r \to 0} \frac{r q_r|_{r+\Delta r} - r q_r|_r}{\Delta r}$$

---

[2]   Turbulent flows are characterized by fluctuations in the local velocity field (recall the familiar breakup of smoke streaming from the tip of a lighted cigarette). Transition to turbulence depends, among other things, on the Reynolds number. For pipe flows, transition from laminar to turbulent flow occurs at a Reynolds number of about 2300.

that is,

$$\rho v_z c_p \frac{\partial T}{\partial z} = -\frac{1}{r}\frac{\partial}{\partial r}(rq_r)$$

Fourier's law for this case is $q_r = -k\,\partial T/\partial r$ and hence

$$\rho v_z c_p \frac{\partial T}{\partial z} = \frac{k}{r}\frac{\partial}{\partial r}\left(r\frac{\partial T}{\partial r}\right) \tag{4.29}$$

Equation (4.29) is a partial differential equation where T is a function of both z and r. However, for fully developed flow the temperature profiles (for q_w = constant) increase linearly in z, and T can be split into the sum of a z-dependent term $T_b(z)$ and an r-dependent one $T_r(r)$. ($T_r(r)$ is not to be confused with the recovery temperature of Section 4.4.3.) Consider a heat balance over an element of tube Δz long in the form

$$q_w \cdot 2\pi R\,\Delta z = \rho\pi R^2 v_{z,\text{avg}} c_p(T_b|_{z+\Delta z} - T_b|_z), \qquad \frac{dT_b}{dz} = \frac{2q_w}{\rho c_p R v_{z,\text{avg}}} \tag{4.30}$$

where T_b is the *bulk temperature* and physically is just the temperature which would result if at a given location z the flow were diverted to an adiabatic mixing chamber and allowed to come to thermal equilibrium. Mathematically speaking, T_b is defined locally by

$$T_b(z) = \frac{1}{\rho c_p v_{z,\text{avg}}\pi R^2}\int_0^R \rho c_p v_z(r) T(r, z) 2\pi r\,dr \tag{4.31}$$

From Eq. (4.30), it is seen that T_b varies linearly with z in known manner, and Eq. (4.29) can be simplified by replacing $\partial T/\partial z$ with dT_b/dz and $\partial T/\partial r$ with dT_r/dr.

$$\frac{k}{r}\frac{d}{dr}\left(r\frac{dT_r}{dr}\right) = \rho c_p v_z \frac{2q_w}{\rho c_p R v_{z,\text{avg}}} = v_z \frac{2q_w}{R v_{z,\text{avg}}}$$

Substituting for v_z (Eqs. (4.23) and (4.25)) gives

$$\frac{1}{r}\frac{d}{dr}\left(r\frac{dT_r}{dr}\right) = 2v_{z,\text{avg}}\left[1 - \left(\frac{r}{R}\right)^2\right]\frac{2q_w}{R v_{z,\text{avg}}k}$$

$$= \frac{4q_w}{kR}\left[1 - \left(\frac{r}{R}\right)^2\right]$$

Integrating gives

$$r\frac{dT_r}{dr} = \frac{4q_w}{kR}\left[\frac{r^2}{2} - \frac{r^4}{4R^2}\right] + C_1$$

$$\frac{dT_r}{dr} = \frac{4q_w}{kR}\left[\frac{r}{2} - \frac{r^3}{4R^2}\right] + \frac{C_1}{r}$$

$$= \frac{4q_w}{kR}\left[\frac{r}{2} - \frac{r^3}{4R^2}\right]$$

since $\partial T/\partial r \ (= dT_r/dr) = 0$ at $r = 0$. Integrating again, we obtain

$$T_r = \frac{4q_w}{kR}\left[\frac{r^2}{4} - \frac{r^4}{16R^2}\right] + C_2, \qquad T = T_b(z) + \frac{4q_w}{kR}\left[\frac{r^2}{4} - \frac{r^4}{16R^2}\right] + C_2$$

At $r = 0$, $T = T_0$, and we can write

$$T = T_0 + \frac{q_w r^2}{kR}\left[1 - \frac{(r/R)^2}{4}\right] \tag{4.32}$$

(Note that T_0 is independent of r but varies with z in the same manner as T_b along the center line of the tube.) Next, we define a heat transfer coefficient in the form

$$h = \frac{q_w}{T_w - T_b} \tag{4.33}$$

where we choose to use the bulk temperature as a reference temperature rather than T_0. From Eq. (4.31)

$$T_b = \frac{\displaystyle\int_0^R \left\{T_0 + \frac{q_w r^2}{kR}\left[1 - \frac{(r/R)^2}{4}\right]\right\}\{2v_{z,\text{avg}}[1 - (r/R)^2]\}2\pi r \, dr}{v_{z,\text{avg}}\pi R^2}$$

and, after some algebra, we find

$$T_b = T_0 + \frac{7}{24}\frac{q_w R}{k} \tag{4.34}$$

From Eq. (4.32), at $r = R$,

$$T_w = T_0 + \frac{3}{4}\frac{q_w R}{k} \tag{4.35}$$

and combining Eqs. (4.33) through (4.35) gives

$$h = q_w\Big/\left(\frac{11}{24}\frac{q_w R}{k}\right) = 1\Big/\left(\frac{11}{48}\frac{2R}{k}\right)$$

or, rearranging,

$$\frac{h(2R)}{k} = \frac{hD}{k} = \frac{48}{11}(= 4.364) \tag{4.36}$$

$$Nu = \frac{48}{11}$$

The dimensionless group hD/k is yet another parameter in convective heat transfer. It is called the **Nusselt number**, and the single numerical value obtained here is a consequence of (1) the simple nature of the fluid flow problem, (2) the thermal condition at the wall, and (3) constant physical properties. (The choice of reference temperature T_b in defining h is appropriate because it is the value which can be easily extracted from a heat balance.) For more complex circumstances (as discussed in Chapter 5), the Nusselt number is a function of other dimensionless groups such as the Reynolds number (Re) and the Prandtl number (Pr).

4.5.3   *EXAMPLE* **Preheater for a Tubular Reactor**

A reactant mixture (essentially water) at 80°F is to be preheated up-stream of a small tubular reactor in a research laboratory. The reactor is to operate at 180°F and requires 10 lb/hr of reactant mixture. Assuming that the preheater is 0.1875 in. inside diameter tubing wound with nichrome heating ribbon over heater tape, estimate (1) the length of preheater required and (2) the maximum temperature at the exit, if the maximum power output of the heater is 50 watts per running foot of preheater tubing.

The physical properties of water at an intermediate temperature (140°F) are (see Appendix B, Table B.7)

$$\rho = 61.4 \text{ lb/ft}^3$$
$$\nu = 5.13 \times 10^{-6} \text{ ft}^2/\text{sec} = 0.01847 \text{ ft}^2/\text{hr}$$
$$c_p = 1.0 \text{ Btu/lb °F}$$
$$k = 0.378 \text{ Btu/ft hr °F}$$

For $D = 0.1875$ in. $= 0.01563$ ft and $\dot{m} = 10$ lb/hr, the Reynolds number for the flow is

$$Re = \frac{\rho D v_{z,\text{avg}}}{\mu} = \frac{4}{\pi D \mu} \rho \frac{\pi D^2}{4} v_{z,\text{avg}} = \frac{4\dot{m}}{\pi D \mu} = \frac{4\dot{m}}{\pi D \rho \nu}$$

$$Re = \frac{(4)(10 \text{ lb/hr})}{(3.14)(0.01563 \text{ ft})(61.4 \text{ lb/ft}^3)(0.01847 \text{ ft}^2/\text{hr})} = 720$$

It is seen that the flow is laminar ($Re < 2300$).

Neglecting heat losses to the surroundings and assuming axial conduction along the preheater is not appreciable, we have

$$q_w = \frac{50 \text{ watts}}{\text{ft}} \frac{3.41 \text{ Btu}}{\text{watt hr}} \frac{1}{(0.01563\pi) \text{ ft}} = 3470 \text{ Btu/ft}^2 \text{ hr}$$

From Eq. (4.30),

$$\frac{dT_b}{dz} = \frac{2q_w}{\rho c_p R v_{z,\text{avg}}} = \frac{2q_w}{c_p} \frac{\pi R}{\rho \pi R^2 v_{z,\text{avg}}} = \frac{\pi D q_w}{c_p \dot{m}}$$

Integrating between $z = 0$ and $z = L$,

$$L = \frac{\dot{m} c_p \Delta T_b}{\pi D q_w} = \frac{(10 \text{ lb/hr})(1 \text{ Btu/lb °F})(100°F)}{\pi (0.01563)(\text{ft})(3470 \text{ Btu/ft}^2 \text{ hr})} = 5.87 \text{ ft}$$

From Eqs. (4.34) and (4.35),

$$T_w = T_b + \frac{11}{48} \frac{q_w D}{k}$$

$$= 180 + \frac{11}{48} \frac{(3470 \text{ Btu/ft}^2 \text{ hr})(0.01563 \text{ ft})}{(0.378 \text{ Btu/ft hr °F})}$$

$$= 212°F$$

So we note the water at the wall just reaches the boiling temperature at the exit.

4.6 CONVECTIVE HEAT TRANSFER WITH CHANGE OF PHASE

4.6.1 The Process of Laminar Film Condensation

In many applications, convective heat transfer is accompanied by a change in phase. Such processes occur in condensation, evaporation, and boiling. Usually, analysis of these physical situations involves complicated coupling between the conservation equations in adjacent phases. Occasionally, however, it is possible to *decouple* the problem, that is, to analyze the transport phenomena in a single phase; and we offer the classic example, the Nusselt condensation problem of laminar film condensation of a pure saturated quiescent vapor. Consider the physical situation illustrated in Figure 4.5, where a saturated vapor is condensing on a vertical wall. As vapor condenses and accumulates at the wall, a film is formed (provided good wetting occurs) which flows down the wall under the action of the gravitational force. If the condensation rate is not too high, the flow behavior is laminar. In addition, it can be demonstrated that convection of momentum and thermal energy, even in the x-direction, can be ignored.

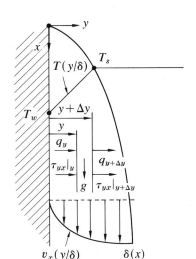

Figure 4.5   Laminar film condensation on a vertical wall.

4.6.2 The Velocity and Temperature Profiles

Applying conservation of momentum to an element of the flow between planes at y and $y + \Delta y$ at position x, the contributions to the momentum balance are just

momentum flux in at y by viscous action: $\tau_{yx} \cdot A|_y$

momentum flux out at $y + \Delta y$ by viscous action: $\tau_{yx} \cdot A|_{y+\Delta y}$

body force action: $(\rho - \rho_v)gA\, \Delta y$

where the third term includes a correction for the buoyant force of the quiescent vapor, that is, the hydrostatic pressure gradient imposed by the vapor. Summing the results and rearranging gives

$$\lim_{\Delta y \to 0} \frac{\tau_{yx}|_y - \tau_{yx}|_{y+\Delta y}}{\Delta y} + (\rho - \rho_v)g = 0$$

that is,

$$-\frac{d\tau_{yx}}{dy} + (\rho - \rho_v)g = 0$$

Substituting for τ_{yx} from Eq. (4.15), we obtain the governing momentum equation

$$\mu \frac{d^2 v_x}{dy^2} + (\rho - \rho_v)g = 0 \tag{4.37}$$

Similarly, the reader may show by using the same procedure used to develop Eq. (2.10) that the defining equation for conservation of thermal energy is just

$$\frac{d^2 T}{dy^2} = 0 \tag{4.38}$$

Equations (4.37) and (4.38) represent the local behavior of the film at any position x. Additionally, there is the requirement that the overall rate at which heat is being transferred to the wall between $x = 0$ and x is just equal to the rate at which heat is released by the condensation process and subsequent subcooling of the condensate below the saturation temperature T_s; that is,

$$\int_0^x k \frac{dT}{dy}\bigg|_{y=0} dx = \int_0^{\delta(x)} \rho v_x [\hat{h}_{fg} + c_p(T_s - T)] \, dy$$

or approximately,

$$\int_0^x \frac{dT}{dy}\bigg|_{y=0} dx \doteq \frac{\rho \hat{h}_{fg}}{k} \int_0^{\delta(x)} v_x \, dy \tag{4.39}$$

where we assume constant properties throughout and neglect sensible heat effects. (Since the latent heat $\hat{h}_{fg}$ is usually much greater than the $c_p(T_s - T_w)$, the latter assumption is sound. For example, $\hat{h}_{fg}$ for steam is about 970 Btu/lb, while $c_p(T_s - T_w)$ is, for most applications, less than 40 Btu/lb.)

Equation (4.39) serves to extract the film thickness $\delta(x)$ from the overall analysis. The procedure is as follows: First, obtain the velocity distribution from Eq. (4.37),

$$\frac{dv_x}{dy} = -\frac{(\rho - \rho_v)g}{\mu} y + C_1$$

$$= \frac{(\rho - \rho_v)g}{\mu}(\delta - y)$$

$$v_x = \frac{(\rho - \rho_v)g}{\mu}\left(\delta y - \frac{y^2}{2}\right) + C_2$$

$$= \frac{(\rho - \rho_v)g}{\mu}\left(\delta y - \frac{y^2}{2}\right) \tag{4.40}$$

where $C_1 = (\rho - \rho_v) g \delta / \mu$ assuming zero shear at $y = \delta$, and $C_2 = 0$ since $v_x = 0$ at $y = 0$. Note that for a given δ, the film surface velocity is

$$v_{x,\delta} = \frac{(\rho - \rho_v) g}{2\mu} \delta^2 \doteq \frac{\rho g}{2\mu} \delta^2 = \frac{g \delta^2}{2\nu} \tag{4.41}$$

while the mass flow rate per unit width (z-direction) of film, Γ, is

$$\Gamma = \int_0^\delta \rho v_x \, dy \doteq \frac{\rho g \delta^3}{3\nu} \tag{4.42}$$

Next, extract the temperature distribution from Eq. (4.38) and the boundary conditions.

$$T = T_w + (T_s - T_w)(y/\delta) \tag{4.43}$$

4.6.3 Film Thickness

Substituting Eqs. (4.40) and (4.43) into (4.39) gives

$$\int_0^x \frac{dx}{\delta} = \frac{\rho \hat{h}_{fg}}{k(T_s - T_w)} \frac{(\rho - \rho_v) g}{\mu} \int_0^\delta \left(\delta y - \frac{y^2}{2} \right) dy$$

$$\int_0^x \frac{dx}{\delta} = \frac{\rho(\rho - \rho_v) g \hat{h}_{fg} \delta^3}{3\mu k(T_s - T_w)} \tag{4.44}$$

Differentiating Eq. (4.44),

$$\frac{dx}{\delta} = \frac{\rho(\rho - \rho_v) g \hat{h}_{fg}}{\mu k(T_s - T_w)} \delta^2 \, d\delta$$

or

$$dx = \frac{\rho(\rho - \rho_v) g \hat{h}_{fg}}{4\mu k(T_s - T_w)} d\delta^4$$

Finally,

$$\delta = \left\{ \frac{4x\mu k(T_s - T_w)}{\rho(\rho - \rho_v) g \hat{h}_{fg}} \right\}^{1/4} \tag{4.45}$$

Thus the film thickness δ is seen to grow like $x^{1/4}$. The heat flux at the wall is

$$q_w = -k \frac{dT}{dy} \bigg|_{y=0}$$

$$= -k \left(\frac{T_s - T_w}{\delta} \right)$$

Defining a condensation heat transfer coefficient with the above sign convention,

$$q_w = h(T_w - T_s)$$

we have simply

$$h = \frac{k}{\delta} \tag{4.46}$$

4.6.4   Expressing the Answer in Dimensionless Form

Rearranging Eq. (4.46) and substituting for δ, Eq. (4.45),

$$\frac{hx}{k} = \frac{x}{\delta}$$

$$\frac{hx}{k} = \left(x^4 \, \frac{\rho(\rho-\rho_v)g\hat{h}_{fg}}{4x\mu k(T_s - T_w)}\right)^{1/4}$$

$$= \left(\frac{x^3(\rho-\rho_v)g}{4\rho\nu^2} \, \frac{\mu c_p}{k} \, \frac{\hat{h}_{fg}}{c_p(T_s - T_w)}\right)^{1/4}$$

Again, we see appearance of a Nusselt number (this time in the context of a convective heat transfer problem with change of phase)

$$Nu_x \equiv \frac{hx}{k} \tag{4.47}$$

a Prandtl number,

$$Pr \equiv \frac{\mu c_p}{k} = \frac{\nu}{\alpha} \tag{4.48}$$

and, for this problem, two additional dimensionless groups. One is the **Grashof number**,

$$Gr_x = \frac{x^3(\rho-\rho_v)g}{\rho\nu^2} = \frac{(\Delta\rho/\rho)gx^3}{\nu^2} \tag{4.49}$$

which will be seen later to be important in all gravity-driven convection problems. The other has, on occasion, been called the **Jakob number**,

$$Ja = \frac{c_p(T_s - T_w)}{\hat{h}_{fg}} \tag{4.50}$$

In terms of the above dimensionless groups, the local Nusselt number for laminar film condensation heat transfer for a saturated vapor on a vertical flat plate is

$$\bullet \qquad Nu_x = \left(\frac{Gr_x Pr}{4Ja}\right)^{1/4} \tag{4.51}$$

4.6.5   *EXAMPLE* Condensation on a Vertical Tube

Steam condenses on the outer surface of a 6.0 in. outside diameter vertical cylinder 1 ft long. Cooling water on the inside of the cylinder maintains the outer surface of the cylinder at 80°F. If the saturation temperature is 100°F, calculate (1) the overall heat transfer coefficient, (2) the total heat transfer rate, and (3) the total condensation rate.

From Eq. (4.51) the local heat transfer coefficient as a function of x/L may be found.

$$Nu_x = h(x)x/k, \qquad h(x) = (k/L)Nu_x(L/x)$$

$$h(x) = (k/L)(Gr_x Pr/4Ja)^{1/4}(L/x)$$

$$Gr_x = \frac{(\Delta\rho/\rho)gx^3}{\nu^2} = \frac{(\Delta\rho/\rho)gL^3}{\nu^2}\left(\frac{x}{L}\right)^3 = Gr_L\left(\frac{x}{L}\right)^3$$

$$h(x) = (k/L)(Gr_L Pr/4Ja)^{1/4}(x/L)^{-1/4} = h_L(x/L)^{-1/4}$$

The average heat transfer coefficient is simply

$$h_{avg} = \frac{1}{L}\int_0^L h(x)\,dx = \int_0^1 h(x)\,d\left(\frac{x}{L}\right)$$

$$h_{avg} = h_L \int_0^1 \left(\frac{x}{L}\right)^{-1/4} d\left(\frac{x}{L}\right)$$

$$h_{avg} = \frac{4}{3}h_L = \frac{4}{3}\left(\frac{k}{L}\right)\left(\frac{Gr_L Pr}{4Ja}\right)^{1/4}$$

For water at the average film temperature (90°F), Table B.7 gives $\rho = 62.1$ lb/ft³, $c_p = 0.997$ Btu/lb °F, $\nu = 0.828 \times 10^{-5}$ ft²/sec, $k = 0.359$ Btu/hr ft °F, and $Pr = 5.12$. For the saturated vapor at 100°F Table B.11 gives $\hat{h}_{fg} = 1036.4$ Btu/lb and $\rho_v = 1/350.8 = 0.00285$ lb/ft³, a negligible value compared to ρ. We find

$$Gr_L = \frac{(62.1/62.1)(32.2)(1^3)}{(0.828 \times 10^{-5})^2} = 0.470 \times 10^{12}$$

$$Pr = 5.12$$

$$Ja = \frac{(0.997)(20)}{1036.4} = 0.01927$$

$$h_{avg} = \frac{4}{3}(0.359/1)[(0.470 \times 10^{12})(5.12)/(4)(0.01927)]^{1/4}$$

$$h_{avg} = \frac{4}{3}(849) = 1131 \text{ Btu/hr ft}^2\,°F$$

Once h_{avg} is found parts (2) and (3) are easy. The total heat transfer is

$$\dot{Q} = h_{avg}\pi DL(T_s - T_w)$$
$$= (1131)(3.14)(0.5)(1.0)(20) = 35{,}500 \text{ Btu/hr}$$

and the total condensation rate is

$$\dot{m} = \frac{\dot{Q}}{\hat{h}_{fg}}$$

$$\dot{m} = \frac{35{,}500}{1036.4} = 34.3 \text{ lb/hr}$$

4.7   DIFFUSION INTO A FALLING FILM

4.7.1   The Process

Gas absorption by a falling film is used widely in industrial practice. Also the mathematical analysis of this problem forms the basis of the so-called *penetration theory* of mass transfer, a concept which has been extensively used in chemical engineering applications. Figure 4.6 depicts a laminar film of liquid flowing down a vertical surface.

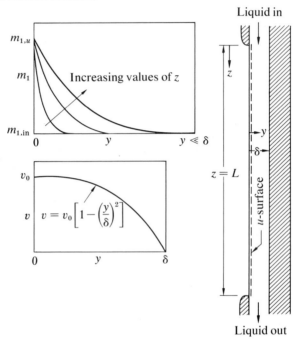

Figure 4.6   Gas absorption by a falling film: velocity profile and dissolved gas concentration profiles.

As in the preceding simple flow which dealt with a falling film, it is assumed that the Reynolds number of the film ($4/\mu \times$ mass flow rate/unit surface width) is low enough to ensure absence of ripples on the liquid surface; this restriction is usually sufficient to preclude the onset of turbulence as well. Further, it is assumed that the rate of absorption is slow enough that gas 1 does not penetrate very far into the film, even at the end L of the contact length. Also, the concentration of gas in the liquid is small enough that its effect on the density and the viscosity of the liquid phase may be neglected. Specification of an isothermal system then enables assumption of constant properties throughout. The transport of gas 1 through the gas phase is assumed to be rapid enough to maintain its concentration at the s-surface, the gas side of the gas-liquid interface, at the constant bulk value; the assumption of thermodynamic equilibrium and Henry's law (to be discussed in Chapter 10) then specifies $m_{1,u}$, the concentration of dissolved gas at the u-surface, the liquid side just under the interface. The drag of the gas phase on the liquid film is assumed negligible so that the velocity gradient in the liquid at the interface is zero. The velocity profile in the liquid is assumed to be fully developed at the start of the contact length, $z = 0$. Since the rate of absorption is low, the mass flow in the y-direction and associated velocity component v_y are negligible.

Combining Eqs. (4.40) and (4.41) of Section 4.6.1, it may easily be shown that the velocity distribution is

$$v_z = v_0 \left[1 - \left(\frac{y}{\delta}\right)^2\right] = \left(\frac{g\delta^2}{2\nu}\right)\left[1 - \left(\frac{y}{\delta}\right)^2\right] \tag{4.52}$$

where now y is measured from the surface of the film, of thickness δ, and v_0 is the velocity at $y = 0$, that is, at the interface.

Although a timewise steady state prevails, the problem is complicated by the fact that mass transport takes place in two spatial directions. Thus we cannot use the conservation of species equation derived in the one-dimensional form of Eq. (3.49); instead we go back to first principles and perform a species mass balance on an elemental volume located between y and $y + dy$, and z and $z + dz$, as shown in Figure 4.7. Species 1 can cross the element boundaries by diffusion and convection. Whenever v_z and z are large, diffusion in the z-direction will be small compared to convection, and may be ignored. Appendix A, Section A.8.3, gives a more rigorous argument for dropping the streamwise diffusion term. Conservation of species 1 requires that the flow of species 1 into the element equals the outflow,

$$\rho m_1 v_z|_z \, dx \, dy - \rho \mathcal{D}_{12} \frac{\partial m_1}{\partial y}\bigg|_y \, dx \, dz = \rho m_1 v_z|_{z+dz} \, dx \, dy - \rho \mathcal{D}_{12} \frac{\partial m_1}{\partial y}\bigg|_{y+dy} \, dx \, dz$$

Expanding the terms evaluated at $y + dy$ and $z + dz$ in Taylor series gives

$$\rho m_1 v_z \, dx \, dy - \rho \mathcal{D}_{12} \frac{\partial m_1}{\partial y} \, dx \, dz = \rho m_1 v_z \, dx \, dy + \rho v_z \frac{\partial m_1}{\partial z} \, dx \, dy \, dz + \cdots$$

$$- \rho \mathcal{D}_{12} \frac{\partial m_1}{\partial y} \, dx \, dz - \rho \mathcal{D}_{12} \frac{\partial^2 m_1}{\partial y^2} \, dx \, dy \, dz + \cdots$$

Canceling terms, dividing by $dx \, dy \, dz$, and then letting dx, dy, and dz approach zero yields

$$v_z \frac{\partial m_1}{\partial z} = \mathcal{D}_{12} \frac{\partial^2 m_1}{\partial y^2} \tag{4.53}$$

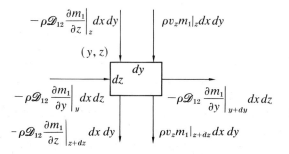

$$-\rho\mathcal{D}_{12}\frac{\partial m_1}{\partial z}\bigg|_z dx\,dy \qquad \rho v_z m_1|_z dx\,dy$$

$$(y, z)$$

$$-\rho\mathcal{D}_{12}\frac{\partial m_1}{\partial y}\bigg|_y dx\,dz \qquad -\rho\mathcal{D}_{12}\frac{\partial m_1}{\partial y}\bigg|_{y+dy} dx\,dz$$

$$-\rho\mathcal{D}_{12}\frac{\partial m_1}{\partial z}\bigg|_{z+dz} dx\,dy \qquad \rho v_z m_1|_{z+dz} dx\,dy$$

Figure 4.7 Species mass balance for an element of a falling film.

Introducing the velocity profile from Eq. (4.52) gives

$$v_0\left[1-\left(\frac{y}{\delta}\right)^2\right]\frac{\partial m_1}{\partial z} = \mathscr{D}_{12}\frac{\partial^2 m_1}{\partial y^2} \tag{4.54}$$

But the small penetration assumption allows the velocity profile to be approximated by $v_z = v_0$, a constant. The differential equation governing the concentration distribution then becomes

$$\frac{\partial m_1}{\partial z} = \frac{\mathscr{D}_{12}}{v_0}\frac{\partial^2 m_1}{\partial y^2} \tag{4.55}$$

subject to the initial condition,

$z = 0$, $m_1 = m_{1,\text{in}}$ where $m_{1,\text{in}}$ is the concentration of the gas in the inlet liquid,

and the boundary conditions,

$y = 0$, $m_1 = m_{1,u}$

$y \to \infty$, $m_1 \to m_{1,\text{in}}$ since the penetration distance is small.

The mathematical problem is identical to that for transient conduction or diffusion in a semi-infinite solid with $\mathscr{D}_{12}/v_0$ replacing α in the former case and $\mathscr{D}_{12}$ in the latter. Thus the solution may be written down immediately as

$$\frac{m_1 - m_{1,\text{in}}}{m_{1,u} - m_{1,\text{in}}} = \text{erfc}\,(\eta), \qquad \eta = \frac{y}{(4\mathscr{D}_{12}z/v_0)^{1/2}} \tag{4.56}$$

4.7.2   The Mass Transfer Coefficient for a Falling Film

In Section 3.10 we introduced the concept of a mass transfer coefficient K and obtained the formula appropriate to the process of diffusion across a *stagnant film*. We are now in a position to obtain a formula for the mass transfer coefficient appropriate to diffusion into a *falling film*. Equation (3.71) defined the mass transfer coefficient in terms of mass units; for a falling film we modify this definition by simply using the appropriate subscripts

$$j_1|_{y=0} = K(m_{1,u} - m_{1,\text{in}}) \tag{4.57}$$

But,

$$j_1|_{y=0} = -\rho\mathscr{D}_{12}\frac{\partial m_1}{\partial y}\bigg|_{y=0}$$

$$= -\rho\mathscr{D}_{12}(m_{1,u} - m_{1,\text{in}})\left[\frac{\partial}{\partial y}\,\text{erfc}\,(\eta)\right]_{y=0}$$

from Eq. (4.56). Therefore

$$K = -\rho\mathscr{D}_{12}\frac{\partial}{\partial y}\left[1 - \frac{2}{\sqrt{\pi}}\int_0^\eta e^{-u^2}\,du\right]_{y=0}$$

Using the chain rule of differentiation gives

$$K = -\rho\mathcal{D}_{12}\left[-\frac{2}{\sqrt{\pi}}\,e^{-\eta^2}\,\frac{1}{(4\mathcal{D}_{12}z/v_0)^{1/2}}\right]_{y=0}$$

or

$$K = \rho\left(\frac{v_0\mathcal{D}_{12}}{\pi z}\right)^{1/2} \tag{4.58}$$

Note that the mass transfer coefficient for this case is position dependent, varying inversely with the square root of distance down the liquid film.

A dimensionless mass transfer coefficient known as the **Stanton number** can be defined as

$$St_z = \frac{j_1}{\rho v_0(m_{1,u}-m_{1,\text{in}})} = \frac{K}{\rho v_0}$$

It is seen that

$$St_z = \frac{1}{\pi^{1/2}}\left(\frac{v_0 z}{\mathcal{D}_{12}}\right)^{-1/2} \tag{4.59}$$

Constant fluid properties and a low absorption rate were assumed in the analysis of the falling film; thus the molar equivalents of Eqs. (4.57) and (4.58) may be written down directly as

$$J_1|_{y=0} = K(x_{1,u}-x_{1,\text{in}}) \tag{4.60}$$

$$K = c\left(\frac{v_0\mathcal{D}_{12}}{\pi z}\right)^{1/2} \tag{4.61}$$

When the molar K is divided by cv_0 to form a molar Stanton number, the result remains identical to Eq. (4.59).

4.7.3   Total Gas Absorption for a Film of Length L

In addition to the local rate of absorption we are also interested in the total gas absorption for a falling film of length L. This may be calculated in a number of ways; since the mass transfer driving force $(x_{1,u}-x_{1,\text{in}})$ is constant with position it is convenient to first derive an expression for the average mass transfer coefficient,

$$K_{\text{avg}} = \frac{1}{L}\int_0^L K\,dz$$

$$= \frac{c}{L}\left(\frac{\mathcal{D}_{12}v_0}{\pi}\right)^{1/2}\int_0^L z^{-1/2}\,dz$$

$$K_{\text{avg}} = 2c\left(\frac{\mathcal{D}_{12}v_0}{\pi L}\right)^{1/2} \tag{4.62}$$

Then the total moles of gas absorbed per unit time is $K_{\text{avg}}\mathcal{P}L(x_{1,u}-x_{1,\text{in}})$ or $\dot{M}$, where $\mathscr{P}$ is the total width (x-direction) of film. Thus,

$$\frac{\dot{M}}{\mathscr{P}} = 2c\left(\frac{\mathscr{D}_{12}v_0 L}{\pi}\right)^{1/2} (x_{1,u} - x_{1,\text{in}}) \text{ lb moles/sec-ft} \qquad (4.63)$$

4.7.4 EXAMPLE Stripping of Ammonia from an Air Stream

A water film 0.012 in. thick and at 80°F runs down the inside wall of a vertical tube of 2 in. inside diameter and 48 in. length. An air stream containing ammonia flows through the tube, the gas phase concentration of ammonia being such that the resulting mole fraction of ammonia in the liquid at the interface is 0.0371. If the entering water is essentially pure, calculate the rate at which ammonia is removed from the air stream.

From Eq. (4.63) the molar rate at which ammonia is absorbed is

$$M = 2\pi Dc\left(\frac{\mathscr{D}_{12}v_0 L}{\pi}\right)^{1/2} (x_{1,u} - x_{1,\text{in}}), \quad \text{since } \mathscr{P} = \pi D$$

Since the ammonia is in dilute solution, we may use properties of pure water from Appendix B, Tables B.7 and B.12c, to obtain

$$c = \frac{\rho}{M} = \frac{(62.2 \text{ lb/ft}^3)}{(18 \text{ lb/mole})} = 3.46 \text{ mole/ft}^3$$

$$\mathscr{D}_{12} = \frac{\nu}{Sc_{12}} = \frac{(0.930 \times 10^{-5} \text{ ft}^2/\text{sec})}{409} = 2.27 \times 10^{-8} \text{ ft}^2/\text{sec}$$

From Eq. (4.41) of Section 4.6.2 the liquid surface velocity v_0 is related to the film thickness by

$$v_0 = \frac{g\delta^2}{2\nu} = \frac{(32.2 \text{ ft/sec}^2)(0.012/12 \text{ ft})^2}{(2)(0.93 \times 10^{-5} \text{ ft}^2/\text{sec})} = 1.73 \text{ ft/sec}$$

Our expression for $\dot{M}$ therefore becomes

$$\dot{M} = (2)(3.14)\left(\frac{2}{12}\right)(3.46)\left(\frac{(2.27 \times 10^{-8})(1.73)(4)}{3.14}\right)^{1/2}(0.0371 - 0)(3600)$$

$$= 0.108 \text{ mole/hr}$$

Hence the mass rate at which ammonia is absorbed is

$$\dot{m} = M_{\text{NH}_3}\dot{M} = (17)(0.108)$$
$$= 1.84 \text{ lb/hr}$$

4.8 SUMMARY

In this chapter we have presented physical situations in which there is conduction or diffusion into a moving fluid. Newton's law of viscosity was introduced together with the idea that the momentum of a moving fluid is *transferred* so that it exerts a force on a wall. Only the simplest of flows were used as examples, established laminar pipe flow and laminar flow down a vertical wall in order to minimize complexity. The concepts of a bulk average temperature, a local heat or mass transfer coefficient, and an average heat or mass transfer coefficient arose in these examples. Certain dimensionless groupings discussed at more length in the next chapter also arose in the examples.

EXERCISES

1. Consider transpiration through a spherical porous shell of inner radius r_1 and outer radius r_2. Assuming steady, radial flow of coolant:

    (a) Show by means of a heat balance over an element of volume $4\pi r^2 \Delta r$ that the differential equation which describes the temperature distribution in the porous shell is

    $$\dot{m}c_p \frac{dT_m}{dr} = 4\pi k_m \frac{d}{dr}\left(r^2 \frac{dT_m}{dr}\right)$$

    where c_p is the heat capacity of the coolant and $\dot{m} = \rho 4\pi r^2 v_r$.
    (b) Show that a solution to the above differential equation is

    $$T_m(r) = C_2 e^{-\beta/r} - C_1$$

    where $\beta \equiv \dot{m}c_p/4\pi k_m$.
    (c) Given that $T_m(r_1) = T_1$, and $T_m(r_2) = T_2$, show that

    $$\frac{T - T_1}{T_2 - T_1} = \frac{e^{-\beta/r} - e^{-\beta/r_1}}{e^{-\beta/r_2} - e^{-\beta/r_1}}$$

    (d) Obtain an expression for the conductive heat flux q_{r_2}.

2. A spacecraft with a nose radius $r_2 = 0.0328$ ft (1 cm) enters Jupiter's atmosphere at a velocity of 156,000 ft/sec and an angle of entry of 6.27°. Under these conditions a heat load q_{r_2}, largely radiative, of 3.17×10^7 Btu/hr ft² (10.0 kw/cm²) is delivered to the outside surface. Molecular hydrogen is used to transpiration-cool the porous carbon nose. If the surface temperature must not rise above 5000°R, what flow rate per unit area at $r = r_2$ of coolant is required? The inside surface, of radius $r_1(r_1/r_2 = \frac{1}{2})$, is at -200°F. Assume that the porous matrix filled with hydrogen has an effective thermal conductivity of 3 Btu/hr ft °F. (*Hint*: Introduce $n_2 = \dot{m}/4\pi r_2^2$ into Exercise 1.)
3. Show that the Eckert number is dimensionless.
4. Estimate the temperature rise at the tip of the main rotor of a helicopter, neglecting heat losses by conduction along the blade or by radiation. The helicopter has an air speed of 200 ft/sec and the speed of the advancing rotor reaches 1100 ft/sec. Consider the limiting cases of zero rotor heat capacity and infinite rotor heat capacity.
5. Parallel the development of Section 4.5 to obtain for steady laminar flow in

a horizontal annulus of inner and outer radii r_1 and r_2, respectively:

    (a) An expression for the velocity profile $v_z(r)$.

    (b) An expression for the temperature profile $T(r)$ in the annulus, assuming that q_{r_1} is constant and that $q_{r_2} = 0$.

    (c) The Nusselt number for the heat transfer.

6. From the results of Exercise 5 and the data of Section 4.5.3, calculate the length of preheater required when

$$2\pi r_1 \cdot q_{r_1} = 50 \text{ watts/ft}$$
$$q_{r_2} = 0$$
$$r_1 = 0.25 \text{ in.}$$
$$r_2 = 0.50 \text{ in.}$$

7. Using Newton's law of viscosity, show that the differential equation for steady laminar flow of a falling film on an inclined plate is

$$\mu \frac{d^2 v_x}{dy^2} = -(\rho - \rho_v)g \sin \beta$$

where y is measured normal to the plate surface, x along the plate, and β is the inclination of the plate from the horizontal.

8. Neglecting vapor drag at the free film surface, show that the thickness δ of the falling film in the previous exercise is

$$\delta = \left(\frac{3\mu\Gamma}{\rho(\rho - \rho_v)g \sin \beta}\right)^{1/3}$$

where Γ is the mass flow rate of film per unit width.

9. Saturated steam at 100°F is condensing on an inclined surface. The surface is 10 cm long, 2 cm wide, and inclined at an angle $\beta = 45°$. Calculate the total condensation rate for a uniform wall temperature of 80°F.

10. Gaseous hydrogen chloride (HCl) is to be stripped from an air stream. The air flows up inside a vertical tube of 1.2 in. inside diameter and 60 in. length. A water film 0.018 in. thick and at 100°F runs down the inside wall of the tube. The gas phase concentration of HCl is such that the resulting mole fraction of HCl in the liquid at the interface is 0.044. If the entering water is free from HCl, determine the rate at which HCl is removed from the air stream. Plot a concentration profile of HCl in the liquid film at the bottom of the tube, that is, plot x_1 versus y at $z = L$.

REFERENCES

Simple Convective Flows

A number of simple problems are solved under the heading of "shell balances" in:

Bird, R. B., W. E. Stewart, and E. N. Lightfoot, *Transport Phenomena.* New York: John Wiley, 1960. (See Chapters 2, 9, 17.)

Laminar Flow in Ducts

A good summary is given by:

> Kays, W. M., *Convective Heat and Mass Transfer*. New York: McGraw-Hill, 1966. Chapter 8.

Flow with Change of Phase

Good descriptions of boiling and condensation are given by:

> Rohsenow, W. M., and H. Y. Choi, *Heat, Mass, and Momentum Transfer*. Englewood Cliffs, N.J.: Prentice-Hall, 1961. Chapters 9 and 10.

Gas-Liquid Transfer

Additional information on gas absorption into liquids is given in:

> Davies, J. T., "Mass Transfer and Interfacial Phenomena", *Advances in Chemical Engineering, Volume 4*, edited by T. B. Drew, J. W. Hoppes, Jr., and T. Vermeulen. New York: Academic Press, 1963.

A more advanced treatise pertinent to the subject is:

> Danckwerts, P. V., *Gas-Liquid Reactions*. New York: McGraw-Hill, 1970.

CHAPTER 5

CONVECTIVE TRANSFER RATES

5.1 INTRODUCTION

As we said in the introduction to the preceding chapter, in the usual case of convective transfer near a wall or near the edge of a jet or plume, the diffusive transfer process acts very nearly at right angles to the convective transfer process. Figure 5.1 illustrates the usual situation for transfer from a wall into a flowing fluid.

For example, consider heat transfer from a hot solid surface into a cold fluid flowing over it. The fluid in contact with the surface is at the surface temperature, while that a short distance away is cooler. It is clear that a temperature gradient exists in the fluid and that Fourier's law then indicates conduction into the fluid. The heat warms the fluid which becomes progressively hotter as it flows along a streamline over the surface. The movement of the fluid brings cold fluid near the surface sweeping the hot fluid away. In this way a high rate of conduction is maintained away from the surface. This process of conduction into a moving fluid is **convective heat transfer**.

In the same way, when the surface imparts a mass concentration of a species to the fluid immediately adjacent to it different from the mass concentration of the species in the oncoming stream (think of dry air blowing over a green leaf which imparts water vapor to the air), **convective mass transfer** occurs. Con-

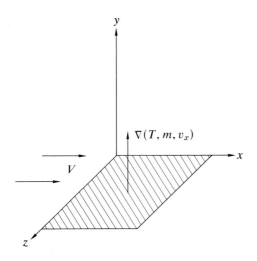

Figure 5.1 Transfer normal to a wall into a parallel flow.

vective mass transfer is mass diffusion into a flowing fluid. A fluid flowing over a wall also exerts a drag on it, by virtue of Newton's law of viscosity. The fluid immediately adjacent to the surface sticks to it, and a velocity gradient is set up. For a Newtonian fluid the shear stress acting on the wall will be the product of the fluid viscosity and the velocity gradient. Such a drag is actually a transfer of momentum from the flowing fluid to the wall and could properly be called **convective momentum transfer**, but is usually just called viscous drag or viscous shear.

Newton's law of cooling is meant to give the heat transfer rate for convective heat transfer from a wall, but where does the heat transfer coefficient come from? What are the relations to be used for the mass and momentum transfer rates? It is the purpose of this chapter to give additional insight into Newton's law of cooling for a fluid flowing near a wall and to present some working relations for estimating convective transfer rates between a wall and such a flowing fluid.

5.2 SURFACE TRANSFER COEFFICIENTS

Newton's law of cooling gives the surface heat flux q_s in terms of a convective heat transfer coefficient h_c, and the temperature difference between the wall at T_s and the average temperature of a fluid some distance away from the wall, T_e:

$$q_s = h_c(T_s - T_e) \tag{5.1}$$

If we were to call the distance measured from the wall y, we could write Fourier's law at the wall as

$$q_s = -k\frac{\partial T}{\partial y}\bigg|_{y=0} \tag{5.2}$$

where k is the fluid conductivity and T the fluid temperature. We see that h_c is identically

$$h_c \equiv \frac{-k\frac{\partial T}{\partial y}\big|_{y=0}}{T_s - T_e} \qquad (5.3)$$

We have seen in Chapter 4 how one can obtain h_c by calculating the temperature distribution in the flowing fluid, finding the gradient at the wall, and substituting it into Eq. (5.3). It is a laborious process, but it is actually what is done in determining h_c analytically. Using a value of h_c, once one is found, is by comparison a very simple process. One simply uses Eq. (5.1).

In a similar manner we can define useful coefficients for mass transfer and momentum transfer. The analogous form of Newton's law of cooling for mass transfer is:

(mass units)

$$j_{1,s} = K(m_{1,s} - m_{1,e}) \qquad (5.4a)$$

(mole units)

$$J_{1,s} = K(x_{1,s} - x_{1,e}) \qquad (5.4b)$$

[Recall that we do not bother to introduce different distinguishing notation for the Ks defined in Eqs. (5.4a) and (5.4b).] The mass transfer coefficient is from Fick's law Eq. (3.43):

$$K = \frac{-\rho\mathscr{D}\frac{\partial m_1}{\partial y}\big|_{y=0}}{(m_{1,s} - m_{1,e})} \qquad \text{(mass units)} \qquad (5.5)$$

As in Chapter 3, we consider either binary mixtures where $\mathscr{D} = \mathscr{D}_{12}$, or situations where species 1 is in small concentration, for which $\mathscr{D}$ is an effective binary diffusion coefficient $\mathscr{D}_{1m}$ for species 1 diffusing in the mixture or solution. To simplify notation the subscripts are left off.

For momentum transfer it is customary to use a skin friction coefficient c_f:

$$\tau_s = c_f \tfrac{1}{2}\rho V^2 \quad \text{(where the sign of } \tau_s \text{ is disregarded)} \qquad (5.6)$$

Again, from Newton's law of viscosity, we define a **skin friction coefficient**:

$$c_f = \frac{\mu\frac{\partial v_x}{\partial y}\big|_{y=0}}{\tfrac{1}{2}\rho V^2} \qquad (5.7)$$

Both local and average values of h_c, K, and c_f are useful. A local value of h_c gives the relation between the value of the heat flux and the temperature difference at a local spot on the wall, for example, at the upstream stagnation point of the nose of an aircraft. An average value will give the average heat flux over the entire surface in terms of the average temperature difference; for example, the average may be taken over the entire nose section of an aircraft.

It is everyday experience that, when the wind blows harder, one's bare face feels cooler. It is obvious that the convective heat and mass transfer coefficients are increasing functions of flow velocity. It is not so obvious, but still apparent, when one thinks about it, that the heat transfer coefficient must be a function of such things as fluid viscosity, density, specific heat, and conductivity as well. The mass transfer coefficient must likewise depend upon fluid viscosity, density, and mass diffusivity. One would suppose also that the size of the object about which or within which the flow occurs would also play a part. Given this size, the flow geometry, the flow velocity and the fluid properties, how does one find a surface transfer coefficient? An example follows.

5.3   AN ANALYTICAL SOLUTION FOR SLUG FLOW

5.3.1   The Physical Situation

To take the simplest possible example illustrating the dependence of transfer coefficient on flow velocity we consider slug flow over a heated isothermal wall. Slug flow denotes the case of uniform flow velocity. Uniform velocity can occur when a solid slab slides over a heated plate or can be a reasonable approximation when a non-Newtonian fluid such as mayonnaise flows in a duct or when a liquid metal enters a heated pipe. Figure 5.2 shows the situation envisioned.

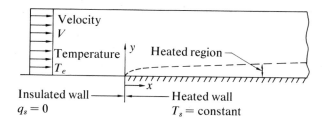

Insulated wall —— | —— Heated wall
$q_s = 0$ $T_s = $ constant

Figure 5.2   Slug flow over a heated wall.

We suppose that the flow is steady and that the thermal conditions have been established for a long time. Upstream from the heated wall the entering medium is isothermal at temperature T_e. Downstream from $x = 0$, the wall is heated and maintained at temperature T_s. A portion of the moving medium adjacent to the heated wall becomes heated by conduction. The temperature gradient in the y-direction perpendicular to the wall is quite steep, for heat flowing upward into a fixed volume of space is quickly swept away downstream by the moving mass with its thermal capacity of ρc_p per unit volume. Only in the immediate vicinity of where the heating commences ($x = 0$) are the streamwise gradients expected to be as steep. Thus the heated region remains thin compared to the distance x, as shown in the figure.

5.3.2   Formulation of the Problem

Appendix A presents a general formulation of the conservation equations. We could take the equation for conservation of energy from there and simplify it appropriately. However, for this rather simple problem, it may readily be derived from first principles. Consider a fixed element of space between x and $x + dx$, and y and $y + dy$, as shown in Figure 5.3.

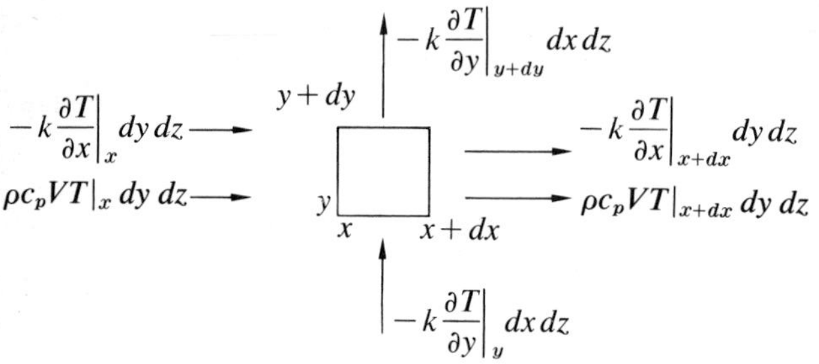

Figure 5.3   Heat balance for slug flow.

Heat can cross the element boundaries by conduction and convection. Whenever V and x are large, conduction in the x-direction will be small compared to convection and may be ignored. Appendix A, Section A.8.3, gives a more rigorous argument for dropping the streamwise conduction term. Conservation of energy requires that the flow of energy into the element equal the outflow,

$$\rho c_p V T|_x \, dy \, dz - k \frac{\partial T}{\partial y}\bigg|_y \, dx \, dz = \rho c_p V T|_{x+dx} \, dy \, dz - k \frac{\partial T}{\partial y}\bigg|_{y+dy} \, dx \, dz$$

As an alternative to the procedure used in, for example, Section 2.5, expand the terms evaluated at $x + dx$ and $y + dy$ in a Taylor series to obtain

$$\rho c_p V T \, dy \, dz - k \frac{\partial T}{\partial y} \, dx \, dz = \rho c_p V T \, dy \, dz + \rho c_p V \frac{\partial T}{\partial x} \, dx \, dy \, dz + \cdots$$

$$- k \frac{\partial T}{\partial y} \, dx \, dz - \frac{\partial}{\partial y}\left(k \frac{\partial T}{\partial y}\right) dy \, dx \, dz + \cdots$$

With the assumption of a constant thermal conductivity k there results

$$\rho c_p V \frac{\partial T}{\partial x} = k \frac{\partial^2 T}{\partial y^2} \qquad\qquad (5.8)$$

The boundary conditions to be satisfied are

$$y = 0: \quad T = T_s$$
$$y \to \infty: \quad T \to T_e$$
$$x = 0: \quad T = T_e$$

5.3.3   Solution

Equation (5.8) may be compared to Eq. (2.41). They are clearly of the same form. In fact by transforming from x to $t = x/V$ they may be made identical. The left-hand side becomes

$$\rho c_p V \frac{\partial T}{\partial x} = \rho c_p V \frac{\partial T}{\partial t} \frac{dt}{dx} = \rho c_p V \frac{\partial T}{\partial t} \frac{1}{V} = \rho c_p \frac{\partial T}{\partial t} \tag{5.9}$$

The previous solution for transient conduction in a semi-infinite slab, Eq. (2.48), thus applies,

$$T^* = \frac{T - T_e}{T_s - T_e} = \text{erfc}\left(\frac{y}{[4\alpha t]^{1/2}}\right)$$

Substituting for t the value x/V yields

$$T^* = \text{erfc}\left(\frac{y}{[4\alpha x/V]^{1/2}}\right) = \text{erfc }\eta \tag{5.10}$$

where, by way of a reminder,

$$\alpha = \frac{k}{\rho c_p}, \qquad \text{erfc }\eta = 1 - \frac{2}{\pi^{1/2}} \int_0^\eta e^{-u^2}\, du$$

Equation (5.10) is plotted in Figure 5.4. From the figure we see that T^* falls very

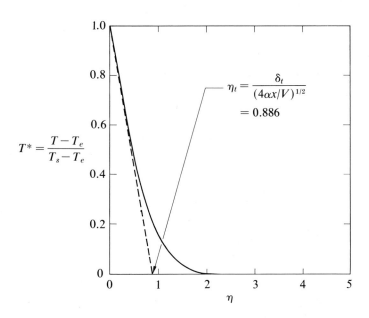

Figure 5.4   Thermal boundary layer for slug flow over a heated wall.

rapidly from the wall value of 1 at $\eta = 0$ to its zero asymptotic value. The tangent to the curve at $\eta = 0$ crosses the η-axis at a value of $\eta_t = 0.886$. This intercept may be used to characterize the thickness of the heated layer as follows:

$$\eta = \frac{y}{(4\alpha x/V)^{1/2}}, \qquad y = \eta\left(\frac{4\alpha x}{V}\right)^{1/2}$$

$$y = \delta_t = 1.772\left(\frac{\alpha x}{V}\right)^{1/2} \tag{5.11}$$

For example, consider a dielectric material such as plastic or oil. Typical property values are

$$k = 0.1 \text{ Btu/hr ft °F}, \qquad \rho = 50 \text{ lb/ft}^3$$

$$c_p = 0.33 \text{ Btu/lb °F}, \qquad \alpha = k/\rho c_p = 6.06 \times 10^{-3} \text{ ft}^2/\text{hr}$$

For a medium moving as slow as 1 ft/min the value of δ_t at $x = 1$ ft is 0.018 ft or less than $\frac{1}{4}$ in. The exact numerical value for the thickness of the heated layer is, of course, rather arbitrary. We could have used the value of η at which T^* crosses 0.10 or 0.01, but as seen in the figure, the order of magnitude is the same for any of the choices.

The heat flux at the wall is given by Fourier's law

$$q = -k\frac{\partial T}{\partial y}\bigg|_{y=0} = k(T_s - T_e)\frac{2}{\pi^{1/2}}\frac{1}{(4\alpha x/V)^{1/2}}$$

$$q = \frac{k}{\pi^{1/2}}(T_s - T_e)\left(\frac{V}{\alpha x}\right)^{1/2} \tag{5.12}$$

We see that the heat flux is linear in $(T_s - T_e)$, just as represented in Newton's law of cooling. The heat transfer coefficient is

$$h_c = \frac{q}{T_s - T_e} = \frac{k}{\pi^{1/2}}\left(\frac{V}{\alpha x}\right)^{1/2} \tag{5.13}$$

This value is the local one at the local value of x.

To obtain the average value over a length L we compute the average,

$$h_{avg} = \frac{1}{L}\int_0^L h_c\, dx \tag{5.14}$$

Substitution of Eq. (5.13) into Eq. (5.14) gives

$$h_{avg} = \frac{2k}{\pi^{1/2}}\left(\frac{V}{\alpha L}\right)^{1/2} \qquad \text{(twice the local value at } x = L) \tag{5.15}$$

For the values just mentioned; $k = 0.1$ Btu/hr ft °F, $\alpha = 6.06 \times 10^{-3}$ ft^2/hr, $V = 1$ ft/min, $L = 1$ ft; we obtain $h_{avg} = 11$ Btu/hr ft^2 °F. Since h_{avg} increases with $V^{1/2}$, a velocity of 1 ft/sec would give $h_{avg} = 11(60)^{1/2} = 85$ Btu/hr ft^2 °F. The solution supports our intuition that higher velocities give larger transfer coefficients.

5.3.4 Mass Transfer

The same problem can be solved for mass transfer into a slug flow. In fact we have solved it in Chapter 4 for the case of gas absorption into a falling film. If the oncoming flow has a mass concentration of $m_{1,e}$ and that at the wall for $x > 0$ is $m_{1,s}$, the solution is

$$\frac{m_1 - m_{1,e}}{m_{1,s} - m_{1,e}} = \text{erfc}\left(\frac{y}{[4\mathcal{D}x/V]^{1/2}}\right) \tag{5.16}$$

The mass transfer rate into the flow, given by Fick's law, is

$$j_1 = -\rho\mathcal{D}\frac{\partial m_1}{\partial y}\bigg|_{y=0}$$

$$j_1 = \frac{\rho\mathcal{D}}{\pi^{1/2}}(m_{1,s} - m_{1,e})\left(\frac{V}{\mathcal{D}x}\right)^{1/2} \tag{5.17}$$

The local mass transfer coefficient we found was

$$K = \frac{\rho\mathcal{D}}{\pi^{1/2}}\left(\frac{V}{\mathcal{D}x}\right)^{1/2} \tag{5.18}$$

As before

$$K_{\text{avg}} = \frac{2\rho\mathcal{D}}{\pi^{1/2}}\left(\frac{V}{\mathcal{D}L}\right)^{1/2} \tag{5.19}$$

5.3.5 The Boundary Layer Concept

In Section 5.3.3 the heated region was found to be thin compared to the length along the wall. For a velocity of only 1 ft/min it was estimated to be $\frac{1}{4}$ of an inch at $L = 1$ ft for a material like oil. A higher velocity results in a thinner region next to the wall. Such a region is known as a **thermal boundary layer**. Similarly, a mass concenetration boundary layer exists in the case of mass transfer. In the case of momentum transfer in a Newtonian fluid, slug flow does not occur. Rather there is a build up of a retarded layer of fluid known as a **momentum boundary layer**.

From a simplistic point of view the heat from the wall must be conducted through a layer of thickness δ_t before it can be carried away by the moving fluid. (Since the boundary layer itself actually carries away all the heat, the thickness δ_t is thinner than the value of y at which the temperature has changed say 99% of the way from the wall value to the free stream value, as shown in Figure 5.4.) From this simplistic boundary layer point of view

$$h_c = \frac{k}{\delta_t}$$

The ratio of L to δ_t is a measure of the effectiveness of convection or fluid movement in promoting transfer. This ratio will be found to be the dimensionless heat flux or surface heat transfer coefficient, the Nusselt number, encountered in Chapter 4 and discussed further in the next section.

5.4   DIMENSIONLESS SURFACE TRANSFER

5.4.1   Heat Transfer

The surface transfer rate can be made dimensionless in one of two ways: (1) by dividing by a conductive flux fabricated from $k/L(T_s - T_e)$ or (2) by dividing by a convective flux $\rho c_p V(T_s - T_e)$. The conductive flux is, of course, that which would occur through a slab of fluid L thick, while the convective one is that which would be imposed upon a porous wall if the fluid entered it at T_e with the full velocity V and emerged at temperature T_s. Since the former flux is quite low, the resulting division would be expected to give a large dimensionless number. On the other hand, the latter flux is large, and a small result would be expected. These two dimensionless fluxes are, respectively, the Nusselt number Nu and the Stanton number, St.

- $$Nu = \frac{qL}{k(T_s - T_e)} = \frac{h_c L}{k} \qquad (5.20)$$

$$St = \frac{q}{\rho c_p V(T_s - T_e)} = \frac{h_c}{\rho c_p V} \qquad (5.21)$$

It may be recalled that these numbers arose in a very natural way in Chapter 4 when we considered pipe flow and transpiration.

It may be seen above that the dimensionless parameters can be regarded equally well as dimensionless fluxes or as dimensionless transfer coefficients. Another advantage to using dimensionless parameters is that the values are independent of the system of units employed.

5.4.2   Mass and Momentum Transfer

The corresponding two parameters for mass transfer are

- $$Nu_m = \frac{j_1 L}{\rho \mathscr{D}(m_{1,s} - m_{1,e})} = \frac{KL}{\rho \mathscr{D}} \qquad (5.22)$$

- $$St_m = \frac{j_1}{\rho V(m_{1,s} - m_{1,e})} = \frac{K}{\rho V} \qquad (5.23)$$

As before, the former is the actual flux divided by the flux that would diffuse through a slab of thickness L. This dimensionless number would usually be large compared to unity. The Stanton number for mass transfer is the actual transfer of species 1 per unit area divided by the mass of species 1 which would be deposited within a porous wall if the fluid entered it at the full velocity V with mass fraction $m_{1,e}$ and emerged with mass fraction $m_{1,s}$, a dimensionless ratio considerably less than one for flows parallel to a wall.

With regard to momentum transfer, the skin friction coefficient is already dimensionless.

5.5 DIMENSIONLESS FLOW FIELD PARAMETERS

5.5.1 Significance of Reynolds Number

In Appendix A it is shown that the dimensionless parameter

$$Re = \frac{VL\rho}{\mu} = \frac{VL}{\nu} \tag{5.24}$$

is the only one governing the constant-property momentum equation. In other words, when the equations for conservation of mass and conservation of momentum are made dimensionless, the only independent parameter characterizing the flow field is Reynolds number. Under steady conditions where property variations are not important and where numerical values of the dimensionless boundary conditions are fixed, all the dependent variables (three dimensionless velocity components and dimensionless pressure) are functions only of the dimensionless position coordinates x^*, y^*, z^* and the dimensionless Reynolds number. In order for density variations to be negligible in a compressible gas, such as air, the velocities must be low compared to the speed of sound (low Mach number subsonic flow). The boundary conditions are simply $v_x^* = v_y^* = v_z^* = 0$ on the walls of some particular geometry.

The gradient of the velocity can be obtained from the solution for the velocity field and multiplied by viscosity to determine the shear stress acting on a wall. When made dimensionless by dividing by $\frac{1}{2}\rho V^2$, this shear stress becomes the skin friction coefficient c_f. This quantity will be a function only of geometry and Reynolds number when property variations are slight. For a fixed geometry, such as a flat plate with a parallel flow over it, we write

$$c_f = c_f(Re) \tag{5.25}$$

The local value will also be a function of $x^* = x/L$; the average value over the length L is only a function of Re.

A practical result of this observation is that we can make a measurement of drag on a small plate in a water tunnel, and apply the results (in the form of a curve of c_f versus Re) for a large plate subjected to drag from air.

Table 5.1, at the end of this chapter, contains c_f versus Re relations for various flow situations. Tables 5.2 and 5.3, also at the end of this chapter, summarize the dimensionless wall fluxes or transfer coefficients and the dimensionless flow field parameters.

5.5.2 Significance of Reynolds and Prandtl Numbers to Heat Transfer

In Appendix A it is shown that the Reynolds and Prandtl numbers are the only dimensionless parameters governing the temperature field in a constant property flow with fixed boundary conditions. By the same line of reasoning used above, it is apparent that, for a given geometry with say isothermal boundary conditions, **the Nusselt number will be a function only of Reynolds number and Prandtl number**.

This conclusion can be tested with the slug flow solution obtained in Section 5.3. Consider Eq. (5.15). Let us make it dimensionless according to Eq. (5.20)

$$Nu_{avg} = \frac{2}{\pi^{1/2}}\left(\frac{VL}{\alpha}\right)^{1/2} \tag{5.26a}$$

To form the Reynolds number given by Eq. (5.24) it is necessary to divide and multiply by $\nu^{1/2}$. The result is

$$Nu_{avg} = \frac{2}{\pi^{1/2}}\left(\frac{VL}{\nu}\right)^{1/2}\left(\frac{\nu}{\alpha}\right)^{1/2}$$

$$Nu_{avg} = \frac{2}{\pi^{1/2}} Re^{1/2}Pr^{1/2} \tag{5.26b}$$

where

$$\bullet \qquad Pr = \frac{\nu}{\alpha} = \frac{\mu c_p}{k} \tag{5.27}$$

The phenomenon of aerodynamic heating was discussed in Chapter 4. The result of relevance in the present context is that a modified form of Newton's law of cooling applies [Eq. (4.21)],

$$q = h(T_s - T_r) \tag{5.28}$$

where $T_r = T_e + r(V^2/2c_p)$ is the recovery temperature, and r is the recovery factor. Recall that for Couette flow $r = Pr$, while for laminar flow over a flat plate $r \doteq Pr^{1/2}$, and for turbulent flow $r \doteq Pr^{1/3}$. Prandtl number is thus seen to affect both h (through Nu) and r.

5.5.3   Significance of Reynolds and Schmidt Numbers to Mass Transfer

Let us repeat the procedure of making the slug flow solution dimensionless; this time we consider mass transfer. Equation (5.19) made dimensionless according to Eq. (5.22) yields

$$Nu_{m,avg} = \frac{2}{\pi^{1/2}}\left(\frac{VL}{\mathscr{D}}\right)^{1/2} = \frac{2}{\pi^{1/2}}\left(\frac{VL}{\nu}\right)^{1/2}\left(\frac{\nu}{\mathscr{D}}\right)^{1/2} \tag{5.29a}$$

$$Nu_{m,avg} = \frac{2}{\pi^{1/2}} Re^{1/2}Sc^{1/2} \tag{5.29b}$$

where

• $$Sc = \frac{\nu}{\mathscr{D}} = \frac{\mu}{\rho \mathscr{D}} \qquad (5.30)$$

Here we find Nusselt number for mass transfer to be a function of Reynolds number and Schmidt number. This result is generally true, as shown in Appendix A. Beyond this observation we note that Eq. (5.29b) is identical to Eq. (5.26b) with Pr replaced by Sc. This result holds true when the mass flux $n = (\rho v)_s$ is not high enough to disturb the momentum flow. In fact, we shall see that

$$St\, Pr^{2/3} \doteq St_m Sc^{2/3} \doteq \frac{C_f}{2}$$

is a good approximation for many flow situations. A notable exception to this last statement, however, is heat transfer to liquid metals for which Pr is much less than unity.

5.5.4   *EXAMPLE* Extrapolation of Heat Transfer Test Data

An experiment was run in a 30 in. square open throat wind tunnel. The heat loss was measured from a 2.75 in. diameter cylinder to an airstream flowing at right angles to the cylinder. Heat loss from the cylinder consisted of both radiation and convection. The measured total heat loss, and a calculation of the radiation contribution for each test condition, are listed below. [For the radiation calculation an internal-total, hemispherical emittance of 0.8 was used in Eq. (6.32).]

Our objective is to demonstrate the utility of dimensionless parameters or, more precisely speaking, the principles of *similarity* for the extrapolation of experimental data. Let us estimate, from the data presented below, the heat loss from a 4.0 in. diameter cylinder at 90°F to a 15 ft/sec air stream at 75°F. The cylinder has an emittance of 0.9.

Test Data for Example 5.5.4

Run	V, ft/sec	T_s, °F	T_e, °F	$q_{radiation}$, Btu/hr ft²	q_{total}, Btu/hr ft²
1	4.5	129.8	88.0	42.2	185.6
2	9.3	97.3	83.1	13.2	82.9
3	11.6	88.8	65.2	20.5	175.3
4	20.0	91.3	82.8	7.4	82.9
5	28.3	85.4	76.6	7.6	127.7
6	27.9	101.0	90.9	10.1	130.1
7	28.9	96.9	88.2	8.3	127.3
8	27.3	79.4	67.2	11.0	175.0
9	37.6	95.0	88.4	5.7	147.9
10	37.4	93.0	83.5	8.1	173.8
11	48.6	91.7	84.6	6.0	173.2
12	57.4	84.2	78.1	5.2	182.8

The first step is to draw a sketch of the physical situation.

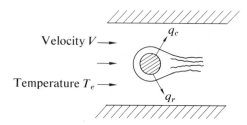

We recognize that heat is lost by convection and radiation (to be explained more fully in Chapter 6),

$$q_{total} = q_c + q_r \tag{5.31}$$

Newton's law of cooling gives

$$q_c = h_c(T_s - T_e) \tag{5.32}$$

while (as we shall see in Chapter 6) for a small object in a large room

$$q_r = \epsilon(\sigma T_s{}^4 - \sigma T_e{}^4) \tag{5.33}$$

Next we must determine h_c for the 4 in. cylinder. We know that $Nu = q_c D/k\,\Delta T = h_c D/k$ is a function of Re and Pr. Since the test fluid was also air, the Prandtl number will be almost the same, approximately 0.7. We can thus use the test data to determine Nu versus Re and interpolate for the value of Re encountered with the 4 in. cylinder.

   For the 4 in. cylinder the following values are applicable at 1 atm total pressure:

$$T_s = 90°F, \quad T_e = 75°F, \quad T_{avg} = 82.5°F$$
$$\nu(T_{avg}) = 17.0 \times 10^{-5} \text{ ft}^2/\text{sec}$$
$$k(T_{avg}) = 0.0154 \text{ Btu/hr ft °F}$$
$$Re = (15)(4/12)/(17.0 \times 10^{-5}) = 29{,}400$$

(The student should check the units in Re to verify that they are consistent.) Since ν for the test fluid is nearly the same as that for the fluid flowing over the 4 in. cylinder, for which $VD = (15)(4/12) = 5$ ft²/sec, we know that for a 2.75 in. test cylinder the velocity which will match the Reynolds number is $V = 5/(2.75/12) = 22$ ft/sec. Test runs 4 and 5 are therefore of most interest, but to get a more complete picture of the behavior indicated by the data, we compute Nu versus Re for all the data. For example, for run 4

$$T_s = 91.3°F, \qquad T_e = 82.8°F, \qquad T_{avg} = 87.0°F$$
$$\nu = 17.4 \times 10^{-5} \text{ ft}^2/\text{sec}$$
$$k = 0.0156 \text{ Btu/hr ft °F}$$
$$q_c = 82.9 - 7.4 = 75.5 \text{ Btu/hr ft}^2$$
$$Nu = \left(\frac{75.5}{8.5}\right)\left(\frac{0.229}{0.0156}\right) = 130.5$$

$$Re = \frac{(20)(0.229)}{17.4 \times 10^{-5}} = 26,300$$

From a graph of Nu versus Re based on the test data we can estimate for the 4 in. cylinder with $Re = 29,400$ that $Nu = 145$. There then results

$$h_c = \frac{0.0154}{0.333}(145) = 6.7 \text{ Btu/hr ft}^2 \text{ °F}$$

$$q_c = 6.7(15) = 100 \text{ Btu/hr ft}^2$$

$$q_r = (0.9)(1.712 \times 10^{-9})[(550)^4 - (535)^4] = 15 \text{ Btu/hr ft}^2$$

$$q_{\text{total}} = q_c + q_r \qquad (5.34)$$

$$q_{\text{total}} = 100 + 15 = 115 \text{ Btu/hr ft}^2$$

5.5.5 *EXAMPLE* **Use of Correlations to Estimate Convective Heat Transfer**

Let us use Table 5.1 to estimate the convective heat loss from a 4 in. diameter cylinder at 90°F to a 15 ft/sec air stream blowing normal to it at 75°F. We suppose that we have no appropriate test data upon which to draw. From Table 5.1 we obtain

$$Nu_D = 0.43 + KRe_D{}^m Pr^{0.31} \qquad (5.35)$$

For $Re_D = 29,400$ (see Section 5.5.4) Table 5.1 gives $K = 0.193$, $m = 0.618$. We obtain in this way

$$Nu_D = 0.43 + 0.193(29,400)^{0.618}(0.7)^{0.31} = 0.43 + 0.193(577)(0.895) = 100$$

The value of h_c then is estimated to be

$$h_c = \frac{0.0154}{0.333}(100) = 4.6 \text{ Btu/hr ft}^2 \text{ °F}$$

and q_c is

$$q_c = (4.6)(15) = 69 \text{ Btu/hr ft}^2$$

The value estimated with Table 5.1 (69 Btu/hr ft²) is 30% smaller than the value estimated in Section 5.5.4 (100 Btu/hr ft²). A reasonable explanation is that the free stream turbulence in the 30 in. wind tunnel used to obtain the data reproduced in Section 5.5.4 was different from that used by the experimenters whose data was used to establish the relation shown in Table 5.1. Complete dynamic similarity would require that the surrounds of the 4 in. cylinder be reproduced in the 2.75 in. cylinder experiment with every length scaled down by the ratio 2.75/4.0. The fact that the 2.75 in. cylinder was in a 30 in. duct with an open throat, which is in no way accounted for by the relations shown in Table 5.1, is apparently significant but does not, of course, affect the order of magnitude of h_c. The relations in Table 5.1 are usually quoted as accurate within ±15%, and idealization of a real system into components which fit the cases covered in Table 5.1 often introduces, as we see above, somewhat greater uncertainty.

5.5.6   *EXAMPLE* Convective Heat Transfer Coefficient in a Pipe

A second example of the use of the correlations in Table 5.1 is taken. Three gallons per minute of hot water at 140°F flow through a schedule 40 $\frac{1}{2}$ in. domestic hot water line. What is the heat transfer coefficient inside the pipe?

The first step is as before to find Reynolds number. For this purpose the velocity is needed. The volume flow rate is

$$VA_c = (3[\text{gal/min}]231[\text{in.}^3/\text{gal}])/(1728[\text{in.}^3/\text{ft}^3]60[\text{sec/min}])$$
$$= 6.69 \times 10^{-3}\ \text{ft}^3/\text{sec}$$

Table B.16 shows the inside diameter to be 0.622 in. The cross-sectional area is

$$A_c = \left(\frac{\pi}{4}\right)D^2 = \left(\frac{\pi}{4}\right)\left(\frac{0.622}{12}\right)^2 = 2.11 \times 10^{-3}\ \text{ft}^2$$

so that the velocity is found to be

$$V = 6.69 \times 10^{-3}/2.11 \times 10^{-3} = 3.17\ \text{ft/sec}$$

Reynolds number is then

$$Re_D = \frac{VD}{\nu} = \frac{(3.17)(0.622/12)}{0.513 \times 10^{-5}} = 32{,}000$$

The second step is to determine whether the flow is laminar or turbulent. From Table 5.1, item 2, we find the transition Reynolds number to be 2300, so the flow is turbulent.

Third, we find Nu. Table 5.1 gives the **Dittus-Boelter equation**,

$$Nu = 0.023\ Re^{0.8}Pr^{0.33}$$
$$Nu = 0.023(32000)^{0.8}(3.01)^{0.33}$$
$$Nu = 0.023(4020)(1.438) = 133$$

Finally, from Nusselt number the convective heat transfer coefficient is found,

$$h_c = Nu\frac{k}{D} = 133\ \frac{0.378}{0.622/12} = 970\ \text{Btu/hr ft}^2\ °\text{F}$$

5.6   NATURAL CONVECTION

5.6.1   The Phenomenon of Natural Convection

There are situations in which we can surmise there is flow, but we know no particular value of flow velocity V. For example, openings are made in the bottom and the top (or top of the back) of electronic packages such as television sets. The reason is to let cold air in at the bottom and warm air out at the top for purposes of cooling the equipment. But often there is no blower forcing the flow at a velocity V which we know. The flow arises naturally in the presence of gravity from density differences caused by the heat transfer process itself. Similar situations arise when mass transfer or combined heat and mass transfer causes a density difference. The transfer process is called **natural** (or **free**) **convection**. Density gradients in body force fields other than gravity may similarly give rise to natural convection.

The distinguishing feature between *natural* convection and *forced* convection is thus the lack of a fan or pump forcing a certain flow independent of the heat or mass transfer rate. *Mixed free and forced convection* may occur when there is a relatively feeble forced convection velocity field so that natural convection velocities of a comparable magnitude also arise.

5.6.2   Order of Magnitude of a Free Convection Velocity

Consider the case of a television set or other piece of electronic gear of height L. Suppose the vertical average temperature of the air inside the case is T_h, and the temperature outside the case is T_c. The loss in pressure of fluid flowing through an orifice is $\frac{1}{2}\rho V_0^2$, because the flow separates at the downstream edge of the orifice and the kinetic energy in the separated stream is merely dissipated. Suppose our electronic package has two equal-area sets of openings top and bottom so that the pressure drop required to cause the flow is ρV_0^2.

The pressure difference causing the flow is the pressure due to a cold column of air $\rho_c g L$ minus that of the hot column $\rho_h g L$. Our estimate of velocity is then simply

$$\rho V_0^2 = (\rho_c - \rho_h)gL = \Delta\rho gL$$

$$V_0^2 = \frac{\Delta\rho}{\rho}gL \qquad\qquad (5.36)$$

For a thermally expansive fluid $\Delta\rho/\rho$ is often written in terms of a volume expansion coefficient β (see Appendix B, Table B.9).

$$\frac{\Delta\rho}{\rho} = \beta(T_h - T_c), \qquad \beta = -\frac{1}{\rho}\frac{\partial\rho}{\partial T}\bigg|_p \qquad\qquad (5.37)$$

For an ideal gas β is simply the reciprocal of the absolute temperature. Consider a height $L = 1$ ft, $g = 32.2$ ft/sec², and a temperature difference of $20°C = 36°F$. Then, if the mean temperature is $30°C$

$$V_0 = \left[\frac{(20)(1.8)}{(273+30)(1.8)}(32.2)(1)\right]^{1/2} = 1.46 \text{ ft/sec}$$

5.6.3   Grashof Number

For forced convection the dimensionless flow field parameter was Reynolds number, while Prandtl number and Schmidt number affected the temperature and mass fields, respectively. In the case of natural convection the appropriate parameter is a Reynolds number based on the natural convection velocity V_0. To avoid square roots we square Re.

$$(Re^2)_{\text{natural convection}} = \frac{V_0^2 L^2}{\nu^2} = \frac{(\Delta\rho/\rho)gL^3}{\nu^2}$$

This quantity is known as **Grashof number**

$$Gr = \frac{(\Delta\rho/\rho)gL^3}{\nu^2} \tag{5.38}$$

The Grashof number is thus a Reynolds number squared for buoyancy-driven convection. For free convection driven by heat transfer, we have Nu as a function of Gr and Pr. For free convection driven by mass transfer, Nu_m is a function of Gr and Sc. For combined heat and mass transfer Nu and Nu_m prove to be functions of Gr, both Pr and Sc, and the ratio of the buoyancy induced by the temperature difference to that induced by the mass concentration difference. This latter ratio is not important if $Pr \doteq Sc$.

For an estimate to determine whether free or forced convection is more important one can compare the associated pressure drops to see which phenomenon governs the flow in our electronic package. Suppose we had a blower producing velocity V. We compare the ratio of pressure drops across two orifices,

$$\frac{\rho V^2}{\rho V_0^2} = \frac{Re^2}{Gr} \tag{5.39}$$

When the ratio is large, forced convection dominates; when it is small, free convection is predominant.

Table 5.1 gives some useful free convection transfer relations.

5.6.4   *EXAMPLE* Simultaneous Heat and Mass Transfer from a Wetted Cylinder

An experiment was run to determine the heat loss from a 2.94 in. diameter horizontal wetted cylinder into still air at 1.0 atm total pressure. Heat loss due to evaporation and total heat loss could be distinguished because the liquid supply rate was metered. As for the example in Section 5.5.4, there was a radiation heat loss as well. Data points are as follows:

Test Data for Example 5.6.4

Run	T_s, °F	T_e, °F	(R.H.)$_e$, %	q_{total}, Btu/hr ft²	q_{evap}, Btu/hr ft²	$q_{\text{total}} - q_{\text{evap}}$, Btu/hr ft²
1	77.9	70.7	56.0	58.4	46.0	12.4
2	78.2	70.2	55.0	58.3	44.6	13.7
3	74.4	65.9	55.0	59.1	44.9	14.2
4	81.9	71.1	57.5	75.6	56.1	19.5
5	79.9	68.5	53.0	81.0	60.4	20.6

Test Data for Example 5.6.4   (continued)

Run	T_s, °F	T_e, °F	(R.H.)$_e$, %	q_{total}, Btu/hr ft²	q_{evap}, Btu/hr ft²	$q_{total} - q_{evap}$, Btu/hr ft²
6	84.1	69.0	53.0	106.4	78.2	28.2
7	85.4	69.1	60.0	107.0	76.3	30.7
8	90.1	72.7	57.5	133.4	99.8	33.6
9	91.8	68.8	54.0	167.9	122.6	45.3
10	97.5	71.0	57.5	162.0	108.0	54.0
11	101.8	70.0	57.0	232.0	165.5	66.5
12	105.1	68.8	57.0	275.0	197.4	77.6
13	109.0	70.9	55.0	322.0	240.5	81.5
14	112.8	71.5	59.0	366.0	274.3	91.7
15	112.5	69.1	55.0	387.0	291.0	96.0

It is proposed to use these data to predict the heat loss from a wetted 4.0 in. diameter horizontal cylinder, with a surface temperature of 90°F, into still air at 75°F, 20% relative humidity (R.H.), and 1.0 atm.

The heat balance on the cylinder may be written as

$$q_{total} = q_{evap} + q_c + q_r$$

where q_c and q_r are, as in the example in Section 5.5.4,

$$q_c = h_c(T_s - T_e), \qquad h_c = \frac{k}{D} Nu$$

$$q_r = \epsilon(\sigma T_s^4 - \sigma T_e^4)$$

and where

$$q_{evap} = n_{1,s}\hat{h}_{fg}$$

Since the evaporation rate is low, we can assume $n_{1,s} = j_{1,s}$ and, using Eqs. (5.4a) and (5.22), write

$$q_{evap} = \frac{\rho \mathscr{D}_{12}}{D}(m_{1,s} - m_{1,e})Nu_m \hat{h}_{fg}$$

For a binary (air-water vapor) ideal gas mixture

$$\rho = \rho_1 + \rho_2 = \frac{P_1 M_1}{\mathscr{R}T} + \frac{P_2 M_2}{\mathscr{R}T} = \frac{P_1 M_1}{\mathscr{R}T} + \frac{(P - P_1)M_2}{\mathscr{R}T}$$

$$m_1 = \frac{\rho_1}{\rho} = \frac{P_1 M_1 / \mathscr{R}T}{\dfrac{P_1 M_1}{\mathscr{R}T} + \dfrac{(P - P_1)M_2}{\mathscr{R}T}} = \frac{P_1 M_1}{P_1 M_1 + (P - P_1)M_2}$$

Using a steam table we can find $P_{1,sat}(T)$, and $P_1/P_{1,sat}$ equals the relative humidity, 20% in the air at the edge of the boundary layer and 100% at the wall.

The Grashof number is $\Delta \rho g D^3/(\rho \nu^2)$. Evaluated for the 4 in. cylinder there results:

$$\rho_{1,e} = 0.00027 \qquad \rho_{1,s} = 0.00213$$
$$\rho_{2,e} = 0.0737 \qquad \rho_{2,s} = 0.0688$$
$$\rho_e = 0.0740 \qquad \rho_s = 0.0709$$
$$m_{1,e} = 0.00365 \qquad m_{1,s} = 0.030$$

Thus

$$\Delta\rho = 0.0031 \text{ lb/ft}^3$$

$$\frac{\Delta\rho}{\rho_{\text{avg}}} = 0.043$$

$$Gr = \frac{(0.043)(32.2)(0.333)^3}{(17.0 \times 10^{-5})^2} = 1.8 \times 10^6$$

We must now check to determine whether or not this value exceeds the largest value of Gr in the data table. Run 15 gives $Gr = 1.8 \times 10^6$, just about an exact match. The values of Pr and Sc are approximately the same for both the 4.0 in. cylinder and the test data. Therefore we have approximate similarity between run 15 and the conditions under which we desire to estimate the transfer for the 4.0 in. cylinder. For Run 15

$$q_{\text{evap}} = 291 \text{ Btu/hr ft}^2$$
$$\hat{h}_{fg} = 1030 \text{ Btu/lb}$$
$$n_{1,s} = 291/1030 = 0.282 \text{ lb/hr ft}^2$$
$$\mathscr{D}_{12} = 0.765(551/460)^{1.685} = 1.04 \text{ ft}^2/\text{hr} \qquad \text{(see Section 3.9.2)}$$
$$Nu_m = \frac{n_{1,s}D}{\rho\mathscr{D}_{12}(m_{1,s} - m_{1,e})} = \frac{(0.282)(2.94/12)}{(0.071)(1.04)(0.060 - 0.008)} = 18$$

To calculate Nu we first determine q_c,

$$q_c = q_{\text{total}} - q_{\text{evap}} - q_r$$

Assuming $\epsilon = 0.96$ for a wetted surface, we obtain

$$q_c = 387 - 291 - (0.96)(1.712 \times 10^{-9})[(572)^4 - (529)^4]$$
$$= 387 - 291 - 47.3$$
$$= 48.7 \text{ Btu/hr ft}^2$$

$$Nu = \frac{q_c D}{k(T_s - T_e)} = \frac{(48.7)(2.94/12)}{(0.0154)(43.4)} = 18$$

The near equality of Nu_m and Nu is encouraging, since for $Sc = Pr$ they should be identical.

Using Nu_m and Nu we can now calculate transfer rates for the 4 in. cylinder,

$$n_{1,s} = \rho\mathscr{D}_{12}(m_{1,s} - m_{1,e})(Nu_m)/D$$
$$= (0.0725)(1.01)(0.0264)(18)/0.333 = 0.104 \text{ lb/hr ft}^2$$

$$q_{evap} = \hat{h}_{fg}n_{1,s} = (1042)(0.104) = 109 \text{ Btu/hr ft}^2$$
$$q_c = (k)(T_s - T_e)(Nu)/D = (0.0156)(15)(18)/0.333 = 13 \text{ Btu/hr ft}^2$$
$$q_r = (0.96)(0.1712)[(5.497)^4 - (5.347)^4] = 16 \text{ Btu/hr ft}^2$$
$$q_{total} = 138 \text{ Btu/hr ft}^2$$

Notice that evaporation contributes 80% to the total heat transfer from the cylinder. This example should give the student an appreciation of what the major heat loss is for a segment of perspiring skin, a swimming pool, or a hot cup of coffee.

5.6.5 *EXAMPLE* Use of Correlations to Estimate Mass Transfer

Let us use Table 5.1 to estimate the mass transfer from a wet 4 in. horizontal cylinder at 90°F to still air at 75°F, 20% relative humidity. In this case Table 5.1 gives, since $\ln(1 + u) = u, u \ll 1$,

$$Nu_m = 0.468(Gr_D Sc)^{1/4}, \qquad 10^3 < Gr_D < 10^7$$

We just calculated $Gr = 1.8 \times 10^6$, and Sc for a water vapor air mixture is approximately 0.61. We thus obtain

$$Nu_m = (0.468)(1.8 \times 10^6 \cdot 0.61)^{1/4} = 15.1$$

This value is in fair agreement with our previous value of 18. The evaporation rate is

$$n_{1,s} = \frac{(0.0725)(1.01)}{0.333}(0.0264)(15.1) = 0.088 \text{ lb/hr ft}^2$$

5.6.6 *EXAMPLE* Heat Loss from a Dip Bath

A second example is taken to be a 6 ft deep, 6 ft wide, 6 ft long un-covered dip bath with a surface temperature of 130°F. What is the convective heat and mass transfer from the water surface to quiet air at 70°F, 30% relative humidity?

As before, we must first recognize what the key physical phenomena are. As in the preceding problem, the air above the air-water interface is saturated with water vapor, molecular weight 18, and is heated to the interface temperature. As a result this warm, moisture-laden air is light and tends to rise, under the influence of gravity, into the surrounding colder, drier, and hence, heavier air. Sensible heat from the surface then transfers by conduction into the colder air, and latent heat loss results from the mass transfer by diffusion into the drier air. We scan Table 5.1 and recognize that item 8 is a reasonable fit to the situation we wish to treat.

Our first computational step is to calculate Gr as in the preceding example. To do so we need $\Delta\rho$. Proceeding as before we use the ideal gas law and Dalton's law of additive pressures for ideal gas mixtures.

$$\rho = \frac{P_1 M_1}{\mathscr{R} T} + \frac{(P - P_1) M_2}{\mathscr{R} T}$$

From Table B.11 we find $P_{1,s} = 2.221$ psia and $P_{1,e} = 0.30(0.3628) = 0.109$ psia.

$$\rho_s = \frac{(2.221)(144)(18)}{(1545)(590)} + \frac{(14.696 - 2.221)(144)(29)}{(1545)(590)}$$

$$\rho_s = 0.00633 + 0.0571 = 0.0634$$

$$m_{1,s} = 0.00633/0.0634 = 0.0998$$

$$\rho_e = \frac{(0.109)(144)(18)}{(1545)(530)} + \frac{(14.696 - 0.109)(144)(29)}{(1545)(530)}$$

$$\rho_e = 0.00035 + 0.0743 = 0.0747$$

$$m_{1,e} = 0.00035/0.0747 = 0.0047$$

$$\frac{\Delta\rho}{\rho_{avg}} = \frac{0.0113}{0.0690} = 0.164$$

$$Gr_L = \frac{g \dfrac{\Delta\rho}{\rho} L^3}{\nu^2} = \frac{(32.2)(0.164)(6)^3}{(17.9 \times 10^{-5})^2} = 3.56 \times 10^{10}$$

Then Gr is found

Our second step is to determine whether laminar or turbulent flow results from the light air near the surface rising into the more dense surrounding air. Figure 5.1, item 8, shows the critical Gr is approximately 7×10^7. Therefore, we believe the flow to be turbulent.

The third step is to determine Nu. The table gives for the heat transfer Nusselt number

$$Nu_L = 0.12(Gr_L Pr)^{1/3}$$
$$Nu_L = 0.12(3.56 \times 10^{10} \cdot 0.69)^{1/3} = 348$$

while for the mass transfer Nusselt number

$$Nu_L = 0.12(Gr_L Sc)^{1/3}$$
$$Nu_L = 0.12(3.56 \times 10^{10} \cdot 0.61)^{1/3} = 335$$

Now, inherent in our method of solution is the assumption that $Pr \doteq Sc$. Therefore we arbitrarily use the mean value for both heat and mass transfer

$$Nu_L \doteq 342$$

The convective (sensible) heat flux can then be found from

$$\frac{q_c L}{k \Delta T} = Nu_L$$

$$q_c = Nu_L \frac{k}{L} \Delta T = 342 \frac{0.0159}{6} (60) = 54.4 \text{ Btu/hr ft}^2$$

The convective mass transfer is found from

$$\frac{j_1 L}{\rho \mathscr{D}_{12} \Delta m_1} = Nu_L$$

$$j_1 = Nu_L \frac{\rho \mathscr{D}_{12}}{L} \Delta m_1 = \frac{Nu_L}{Sc} \frac{\nu}{L} \rho \, \Delta m_1$$

We write j_1 in the above form, because it is easier to find values of ν in Table B.5 than $\mathscr{D}_{12}$ in Table B.12. We find

$$j_1 = \frac{342}{0.61} \frac{(17.9 \times 10^{-5})(3600)}{6} (0.0690)(0.0998 - 0.0047)$$

$$j_1 = 0.395 \text{ lb/hr ft}^2$$

The latent heat loss is $\hat{h}_{fg} = 1019.5$ Btu/lb. The heat flux lost by evaporation is then

$$q_{\text{evap}} = (0.395)(1019.5) = 403 \text{ Btu/hr ft}^2$$

The total convective heat loss from sensible heat transfer and evaporative cooling is then

$$q_{\text{conv}} = 403 + 54 = 457 \text{ Btu/hr ft}^2$$

In addition there will be a radiative heat loss as explained in the next chapter.

5.7   SUMMARY

In this chapter we have seen that dimensionless surface fluxes of heat (Nu or St), mass (Nu_m or St_m), and momentum (c_f) into a flowing fluid with constant properties are related to dimensionless flow field parameters (Re, Pr, Sc, Gr) and geometry. Table 5.1 gives a few such relations. Tables 5.2 and 5.3 summarize the dimensionless surface fluxes and flow field parameters commonly encountered. As shown in the examples, these tables can be used to make engineering estimates of transfer coefficients and transfer rates for many technologically important processes.

Table 5.1 Illustrative Relations for Estimating Convective Transfer Coefficients (Isothermal Walls, Properties at Reference Temperature)[a]. For Mass Transfer Replace Pr with Sc and Nu with Nu_m (excepting Flow Geometry No. 9.)

Flow Geometry	Laminar Flow	Transition	Turbulent Flow
1. Flow parallel to a flat plate	$c_f(x) = 0.664Re_x^{-1/2}$ $Nu_x = 0.332Re_x^{1/2}Pr^{0.33}$ $(Pr \geqslant 0.5)$ $Nu_x = 0.565Re_x^{1/2}Pr^{1/2}$ $(Pr \leqslant 0.025)$	$Re_x \doteq 5 \times 10^5$	$\dfrac{1}{\sqrt{c_f(x)}} = 4.13 \ln \left[Re_x c_f(x) \right]$ $Nu_x = 0.0296Re^{0.8}Pr^{0.6}$ $(10.0 \geqslant Pr \geqslant 0.5)$
2. Flow in a duct (a) A straight pipe	$c_f = 16/Re_D$ $f = 64/Re_D$ $Nu_D = 3.65 + \dfrac{0.0668(D/L)Re_D Pr}{1 + 0.04[(D/L)Re_D Pr]^{2/3}}$	$Re_D \doteq 2300$	For f see Figure 8.3 $c_f = f/4$ $Nu_D = 0.023Re^{0.8}Pr^{0.33}, Pr > 0.5$ (See Table 8.1v)
(b) Parallel planes	$c_f = 24/Re_{D_h}$ $f = 96/Re_{D_h}$ $Nu_{D_h} = 7.54 + 0.0234 \dfrac{Re_{D_h} Pr}{(L/D_h)}$	$Re_{D_h} = 2800$	For f see Figure 8.3 $c_f = f/4$ Nu_{D_h} same as 2a
3. Flow across a circular cylinder	$Nu_D = 0.43 + K Re_D^m Pr^{0.31}$ $1000 \leqslant Re_D < 4000$ $(K = 0.53, m = 0.5)$ $4000 \leqslant Re_D < 40000$ $(K = 0.193, m = 0.618)$	$Re_D = 40000$	$Nu_D = 0.43 + 0.0265Re_D^{0.8}Pr^{0.31}$

4. Flow across a sphere

$$Nu_D = 2 + 0.37 Re_D^{0.6} Pr^{0.33}$$

$Re_D \doteq 150,000$

Use the correlation for Nu_L for a vertical wall with $L \doteq 2.5D$

5. Free convection on a horizontal cylinder

$$Nu_D = \frac{2}{\ln\{1 + 2/[0.468(Gr_DPr)^{1/4}]\}}$$

$Gr_D \doteq 10^7$

6. Free convection on a sphere

$$Nu_D = 2 + K(Gr_DPr)^{1/4}$$
$$0 < Gr_DPr < 50, \quad K = 0.3$$
$$50 < Gr_DPr < 200, \quad K = 0.4$$
$$200 < Gr_DPr < 10^6, \quad K = 0.5$$
$$10^6 < Gr_DPr < 10^8, \quad K = 0.6$$

$Gr_D \doteq 10^8$

7. Free convection on a vertical wall

$$Nu_x = 0.508 \frac{Gr_x^{1/4}Pr^{1/2}}{[0.952 + Pr]^{1/4}} \quad (x = L)$$

$$Nu_L = \frac{4}{3}Nu_x(x=L)$$

$Gr_x \doteq 10^8$

$$Nu_x = 0.0296 \frac{Gr_x^{0.4}Pr^{0.47}}{[1 + 0.494Pr^{0.67}]^{0.4}}$$

$$\bar{h} = \frac{k}{L}Nu_L = \frac{1}{L}\int_0^{x_{\text{trans}}} h(x)_{\text{lam}}\, dx + \frac{1}{L}\int_{x_{\text{trans}}}^{L} h(x)_{\text{turb}}\, dx$$

^{a} $T_{\text{ref}} \doteq \dfrac{T_s + T_e}{2}$ for low speed external flows; $T_{\text{ref}} \doteq \dfrac{T_s + T_b}{2}$ for duct flows.

Table 5.1 (Continued)

Flow Geometry	Laminar Flow	Transition	Turbulent Flow
8. Free convection on a horizontal square (a) Facing up	$Nu_L = 0.54(Gr_LPr)^{1/4}$ $10^5 < Gr_L < 7 \times 10^7$	$Gr_L \doteq 7 \times 10^7$	$Nu_L = 0.12(Gr_LPr)^{1/3}$ (Note that L is of no importance.)
(b) Facing down	$Nu_L = 0.27(Gr_LPr)^{1/4}$ $3 \times 10^5 < Gr_L < 3 \times 10^{10}$		
9. Film condensation of pure saturated vapor on a vertical wall	$Nu_x = \{Gr_xPr_l/4Ja\}^{1/4}$ $Gr_x = \dfrac{g[(\rho_l - \rho_v)/\rho_l]x^3}{\nu_l^2}$ $Ja = c_p\Delta T/h_{fg}$ $(\Delta T = T_{sat} - T_w)$	$Re_F \doteq 1800$ $Re_F = \dfrac{4\Gamma}{\mu_l}$ $\Gamma = \int_0^\delta \rho u\,dy$ $= \int_0^x (q(x)/\hat{h}_{fg})\,dx$	
10. Flow through a packed bed of spheres	$StPr^{2/3} = 1.625Re_D^{-1/2}$ $15 < Re_D < 120$ $Re_D = \dfrac{\dot{m}D}{A_c\mu}$	$Re_D = 120$ $\dfrac{\mathscr{P}}{A_c} = \dfrac{6(1-\epsilon_v)}{D}$	$StPr^{2/3} = 0.687Re^{-0.327}$ $120 < Re < 2000$ ϵ_v volume void fraction

Additional relations for turbulent flow may be found in Chapter 8.

Table 5.2 Summary of Dimensionless Wall Fluxes and Wall Transfer Coefficients

Type of Surface Flux	Transfer Coefficient	Nusselt Number or Friction Factor	Stanton Number or Skin Friction Coefficient
Heat q	$h_c = \dfrac{q_s}{(T_s - T_e)}$	$Nu = \dfrac{h_c L}{k} = \dfrac{q_s L}{k(T_s - T_e)}$ For pipe flow use D instead of L	$St = \dfrac{h_c}{\rho c_p V} = \dfrac{q_s}{\rho c_p V (T_s - T_e)}$
Mass j_i	$K = \dfrac{j_{i,s}}{(m_{i,s} - m_{i,e})}$	$Nu_m = \dfrac{KL}{\rho \mathcal{D}} = \dfrac{j_{i,s} L}{\rho \mathcal{D}(m_{i,s} - m_{i,e})}$	$St_m = \dfrac{K}{\rho V} = \dfrac{j_{i,s}}{\rho V (m_{i,s} - m_{i,e})}$
Shear τ		$f = \dfrac{\Delta P}{\left(\dfrac{L}{D}\right)(\frac{1}{2}\rho V^2)}$ (pipe flow)	$c_f = \dfrac{\tau}{\frac{1}{2}\rho V^2}$

Table 5.3 Summary of the Chief Dimensionless Parameters Influencing Heat, Mass, and Momentum Flow Fields

Number	Definitions	Area of Significance
Reynolds	$Re = VL/\nu$	Forced Convection. For plates L is the length. For ducts L is replaced by the hydraulic diameter, $D_h = 4A_c/\rho$.
Prandtl	$Pr = \nu/\alpha$	Convective heat transfer, both free and forced
Schmidt	$Sc = \nu/\mathcal{D}$	Convective mass transfer, both free and forced
Grashof	$Gr = \dfrac{(\Delta\rho/\rho)gL^3}{\nu^2}$	Free convection
Froude	$Fr = Re^2/Gr = \rho V^2/\Delta\rho gL$	Mixed free and forced convection
Eckert	$Ec = V^2/2c_p(T_s - T_e)$	Convective heat transfer in high speed flow

EXERCISES

1. Consider a length of horizontal 1 in. schedule 40 pipe (see Appendix B, Table B.16, for actual dimensions). What is the convective heat transfer coefficient on the outside of the pipe when the pipe is 110°F and when

    (a) it is immersed in still air at 90°F (1 atm pressure), and
    (b) it is immersed in still water at 90°F?

    (See Tables B.5, B.7, and B.9 for properties.)

2. Consider a 21 ft length of 1 in. schedule 40 pipe. What is the convective heat transfer coefficient and the friction factor f inside the pipe when the pipe is 110°F and

    (a) water at 90°F flows at 10 ft/sec,
    (b) oil (SAE 50) at 90°F flows at 10 ft/sec,
    (c) air at 90°F and 1 atm flows at 10 ft/sec?

3. Estimate the heat loss in Btu/hr from five sides of a 250°F oil pan 1 ft wide, 6 in. deep, and 30 in. long when 70°F air flows over it at 88 ft/sec.

4. Air at 0°F blows at 30 ft/sec over the exterior of a 12 in. diameter pipeline spanning a river gorge. The line is heated to 100°F to reduce the viscosity of the fluid inside. What would be the heat loss per 100 ft of pipe if the line were uninsulated?

5. Insulating "wet" suits worn by scuba divers are usually made of $\frac{1}{8}$ in. thick foam neoprene ($k = 0.025$ Btu/hr ft °F) which traps next to the skin a layer of stagnant water, perhaps $\frac{1}{8}$ in. thick. Determine the rate of heat loss from a 6 ft tall diver swimming at 5 miles per hour in 55°F sea water if his skin temperature does not fall below 74°F. Express your answer in kcal/hr.

6. Air at 60°F, 50% relative humidity, flows at 5 ft/sec over a 40 ft long swimming pool at 80°F. Estimate the average heat loss per square foot due to:

    (a) sensible heat transfer
    (b) evaporation

7. Estimate the convective and evaporative heat losses from a 135°F water bath into still air at 80°F, 15% relative humidity. The bath is in a chemical dip tank 3 ft wide.

The student, before attacking Exercises 8 and 9, would benefit from a reading of Appendix A.

8. Pump performance tests involve measuring the flow rate pumped Q_f and horsepower consumed P_{hp} versus pump rpm N_{rpm} and pressure difference between the pump inlet and outlet. Pump performance curves are usually plotted as pressure difference (or pressure difference divided by ρg, called *head, H*), power consumption, and efficiency η versus volume flow rate for a constant pump speed. A family of pumps may have different-sized rotors which are otherwise geometrically similar. At times data may be available only for a small member of the family, with rotor size D_1, and those data for water, when it is intended to use a large unit with rotor D_2 to pump, say, jet aircraft fuel. An observation which then assumes some importance is that at large values of Reynolds number the skin friction coefficient is insensitive to Re (nearly constant). Explain how you would plot pump performance data in the most useful fashion. Engineers characterize a family of pumps by its specific speed, a value of

$$N_S = N_{rpm} Q_f^{1/2} / H^{3/4}$$

defined at the point of maximum efficiency when pumping water. Why do all members of a family have the same value of specific speed?

9. Dr. Ernest Schmidt of the Technische Höchschule, Munchen, once attempted to persuade his wife not to leave windows open too long during winter to get fresh air. He ran an experiment in a 23 cm high, 23 cm wide box filled with carbon dioxide gas. He found that 11 sec after opening a model window on the upside down box, the model room had ceased to transfer mass by convection. Explain how the model test results can be used to estimate how long to open a window in a 4 meter wide room with a 4 meter ceiling when the room is 10°C hotter than the outside air.

10. A thermometer covered with wet cotton is known as a **wet-bulb** thermometer. A dry bulb and wet bulb pair of thermometers constitute a psychrometer, a device used to measure the humidity of air which is blown over the thermometers. When the wet bulb reads the same as the dry bulb, the air has 100% relative humidity and is saturated. What is the principle of the device? Use Table B.11 and values of the Schmidt number and Prandtl number from Tables B.5 and B.12 to construct a chart of wet bulb temperature versus dry bulb temperature for 50% relative humidity. Compare your results with a psychrometric chart, available in any good engineering handbook.

REFERENCES

Surface Transfer Coefficients

For heat transfer a good handbook is:

> W. H. McAdams, *Heat Transmission*. New York: McGraw-Hill, 1954.

For mass transfer useful relations are given by:

> Treybal, R. E., *Mass Transfer Operations*, New York: McGraw-Hill, 1968. (See Table 3.3, page 63.)
> Spalding, D. B., *Convective Mass Transfer*, New York: McGraw-Hill, 1963. (See particularly pp. 48–56.)

Similarity and Dimensional Analysis

Virtually every book written on the subject of transfer processes contains a section on similarity and dimensional analysis. See, for example:

> Boelter, L. M. K., V. H. Cherry, H. A. Johnson, and R. C. Martinelli, *Heat Transfer Notes*. New York: McGraw-Hill, 1965. (See Chapter 11.)
> Chapman, A. J., *Heat Transfer*, 2nd ed. New York: Macmillan, 1967. (See Sections 6.7 and 7.10.)

CHAPTER 6

RADIATION AND

FREE MOLECULE TRANSFER

6.1 INTRODUCTION

We have considered thus far transport by diffusion. For example, in heat conduction in a gas at atmospheric pressure and room temperature, an energy carrier, a molecule, is able to travel only a very short distance l before colliding with another molecule. The result is that the energy must diffuse through the gas, taking a very tortuous course. This diffusive process, where the system characteristic size L is large compared with the mean free path l, results in transport down a temperature gradient at a rate given by Fourier's law,

$$\mathbf{q} = -k\boldsymbol{\nabla}T, \qquad l \ll L$$

When mean free paths are long, however, as is the case for a molecule in a vacuum chamber or a photon in an ordinary room, the energy, mass, and momentum travel with the carrier in a direct straight line path from one wall to another. We do not use Fourier's law of conduction, Fick's law of diffusion, or Newton's law of viscosity. Instead, we take account of the fact that molecules or photons travel in straight lines from one wall to another.

We will consider transport when the distances between walls are short compared to the distance between collisions so that molecule-molecule interactions or photon-molecule interactions can be ignored. A photon from a room temperature source has a characteristic wavelength of roughly 10μ (10^{-5} meter) which is short compared to distances between walls in the usual system. The mean free path of such a photon in ordinary air is approximately 20 km, the exact value depending upon the wavelength, amount of dust, water vapor, CO_2, and other trace impurities. This value is very large compared to the distances between walls. An air molecule at room temperature has a size (also a de Broglie wavelength) of the order of magnitude 10^{-10} meter, but the mean free path l at room temperature is on the order of 5 cm divided by the pressure in microns of mercury, or, stated in another way, $l = 5 \times 10^{-5}$ (m-torr)/P. At a vacuum chamber pressure of 10^{-7} torr, the mean free path is on the order of 500 meters, and is consequently longer than the largest of vacuum chambers.

Applications in which the mean free paths are long are too numerous to catalog. Detection of ocean currents, breast cancers or circulatory ailments, fires, and hot bearings is accomplished by detectors responding to thermal photons emitted by water, human skin, burning substances, and passing freight cars, respectively. Spacecraft are heated by photons from the sun and, in low altitude orbits, by encounters with molecules as well, while they are cooled by emission of photons into space. The spacecraft also have forces exerted on them by fluxes of these carriers. Radiation heating of trees and fruit by orchard heaters or occupants of rooms by radiant panels is common. Industrial furnaces are used for melting, annealing, and other processing of materials. Vacuum systems are used in evaporative deposition of coatings, in insulating cryogens, and in the refining of metals. The engineer wishes to compute molecular fluxes to vacuum pump inlets and cryogenic walls in such systems.

In the first half of the chapter are presented the macroscopic descriptions and concepts which are used by engineers in practice. The second half gives a brief introduction to the underlying microscopic concepts and shows how these concepts have practical macroscopic consequences.

6.2 TOTAL EMISSION FROM BLACK SURFACES

6.2.1 Black Surfaces

Many surfaces encountered in technology can be idealized as being black to photons and/or molecules. The word *black* merely denotes that no photons or molecules are reflected from the surface; all photons are *absorbed* and all molecules are either *adsorbed* or *condense*. Whether or not a surface is black or nearly so can be determined, for example, by directing a stream of molecules from a source in a vacuum chamber onto a surface and determining whether or not they are promptly reflected with the same energies they possessed before striking the surface. Much the same can be done with photons from a radiant source. An obvious consequence of our definition of a black surface is that all photons, or molecules, leaving the surface are *emitted* by the surface.

6.2.2   Radiant Fluxes

All have experienced feeling the warmth of the sun on the skin, feeling heat radiated across a room by a fire or radiant heater, and feeling the cooling of the forehead when gazing at the stars in the black sky on a clear night. Quantitative measurements of emission from black surfaces led J. Stefan, in 1879, to postulate the law

● $$q_b{}^+ = \sigma T_1{}^4 \tag{6.1}$$

where $q_b{}^+$ is the *outgoing* radiant heat flux leaving a black surface at *absolute* temperature T_1 and σ is the Stefan-Boltzmann constant,

$$\sigma = 5.6697 \times 10^{-8} \text{ w/m}^2 \, {}^\circ\text{K}^4$$
$$= 1.712 \times 10^{-9} \text{ Btu/hr ft}^2 \, {}^\circ\text{R}^4$$

The outgoing heat flux is called the **radiosity**. When a surface is surrounded by a black source at temperature T_2, the surface experiences an *incoming* flux $q_b{}^-$, given by

$$q_b{}^- = \sigma T_2{}^4 \tag{6.2}$$

The incoming flux is called the **irradiation**.

The *net* outward radiant flux from a surface is simply the radiosity minus the irradiation,

● $$q = q^+ - q^- \tag{6.3}$$

Thus if a surface is black at temperature T_1 and is completely surrounded by a black source at temperature T_2, the net flux from surface 1 is

$$q_1 = \sigma T_1{}^4 - \sigma T_2{}^4 \tag{6.4}$$

Notice that, when the surface and source are at the same temperature T_1, Eq. (6.4) gives $q_1 = 0$, and the system can attain a state of thermodynamic equilibrium.

For the present, Stefan's law, Eq. (6.1), is best regarded as based on experimental observation. Later, in Section 6.6.7, we will regard this law from a microscopic viewpoint in which each photon leaving the surface is an energy carrier, and, by summing over all such photons, we will find the required expression for the outgoing radiant heat flux. Notice also that, in writing down Eqs. (6.3) and (6.4), we have *assumed* that an isothermal black surface radiates at the rate $q_b{}^+ = \sigma T_1{}^4$ regardless of the irradiation falling upon the surface; we suppose that $q_b{}^+$ is $\sigma T_1{}^4$ even when the irradiation is zero. No experiment yet performed has shown this assumption to be in error.

The outgoing flux, when it comes from a black body so that no reflected photons make up part of it, is due to emission only. Thus the radiosity $q_b{}^+$ of a black body is also called the **black body emissive power**. Table 6.1 shows the black body emissive power for a few temperatures to emphasize that radiation is particularly important at elevated temperatures. Even at room temperature the flux is of a magnitude comparable to natural convection heat transfer fluxes or fluxes due to conduction in a gas.

Table 6.1   Black Body Emission

Source Temperature	Emission	
	w/m²	Btu/hr ft²
300°K (room temperature)	459	146
1000°K (cherry-red hot)	56,700	18,000
3000°K (lamp filament)	4,590,000	1,460,000
5700°K (sun temperature)	60,000,000	19,000,000

6.2.3   EXAMPLE Radiative Cooling of a Catalytic Afterburner

A catalytic afterburner is part of an automobile exhaust system. The incoming stream of 300 lb/hr exhaust gas contains a mass fraction m_1 of CO equal to 0.04, and preheated air is added to this stream to oxidize the CO to CO_2 inside the device. A system design requirement is that 60% of the heat liberated by the combustion must be radiated away by the afterburner case. How much radiator area is required if the case is allowed to attain a temperature of 1400°F? Assume the case surface to be black, the surrounds to be black and at 140°F, and the heat of combustion of CO to CO_2 to be 4350 Btu/lb CO.

The total heat release is

$$\dot{Q} = \dot{m} m_1 \, \Delta \hat{h}_{comb} = (300)(0.04)(4350)$$
$$\dot{Q} = 52{,}200 \text{ Btu/hr}$$

The radiosity of the black radiator would be

$$q^+ = \sigma T_1^4 = (1.712 \times 10^{-9})(1860)^4 = 2.05 \times 10^4 \text{ Btu/hr ft}^2$$

The irradiation is almost negligible by comparison;

$$q^- = \sigma T_2^4 = (1.712 \times 10^{-9})(600)^4 = 222 \text{ Btu/hr ft}^2$$

The net flux is therefore

$$q = q^+ - q^- = 20{,}500 - 222 = 20{,}300 \text{ Btu/hr ft}^2$$

A radiating area is then found from

$$\dot{Q}_r = qA, \quad A = \dot{Q}_r/q = (0.60)(52{,}200)/20{,}300$$
$$A = 1.54 \text{ ft}^2$$

6.2.4   Molecular Fluxes

The molecular flux bombarding a surface is denoted as J^- (molecules/m² sec). When a gas is in thermodynamic equilibrium at temperature T,

$$\bullet \qquad\qquad J^- = \tfrac{1}{4}\mathcal{N}\bar{v} \qquad\qquad\qquad \textbf{(6.5a)}$$

where $\mathcal{N}$ is the number of molecules per unit volume given by the ideal gas law and $\bar{v}$ is the mean molecular speed.

$$\mathcal{N} = \frac{P}{kT} \qquad \bar{v} = \left(\frac{8kT}{\pi m}\right)^{1/2}$$

The quantity k, Boltzmann's constant, is the universal gas constant divided by Avogadro's number, that is, the gas constant per molecule. Table 6.2 gives numerical values for k and Avogadro's number. The molecular mass m is obtained by multiplying the value for the atomic mass unit by the molecular weight.

Table 6.2 Selected Physical Constants and Conversion Factors

Quantity	Symbol	
Atomic mass unit	amu	1.66043×10^{-27} kg
Velocity of light	c	$2.997925 \times 10^{+8}$ meter/sec
Planck constant	h	6.6256×10^{-34} joule sec
Boltzmann constant	k	1.38054×10^{-23} joule/°K
Avogadro number	N_{Av}	6.02252×10^{26} molecules/kg mole
Length		$10^6 \, \mu$/meter
		10^{10} Å/meter
Energy		10^7 erg/joule
		6.2418×10^{18} ev/joule
Pressure		1.01325×10^5 newtons/m² atm
		760 torr/atm

Now picture the gas (say water vapor) contained in an enclosure, the walls of which are coated with the condensed phase (water or ice). At thermodynamic equilibrium the system is isothermal, and molecules must leave the walls at the same rate at which they arrive. The flux of molecules emitted by the condensed phase on the surface (assumed to be black) is

$$J_b^+ = \tfrac{1}{4}\mathcal{N}\bar{v} \tag{6.5b}$$

Observe that Eq. (6.5b) has thus been derived in a *thermodynamic* manner; no physical model for the emission process has been postulated. Notice also that Eq. (6.5b) is the molecular analog to Eq. (6.1) which applied to photon emission, and it too may be regarded for now as a law based on experiment. As was the case for photons, experiments on the evaporation of various substances into a vacuum have shown that the law is valid even when the incident flux is zero. In Section 6.6.9 this law will also be derived from microscopic concepts.

The flux of energy striking a black surface being bombarded by molecules is obtained by multiplying the flux of molecules by the energy that they bring to the surface. As shown in Section 6.6.9, the result for a monatomic gas is

$$q^- = \tfrac{1}{4}\mathcal{N}v(2kT)$$

A polyatomic gas carries additional energy so we can write more generally

$$q^- = \tfrac{1}{4}\mathcal{N}\bar{v}(m\hat{u} + \tfrac{1}{2}kT) \tag{6.6}$$

where $\hat{u}$ is the internal energy per unit mass. For a monatomic gas $m\hat{u} = 3kT/2$.

6.2.5   *EXAMPLE* Evaporation of Aluminum into a Vacuum

Determine the rate of evaporation from an aluminum surface into a vacuum if its surface temperature is 889°C. *The Handbook of Chemistry and Physics* gives the saturation vapor pressure of aluminum at 889°C as 1.00×10^{-3} mm Hg (1 mm Hg = 1 torr).

The saturation vapor pressure is precisely the pressure of pure vapor which would be in equilibrium with the condensed phase surface at the specified temperature. Therefore we may use Eq. (6.5b) directly

$$J_b{}^+ = \tfrac{1}{4}\mathcal{N}\bar{v}$$

Assuming a monatomic gas ($M = 27$) we obtain, in SI units,

$$\bar{v} = \left(\frac{8kT}{\pi m}\right)^{1/2}$$

$$\bar{v} = \left[\frac{(8)\,(1.38 \times 10^{-23})\,(1162)}{(3.14)\,(1.66 \times 10^{-27})\,(27)}\right]^{1/2} = 0.95 \times 10^3 \text{ m/sec}$$

The molar concentration is

$$c = \frac{\mathcal{N}}{N_{Av}} = \frac{P}{\mathscr{R}T} = \frac{(10^{-3}/760)\,(1.013 \times 10^5)}{(8.3 \times 10^3)\,(1162)} = 1.38 \times 10^{-8} \text{ kg mole/m}^3$$

Upon substitution there results

$$\frac{J_b{}^+}{N_{Av}} = \tfrac{1}{4}(0.95 \times 10^3)\,(1.38 \times 10^{-8}) = 3.28 \times 10^{-6} \text{ kg mole/m}^2 \text{ sec}$$

6.3   TRANSPORT BETWEEN INFINITE PARALLEL BLACK WALLS

6.3.1   Radiant Energy Transport

We have already noted in Eq. (6.4) that the net flux from one black surface entirely surrounded by another is simply $\sigma T_1{}^4 - \sigma T_2{}^4$. Thus the net flux of radiation between infinite parallel black walls is also given by this simple relation.

6.3.2   Free Molecule Conduction

Exactly the same can be written for two parallel plates exchanging mass. The net flux is simply

$$\tfrac{1}{4}\mathcal{N}_1\bar{v}_1 - \tfrac{1}{4}\mathcal{N}_2\bar{v}_2 = J^+ - J^- \tag{6.7}$$

where $\mathcal{N}_1$ is computed using the saturation vapor pressure at surface temperature T_1 of the one plate, and $\mathcal{N}_2$ is similarly computed for the second plate. But Eq. (6.7) is not particularly useful. For example, it describes transfer of mass across the gap between two plates when both plates are of the same substance; the one at the higher temperature undergoes net evaporation and the other net condensation. Such a situation is not of much practical importance.

The more usual engineering situation is that of two parallel walls containing a superheated gas with no net transfer of mass. Suppose that the pressure is low enough for the distance between the plates to be small compared to the mean free path so that "free molecule" conduction will take place. One wall, the left one, is heated to temperature T_1, while the other wall, the right one, is colder at temperature T_2. We assume that the surfaces are black to incident molecules (though not necessarily to photons). Then the molecules are adsorbed on the walls long enough for those which are reemitted to have acquired the statistical nature of a gas in thermodynamic equilibrium with the wall.

The situation described above occurs quite commonly in the vacuum jackets around pipes and containers used to hold cryogenic liquids such as liquid helium, hydrogen, and nitrogen. An evacuated space around the container is used to reduce the boil-off losses of the cryogens. The space is evacuated to reduce the number density so that there are not too many energy carriers, and layers of shielding with spacing less than the mean free path are used to prevent the energy carriers from crossing directly to the cold inner wall in contact with the cryogen from the relatively hot outer wall heated by the surrounds at room temperature.

Once one wall is heated and the other cooled, the number of molecules per unit volume in each space divides itself into two populations, one consisting of molecules going from left to right $\mathcal{N}_1^+$ and one going from right to left $\mathcal{N}_2^-$. The sum of the two parts equals the whole population

$$\mathcal{N} = \mathcal{N}_1^+ + \mathcal{N}_2^- \tag{6.8}$$

The number of molecules crossing per unit area per unit time from left to right J^+ equals the number crossing from right to left J^-, because we are supposing that steady state prevails with no net mass transfer. We assume that the wall material has a very low vapor pressure so that molecules of the wall material are not subliming from one wall and condensing on the other. (High vacuum systems are constructed from stainless steel rather than a metal such as zinc.) We therefore write

$$J^- = J^+ \tag{6.9}$$

In Eqs. (6.5a,b) the quantity $\mathcal{N}$ was the total population. However, in the case when T_1 was equal to T_2, the quantity $\mathcal{N}$ was twice the population $\mathcal{N}_1^+$ of the molecules going one way. Thus Eqs. (6.5a,b) may be rewritten generally as

$$J^- = \tfrac{1}{2}\mathcal{N}_2^- \bar{v}_2 \tag{6.10a}$$

$$J^+ = \tfrac{1}{2}\mathcal{N}_1^+ \bar{v}_1 \tag{6.10b}$$

We now assume that, because the surfaces are black, the emitted molecules from each have a velocity distribution identical to that characterizing equilibrium emission. Therefore

$$\bar{v}_1 = \left(\frac{8kT_1}{\pi m}\right)^{1/2}, \qquad \bar{v}_2 = \left(\frac{8kT_2}{\pi m}\right)^{1/2}$$

Equations (6.8) to (6.10b) are a set of four simultaneous algebraic equations in four unknowns which must be solved. Substituting Eqs. (6.10a) and (6.10b) into (6.9) and using Eq. (6.8) to eliminate $\mathcal{N}_2^-$ gives

$$\mathcal{N}_1^+ = \mathcal{N}\frac{\bar{v}_2}{\bar{v}_1 + \bar{v}_2} = \mathcal{N}\frac{T_2^{1/2}}{T_1^{1/2} + T_2^{1/2}} \tag{6.11}$$

Equation (6.10b) then gives

$$J^+ = \tfrac{1}{4}\mathcal{N}\left(\frac{8kT_M}{\pi m}\right)^{1/2}, \qquad T_M^{1/2} = \frac{2T_1^{1/2}T_2^{1/2}}{T_1^{1/2} + T_2^{1/2}} \tag{6.12}$$

The *net* flux transferred is then the flux crossing from left to right q_1^+ minus that crossing from right to left. From Eq. (6.6), assuming c_v constant, we obtain

$$q_{net} = J^+(mc_vT_1 + \tfrac{1}{2}kT_1) - J^-(mc_vT_2 + \tfrac{1}{2}kT_2)$$

$$q_{net} = \tfrac{1}{4}\mathcal{N}\left(\frac{8kT_M}{\pi m}\right)^{1/2}(mc_v + \tfrac{1}{2}k)(T_1 - T_2)$$

Introducing the ratio of specific heats and the relation $c_p = c_v + k/m$, we can write

$$q_{net} = \tfrac{1}{4}\mathcal{N}\left(\frac{8kT_M}{\pi m}\right)^{1/2}\left(\frac{\gamma + 1}{2\gamma}mc_p\right)(T_1 - T_2) \tag{6.13}$$

This result gives the steady-state power transfer between two parallel plates when there is no net mass transfer. Note that k/m is $\mathcal{R}/M$ and $\mathcal{N}m$ is density ρ.

6.3.3 *EXAMPLE* Superinsulation

A vacuum system is evacuated to 10^{-5} mm Hg pressure. Calculate the free molecule conduction between two layers of superinsulation 0.1 in. apart when one layer is at 70°F and the other is at 69°F.

Again our first step is to compute $\bar{v}$, this time for $T_M = 69.5°F = 294°K$.

$$\bar{v} = \left(\frac{8kT_M}{\pi m}\right)^{1/2} = \left(\frac{(8)(1.38 \times 10^{-23})(294)}{(3.14)(1.66 \times 10^{-27})(29)}\right)^{1/2} = 4.63 \times 10^2 \text{ m/sec}$$

or

$$\bar{v} = 1520 \text{ ft/sec} = 5.47 \times 10^6 \text{ ft/hr}$$

As we have said, in Eq. (6.13) the grouping $\mathcal{N}m$ is just the density ρ,

$$\rho = \frac{PM}{\mathcal{R}T} = \frac{(10^{-5})(14.7)(144)(29)}{(760)(1545)(529)} = 0.987 \times 10^{-9} \text{ lb/ft}^3$$

For air $c_p = 0.24$ and $\gamma = 1.4$. Substituting these numbers in Eq. (6.13) we obtain

$$q = \tfrac{1}{4}(0.987 \times 10^{-9})(5.47 \times 10^{6})\left(\frac{2.4}{2.8}\right)(0.24)(70\text{--}69)$$

$$= 2.78 \times 10^{-4} \text{ Btu/hr ft}^2$$

We see that evacuation to 10^{-5} mm Hg effectively eliminates conduction at room temperature, for if the mean free path were short compared to the gap we would have

$$q = \frac{k\,\Delta T}{L} = \frac{(0.015)(70\text{--}69)}{(0.1)/(12)} = 1.8 \text{ Btu/hr ft}^2$$

But, in fact, the mean free path is about 5 m for this situation, and q has the much lower value of 2.78×10^{-4} Btu/hr ft².

6.4   TRANSFER BETWEEN FINITE BLACK WALLS

6.4.1   The Shape Factor

In the development which follows, the shape factor concept is presented in the context of radiation (photon) transport. However, the concept is equally applicable to molecular transport in, for example, a vacuum chamber.

Consider two finite walls such as those shown in Figure 6.1. To find the one-way rate of energy (power) transfer by thermal radiation from surface 1 to surface 2 we introduce the shape factor concept. The shape factor F_{1-2} is the fraction of power leaving surface 1 which is intercepted by surface 2. We can therefore write simply

$$Q_{1-2}^{+} = A_1 \sigma T_1{}^4 F_{1-2} = A_1 F_{1-2} \sigma T_1{}^4 \tag{6.14}$$

It will be shown in Section 6.8.3 that $A_1 F_{1-2}$ is given by the double area-integral

$$A_1 F_{1-2} = \int_{A_1}\int_{A_2} \frac{\cos\theta_1 \cos\theta_2\, dA_2\, dA_1}{\pi r_{1-2}^2} \tag{6.15}$$

where dA_1 and dA_2 are elements of area as shown in Figure 6.1. Figures 6.2 and

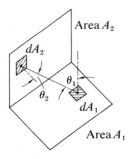

Area A_2

dA_2

θ_1

θ_2

dA_1

Area A_1

Figure 6.1   Two finite areas.

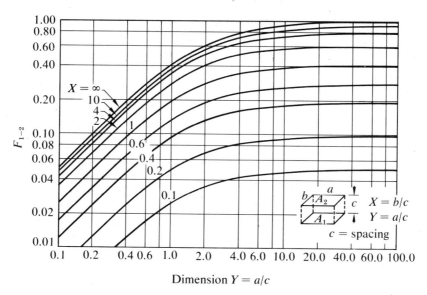

Figure 6.2 Shape factors for opposite rectangles.

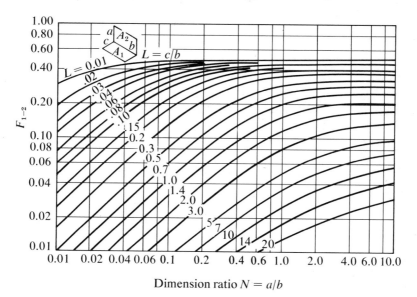

Figure 6.3 Shape factors for adjacent rectangles.

6.3 present values for adjacent and opposite rectangles. The power transfer from 2 to 1 is

$$\dot{Q}_{2-1}^+ = A_2 F_{2-1} \sigma T_2^4$$

The net transfer is consequently

$$\dot{Q}_{1-2} = A_1 F_{1-2} \sigma T_1^4 - A_2 F_{2-1} \sigma T_2^4 \tag{6.16}$$

Note the symmetry in the double integral of Eq. (6.15). We see that the 1 and 2 subscripts are interchangeable; therefore

$$A_1 F_{1-2} = A_2 F_{2-1} \tag{6.17}$$

and Eq. (6.16) may be written

$$\dot{Q}_{1-2} = (\sigma T_1{}^4 - \sigma T_2{}^4) A_1 F_{1-2} \tag{6.18}$$

Imagine a complete enclosure such as an empty room. Suppose that the surfaces of the enclosure are m in number, each isothermal and black. Consider the ith surface, for example, the left-hand wall. The net power transfer by radiation between that surface and all the others is

$$\dot{Q}_i = \sum_{j=1}^{m} \dot{Q}_{i-j} = \sum_{j=1}^{m} (\sigma T_i{}^4 - \sigma T_j{}^4) A_i F_{i-j} \tag{6.19}$$

The sum may be imagined to include surface i itself, for F_{i-i} is not zero for a concave surface, but the difference in σT^4 is zero for this term in any event.

We now have in Eq. (6.19) a general result, but let us think a bit about the meaning of F_{i-j}. First, suppose all surfaces j other than i itself were at zero degrees absolute temperature. Then the net power loss by convex surface i would be $A_i \sigma T_i{}^4$. Equation (6.19) for this case shows

$$\sum_{j=1}^{m} F_{i-j} = 1 \tag{6.20}$$

We see that F_{i-j}, a positive number, cannot be greater than unity. Now consider each term separately in Eq. (6.18) or (6.19). The term $\sigma T_i{}^4 A_i F_{i-j}$ is the one-way power from surface i to j. The total power leaving the surface is $A_i \sigma T_i{}^4$. The shape factor F_{i-j} is thus the fraction of the power radiated by surface i which is intercepted by surface j. Seen in this light it is obvious why the shape factors in a complete enclosure must sum to unity.

As we have said, the shape factor F_{i-j} is the fraction of power radiated by surface i which is intercepted by surface j. But another way, and perhaps a better way, of saying the same thing is to say that the shape factor F_{i-j} is the fraction of the surrounds of surface i taken up by surface j. Consider the one-way power from j to i, $A_j F_{j-i} \sigma T_j{}^4$, which by Eq. (6.17) is equal to $A_i F_{i-j} \sigma T_j{}^4$. The average power per unit area of surface i *received* from surface j is consequently $F_{i-j} \sigma T_j{}^4$. This average flux incident on i is seen to be the flux leaving j times the fraction of the surrounds of i taken up by j, F_{i-j}.

6.4.2   Shape Factor Values

Some shape factors can be found from inspection. For example, consider a convex surface 1, such as the outside of a long tube, completely enclosed by a second surface 2, for example, the inner surface of a larger concentric cylinder. The shape factor F_{1-2} is unity, and from Eq. (6.17) the shape factor F_{2-1} is A_1/A_2. For a second example, consider the sides of a long equilateral triangle (perhaps an A-frame cabin). The shape factor from one side to another is, by virtue of Eq. (6.20), one-half. By inspection, one-third is the shape factor between two sides of a regular tetrahedron. Also by inspection the shape factor F_{1-2} from a small area 1 adjacent to a large perpendicular area 2 is one-half.

   In some other situations, Eq. (6.15) can be analytically integrated to yield the shape factor. It is easily shown that the shape factor from a small area to any surface of revolution, whose axis is coincident with the normal of the small area, is simply $\sin^2 \theta$, where θ is the angle from the normal of the area to the limiting ray tangent to or at the edge of the surface. For example, the shape factor from a small detector to a telescope mirror or lens can be found from this relation, as can also the shape factor from a downward facing area on a satellite to the planet about which it orbits. The shape factor from a small area to an infinitely long object parallel to the area is equal to $\frac{1}{2}(\sin \theta_a - \sin \theta_b)$, where θ_a and θ_b are the angles between the normal and the two limiting rays. The shape factor from an area A_1 on the inside of a sphere wall of radius R to another area A_2 also on the inside of the sphere wall is simply $A_2/4\pi R^2$.

   In many other cases of practical interest, analytical or numerical results may be obtained from the definition Eq. (6.15) or its equivalent

$$A_1 F_{1-2} = \int_{A_1} \int_{2\pi} \cos \theta_1 \frac{d\omega_{1-2}}{\pi} \, dA_1$$

where ω denotes solid angle and is discussed in Section 6.6.5. The values in Figures 6.2 and 6.3 were obtained in this manner.

6.4.3   Shape Factor Algebra

   The information stored in graphs, such as Figures 6.2 and 6.3, is more useful than might be thought at first. The applicability of the data can be extended by the use of shape factor algebra. Figure 6.4 shows a simple example.

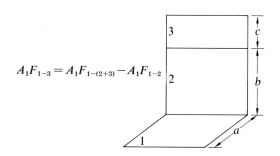

$$A_1 F_{1-3} = A_1 F_{1-(2+3)} - A_1 F_{1-2}$$

Figure 6.4   A simple case of shape factor algebra.

   By clever manipulation of the limits of integration, a much more powerful theorem of shape factor algebra may be proven. For example, referring to Figure 6.5, for two adjacent or opposite rectangles, subdivided by a plane perpendicular to them, the theorem states

$$A_1 F_{1-4} = A_2 F_{2-3}$$

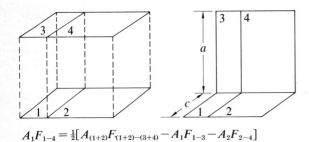

$$A_1F_{1-4} = \tfrac{1}{2}[A_{(1+2)}F_{(1+2)-(3+4)} - A_1F_{1-3} - A_2F_{2-4}]$$

Figure 6.5 A more complex example of shape factor algebra.

To illustrate the use of the above relation consider how we would determine F_{1-4} for the geometry of Figure 6.5. We start by writing

$$A_{(1+2)}F_{(1+2)-(3+4)} = A_1F_{1-3} + A_1F_{1-4} + A_2F_{2-3} + A_2F_{2-4}$$

Then we introduce $A_1F_{1-4} = A_2F_{2-3}$

$$A_{(1+2)}F_{(1+2)-(3+4)} = A_1F_{1-3} + 2A_1F_{1-4} + A_2F_{2-4}$$

so that we may solve for the desired quantity,

$$A_1F_{1-4} = \tfrac{1}{2}[A_{(1+2)}F_{(1+2)-(3+4)} - A_1F_{1-3} - A_2F_{2-4}]$$

in terms of values obtainable from Figure 6.3.

6.4.4 *EXAMPLE* Radiation Interchange on a Space Vehicle

Calculate the irradiation on a 1 meter square plane area on a space vehicle immediately adjacent and perpendicular to a solar cell array 1.5 meters long. The solar cells are at 120°F and may be considered black.

In this case we can use Figure 6.3 directly. The length b is 1 meter as is the length c. Length a is 1.5 meters. Consequently, $L = 1$ and $N = 1.5$. The shape factor is seen to be approximately 0.225. The flux incident on surface 1, the 1×1 area, is then

$$q_1^- = \frac{\dot{Q}_{2-1}^+}{A_1} = \frac{A_2F_{2-1}\sigma T_2^4}{A_1} = F_{1-2}\sigma T_2^4$$

$$q_1^- = (0.225)(1.712 \times 10^{-9})(580)^4 = 44 \text{ Btu/hr ft}^2$$

6.4.5 *EXAMPLE* Design of a Radiant Heating Panel

We wish to design a radiant heating panel on the ceiling of a hospital room so that the face of a bedridden patient would not be uncomfortably cold in 60°F air. The ceiling is 6 ft above the bed and the panel operates at 180°F. Even though painted a pleasant off-white matt color, the infrared emissivity can be 0.85 or 0.9 by virtue of the low temperature (see Tables 6.3 and 6.6 and Table B.15, Appendix B), so we can reasonably assume the panel to be black for heat radiation. As a rough design criterion we choose to size the panel so that the patient's face, modeled as a dry adiabatic black surface, attains a 90°F equilibrium temperature.

We first make a heat balance on the patient's face; for a dry adiabatic surface

$$\dot{Q}_{in} = \dot{Q}_{out}$$

$$A_{\text{Panel}}F_{P-F}\sigma T_P^4 + A_{\text{Room}}F_{R-F}\sigma T_R^4 = A_{\text{Face}}\sigma T_F^4 + h_c A_{\text{Face}}(T_F - T_R)$$

We note

$$A_{\text{Panel}}F_{P-F} = A_{\text{Face}}F_{F-P}, \qquad A_{\text{Room}}F_{R-F} = A_{\text{Face}}F_{F-R}$$

$$F_{F-P} + F_{F-R} = 1$$

Substituting and solving for F_{F-P} we find

$$F_{F-P} = \frac{\sigma T_F^4 - \sigma T_R^4 + h_c(T_F - T_R)}{\sigma T_P^4 - \sigma T_R^4}$$

We estimate h_c from Table 5.1 to be roughly 0.7 Btu/ft² hr °F.

$$F_{F-P} = \frac{(157 - 126) + 0.7(30)}{(287 - 126)} = \frac{52}{161} = 0.323$$

This shape factor requires say a circular disk subtending a half angle of $\sin^2 \theta = 0.323$, $\theta = 34.7°$. A circular area would therefore have to have radius

$$R = 6 \tan (34.7°) = 4.15 \text{ ft}$$

and the resulting panel area is about 54 ft².

6.5 TRANSFER BETWEEN FINITE GRAY SURFACES

6.5.1 Nonblack Opaque Surfaces

Although many surfaces are nearly black, others are not. Bare metals, in particular, reflect 90–98% of the irradiation. Ceramics, such as aluminum, beryllium, and magnesium oxides, reflect appreciably when irradiated with high temperature radiation. The fraction of the irradiation which is reflected is called the hemispherical reflectance ρ, and, if the surface is opaque, the remainder α is absorbed. These two fractions sum to unity.

$$\alpha + \rho = 1 \tag{6.21}$$

The power absorbed per unit area when a nonblack surface is irradiated by a surrounding black surface at T_2 is then

$$q_{\text{absorbed}} = \alpha q^- = \alpha \sigma T_2^4$$

Experimental observations indicate that a nonblack surface emits radiation at a rate less than a black surface would at the same temperature. This fraction of the black body emissive power emitted by a nonblack surface is called the hemispherical emittance ϵ. The radiation leaving surface 1 is thus composed of an emitted flux $\epsilon_1 \sigma T_1^4$ and a reflected flux $\rho_1 q_1^- = \rho_1 \sigma T_2^4$,

$$q_1^+ = \epsilon_1 \sigma T_1^4 + \rho_1 \sigma T_2^4$$

The net radiative flux leaving surface 1 is $q_1 = q_1^+ - q_1^-$,

$$q_1 = (\epsilon_1 \sigma T_1^4 + \rho_1 \sigma T_2^4) - \sigma T_2^4 \tag{6.22a}$$

or, rearranging,

$$q_1 = \epsilon_1 \sigma T_1^4 - (1 - \rho_1) \sigma T_2^4$$

But from Eq. (6.21) $\alpha = 1 - \rho$; therefore

$$q_1 = \epsilon_1 \sigma T_1^4 - \alpha_1 \sigma T_2^4 \tag{6.22b}$$

We see that the net radiative flux leaving a surface can be regarded as either the difference between the radiosity q^+ and the irradiation q^-, Eq. (6.22a), or as the difference between the emitted flux $\epsilon_1 \sigma T_1^4$ and the absorbed flux $\alpha_1 \sigma T_2^4$, Eq. (6.22b).

The second law of thermodynamics states that no energy can flow from a colder body to a warmer one. Thus ϵ can be no larger than unity, for if ϵ were to exceed unity, $\epsilon \sigma T_1^4 - \alpha \sigma T_2^4$ could be greater than zero for $T_1 < T_2$. Recall that the absorptivity α cannot exceed unity.

An important special case, which has been exploited widely in engineering practice, is that of a *gray* surface. A surface is said to be gray when its absorptivity equals its emittance, that is, $\alpha = \epsilon$. Section 6.7 will show that, in general, surfaces have properties which are a function of the energy of the photons being emitted or absorbed. Gray surfaces are special ones whose properties are independent of photon energy. For the present we will restrict our attention to systems having gray surfaces and develop some useful engineering relations.

6.5.2   Diffuse Reflection

A further restriction on the development that follows is that the reflection is perfectly diffuse. Reflection from rough ceramics or sintered metals may often be regarded as perfectly diffuse to a good approximation. However, even mirrorlike surfaces, called **specular** surfaces, may be regarded as perfectly diffuse when they are diffusely irradiated. A perfectly diffuse surface has no preferred direction of reflection; its *brightness* does not change with direction of view. Again detailed explanations are deferred until Sections 6.6 and 6.7. The important practical consequence of diffuseness is that the same shape factor which applies to radiation transfer between black surfaces also applies to radiation transfer between nonblack perfectly diffuse surfaces.

6.5.3   Transfer in an Enclosure

Consider an opaque surface on the inside of a space vehicle, a room, an industrial furnace, or other such enclosure. Let the surface be subdivided and the subdivisions assigned numbers, $i = 1, 2, \ldots, m$. The total incoming radiation for the ith surface from all the other surfaces, including i itself, if it is concave, is

$$A_i q_i^- = \sum_{j=1}^m q_j^+ A_j F_{j-i}$$

Introducing the shape factor reciprocal relation, Eq. (6.17), gives

$$A_i q_i^- = \sum_{j=1}^{m} A_i F_{i-j} q_j^+$$

or

$$q_i^- = \sum_{j=1}^{m} F_{i-j} q_j^+ \qquad (6.23)$$

As before, the radiosity q_j^+ can be written as a sum of emitted and reflected fluxes,

$$q_j^+ = \epsilon_j \sigma T_j^4 + \rho_j q_j^- \qquad (6.24)$$

Substituting Eq. (6.24) into Eq. (6.23) and rearranging yields

$$q_i^- - \sum_{j=1}^{m} F_{i-j} \rho_j q_j^- = \sum_{j=1}^{m} F_{i-j} \epsilon_j \sigma T_j^4 \qquad (6.25a)$$

Using the delta function δ_{ij}, which is zero when $i \neq j$ and unity when $i = j$, permits us to write

$$\sum_{j=1}^{m} (\delta_{ij} - F_{i-j} \rho_j) q_j^- = \sum_{j=1}^{m} F_{i-j} \epsilon_j \sigma T_j^4 \qquad (6.25b)$$

This set of m simultaneous linear algebraic equations in the m unknowns q_j^-, $j = 1, 2, \ldots, m$ can be solved by any of the methods of linear algebra, for example, Cramer's rule, successive elimination, or matrix inversion. Alternatively iteration can be used in a numerical solution procedure.

Any set of linear algebraic equations can be interpreted in terms of a linear electrical network. The Oppenheim radiation network interprets heat flux $\dot{Q}$ as current and radiosity q^+ as voltage. The net radiative power leaving surface i is

$$Q_i = A_i q_i = A_i q_i^+ - A_i q_i^-$$

Substituting for q_i^- from Eq. (6.23), and using the property of shape factors summing to unity, yields

$$\dot{Q}_i = A_i q_i^+ - A_i \sum_{j=1}^{m} F_{i-j} q_j^+$$

$$\dot{Q}_i = \sum_{j=1}^{m} A_i F_{i-j} (q_i^+ - q_j^+) \qquad (6.26)$$

To relate the heat flow $\dot{Q}_i$ to the surface temperature T_i, we obtain q_i^- from Eq. (6.24), replacing the subscript j by subscript i, and substitute it into

$$\dot{Q}_i = A_i q_i^+ - A_i q_i^-$$

to obtain

$$\dot{Q}_i = A_i q_i^+ - A_i \frac{q_i^+ - \epsilon_i \sigma T_i^4}{\rho_i}$$

$$= \frac{(1 - \rho_i) A_i}{\rho_i} \left[\frac{\epsilon_i}{1 - \rho_i} \sigma T_i^4 - q_i^+ \right]$$

But $1 - \rho_i = \alpha_i$, and $\alpha_i = \epsilon_i$ for a gray surface; thus

$$\dot{Q}_i = \frac{(1-\rho_i)A_i}{\rho_i}[\sigma T_i^4 - q_i^+] \tag{6.27}$$

Equations (6.26) and (6.27) then permit a network to be constructed as shown in Figure 6.6.

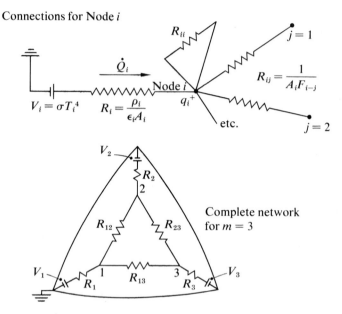

Figure 6.6   The radiation network.

The source voltage V_i is

$$V_i = \sigma T_i^4 \tag{6.28}$$

The total radiant power $\dot{Q}_i$ is represented by the current flowing from source voltage V_i through the *surface* resistance R_i to the surface voltage q_i^+, where

$$R_i = \frac{\rho_i}{(1-\rho_i)A_i} = \frac{1-\epsilon_i}{\epsilon_i A_i} \tag{6.29}$$

Likewise, the network interpretation of Eq. (6.26) is that of a shape factor resistance R_{ij} connecting two surface nodes with voltages q_i^+ and q_j^+, respectively

$$R_{ij} = \frac{1}{A_i F_{ij}} \tag{6.30}$$

As an example, consider the radiant power transfer between two long concentric cylinders, the inner one at temperature T_1 with area $A_1 = \pi D_1 L$. By inspection $F_{1-2} = 1$. The network is shown in Figure 6.7.

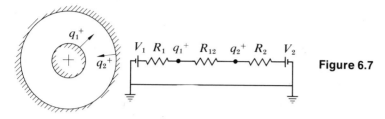

Figure 6.7 The radiation network for two concentric cylinders.

The net power transfer is the current flowing in the network

$$\dot{Q} = \frac{1}{R_1 + R_{12} + R_2} (\sigma T_1{}^4 - \sigma T_2{}^4)$$

$$\dot{Q} = \frac{1}{\dfrac{1-\epsilon_1}{\epsilon_1 A_1} + \dfrac{1}{A_1} + \dfrac{1-\epsilon_2}{\epsilon_2 A_2}} (\sigma T_1{}^4 - \sigma T_2{}^4)$$

$$\dot{Q} = \frac{\epsilon_1 A_1}{1 + \dfrac{\epsilon_1 A_1}{\epsilon_2 A_2}(1-\epsilon_2)} (\sigma T_1{}^4 - \sigma T_2{}^4) \tag{6.31}$$

This result reduces to a particularly simple one when

$$(\epsilon_1 A_1 / \epsilon_2 A_2)(1 - \epsilon_2)$$

is very small, perhaps because $\epsilon_1 A_1 / \epsilon_2 A_2$ is small or because $1 - \epsilon_2$ is small. Then we obtain

$$\dot{Q} = \epsilon_1 A_1 (\sigma T_1{}^4 - \sigma T_2{}^4) \tag{6.32}$$

This equation is of great practical importance, for it applies to a common situation. Recall that it is the relation used in Examples 5.5.4 and 5.6.4.

6.5.4   Further Simplifications for an Enclosure

Radiant transfer between diffuse surfaces which lie upon the inside wall of a sphere is particularly simple to calculate because the shape factor F_{i-j} degenerates to a function of surface j only.

$$F_{i-j} = \frac{A_j}{A_{\text{total}}}, \qquad A_{\text{total}} = 4\pi R^2 \tag{6.33}$$

In this case Eq. (6.23) yields a constant q^- independent of i. Equation (6.25a) can thus be solved for q^-

$$q^- = \frac{\displaystyle\sum_{j=1}^{m} \epsilon_j A_j \sigma T_j{}^4}{A_{\text{total}} - \displaystyle\sum_{j=1}^{m} A_j \rho_j} = \frac{\displaystyle\sum_{j=1}^{m} \epsilon_j A_j \sigma T_j{}^4}{\displaystyle\sum_{j=1}^{m} \epsilon_j A_j} \tag{6.34}$$

The net radiative flux leaving surface i is

$$q_i = q_i{}^+ - q_i{}^-$$

Substituting from Eq. (6.24)

$$q_i = \epsilon_i \sigma T_i^4 + (1 - \epsilon_i) q^- - q^-$$

and from Eq. (6.34)

$$q_i = \epsilon_i \left[\sigma T_i^4 - \frac{\sum_j \epsilon_j A_j \sigma T_j^4}{\sum_j \epsilon_j A_j} \right] \tag{6.35}$$

Multiplying both sides by A_i, changing the index in the numerator from j to k, and rearranging gives

$$\dot{Q}_i = \sum_k A_i \mathscr{F}_{i-k} (\sigma T_i^4 - \sigma T_k^4) \equiv \sum_k \dot{Q}_{i-k}$$

where $\mathscr{F}_{i-k} \equiv \epsilon_i \epsilon_k A_k / \sum_j \epsilon_j A_j$. The quantity $\dot{Q}_{i-k}$ may be interpreted as the net transfer between i and k. Replacing A_k by $F_{i-k} A_{\text{total}}$ gives

$$\dot{Q}_{i-k} = A_i F_{i-k} \frac{\epsilon_i \epsilon_k}{\dfrac{1}{A_{\text{total}}} \sum_j \epsilon_j A_j} (\sigma T_i^4 - \sigma T_k^4) \tag{6.36a}$$

where the denominator is ϵ_{avg}, the area weighted average emittance of the enclosure.

$$\dot{Q}_{i-k} = A_i F_{i-k} (\sigma T_i^4 - \sigma T_k^4) \left(\frac{\epsilon_i \epsilon_k}{\epsilon_{\text{avg}}} \right) \tag{6.36b}$$

Another simple result is obtained when ϵ_k coincides with ϵ_{avg} or when ϵ_i coincides with ϵ_{avg}. In the former case

$$\dot{Q}_{i-k} = A_i F_{i-k} \epsilon_i (\sigma T_i^4 - \sigma T_k^4), \qquad \epsilon_k = \epsilon_{\text{avg}} \tag{6.37}$$

while in the latter case

$$\dot{Q}_{i-k} = A_i F_{i-k} \epsilon_k (\sigma T_i^4 - \sigma T_k^4), \qquad \epsilon_i = \epsilon_{\text{avg}} \tag{6.38}$$

In summary, we have in Section 6.5.3 a fairly powerful general method for obtaining an engineering answer for radiant power transfer in an enclosure. When the enclosure is not too elongated, Eq. (6.36b) or one of two simpler ones found from it, Eqs. (6.37) and (6.38), may be used. In each case the total power from any surface i is

$$\dot{Q}_i = \sum_{k=1}^{m} \dot{Q}_{i-k} \tag{6.39}$$

6.5.5   *EXAMPLE*   Radiant Transfer inside a Furnace

Let us estimate the net heat flux on the floor of a long furnace 10 ft by 10 ft with side walls at 2500°F and a roof at 2000°F. Such furnaces are commonly used in stress relieving, annealing, enameling, melting, and other heating processes. The floor is 500°F and all surfaces are gray and diffuse and have an emittance 0.5. We superimpose the network on a sketch of the furnace in Figure 6.8.

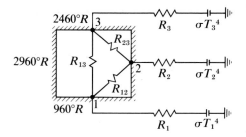

Figure 6.8 Radiation network super-posed on a furnace.

For a rough estimate we take each wall to be a single node. By sym-metry the side walls can be treated together. The shape factor F_{1-3} is 0.414 from Figure 6.2.

$$R_3 = R_1 = \frac{1-\epsilon_1}{\epsilon_1 A_1} = \frac{0.50}{(0.50)(10)} = 0.1$$

$$R_2 = \frac{1-\epsilon_2}{\epsilon_2 A_2} = \frac{0.50}{(0.50)(20)} = 0.05$$

$$R_{13} = \frac{1}{A_1 F_{1-3}} = \frac{1}{(10)(0.414)} = 0.2415$$

$$R_{12} = R_{23} = \frac{1}{A_1 F_{1-2}} = \frac{1}{10(1-0.414)} = 0.1707$$

The network is simplified by means of the delta-wye transformation shown in Figure 6.9.

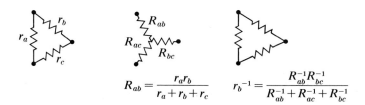

$$R_{ab} = \frac{r_a r_b}{r_a + r_b + r_c} \qquad r_b^{-1} = \frac{R_{ab}^{-1} R_{bc}^{-1}}{R_{ab}^{-1} + R_{ac}^{-1} + R_{bc}^{-1}}$$

Figure 6.9 The delta-wye and wye-delta transformations.

First a delta-wye transformation gives

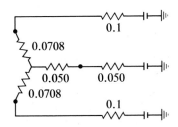

and then a wye-delta transformation gives

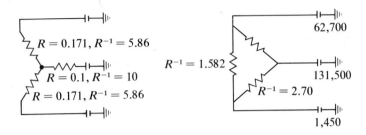

The rate of heat transfer into node 1 per foot of furnace length is:

$$\dot{Q} = 1.582(\sigma T_3^4 - \sigma T_1^4) + 2.70(\sigma T_2^4 - \sigma T_1^4)$$
$$\dot{Q} = 1.582(62,700 - 1,450) + 2.70(131,500 - 1,450)$$
$$\dot{Q} = 448,000 \text{ Btu/hr ft}$$

The quantity 100,000 Btu is sometimes called a **therm**; thus $\dot{Q} = 4.48$ therms/hr ft.

The student should rework this problem using Eq. (6.25) and solving by matrix inversion or Cramer's rule. Even when matrix inversion is used for solution it is advisable to draw the appropriate network as a conceptual aid. Drawing a simple network greatly facilitates conceptual design of a spacecraft thermal control system, for example.

6.6   SPECTRAL DISTRIBUTIONS

6.6.1   The Distribution Concept

In Sections 6.2 to 6.5 we have considered radiation and, to a lesser extent, free molecule transfer only from a total *macroscopic* point of view, since most engineering is based upon such a view. In the remainder of the chapter a *second pass* is made through the subject matter from a more *microscopic* point of view so that nongray walls can be treated. The student wishing an early introduction to the whole field of transfer processes may defer studying this portion of the text and proceed directly to Chapter 7.

Not all the molecules emitted by a wall have the same velocity nor do all the photons emitted have the same energy. The differences in behavior of molecules or photons due to differences in energy are used to good advantage by engineers. For example, a photon with an energy of less than 1 ev (the work done on a charge equal to the electron's charge in forcing it through a potential difference of 1 volt, 1 joule $= 6.2418 \times 10^{18}$ ev) produces no useful energy in a silicon solar cell but merely raises its temperature and thus decreases its efficiency. The engineer employs a filter which reflects the photons of undesirable energies. Ordinary window glass transmits photons very well; approximately 90% pass through, and only 2% are absorbed, when the photons have energies between approximately 0.45 and 4 ev; but the glass absorbs photons, again approximately 90% of them when the photons have somewhat lower energies, and essentially none are transmitted. In a hothouse the glass transmits during the day solar radiation needed by the plants for photosynthesis and both during the day and night prevents cooling of the plants by radiation to the cold sky, because the glass is opaque to thermal photons and emits nearly the same energy back to the plants as is emitted by the plants to the glass. These phenomena which depend upon different behaviors for photons of different energies are said to be nongray or *spectrally selective*. We will show how such phenomena can be treated using the concept of a **distribution function**.

6.6.2 Velocity, Energy, and Momentum Relations

We are concerned primarily with two types of carriers in this introductory treatment, photons and monatomic ideal gas molecules. In special circumstances we will treat polyatomic molecules. For other carriers such as electrons, polyatomic molecules in the general case, and ions, the reader should consult more advanced texts. The monatomic molecule in a large container may have any velocity v (we exclude velocities approaching the speed of light) and at that velocity has momentum mv and energy $\frac{1}{2}mv^2$ (neglecting electronic excitation which occurs at high temperatures), where m is the mass of the molecule (1 amu equals 1.66043×10^{-27} kg, $m = M \cdot$ amu).

A photon in vacuum has velocity $c = 2.997925 \times 10^8$ m/sec, momentum $h\nu_f/c$, and energy $h\nu_f$, where h is Planck's constant, $h = 6.6256 \times 10^{-34}$ joule sec, and ν_f is the frequency in cycles per second, which may have any value. The relation between energy E and frequency is known as Einstein's photoelectric law. Albert Einstein was able to show that the action of light on a surface is to cause electrons to be emitted only when the photons have more energy than E, the surface work function.

The frequency of a 1 ev photon is enormous. For this reason, wavenumber ν is often used instead of ν_f; wavenumber is simply frequency divided by the velocity of light. It consequently has units of reciprocal length; a commonly used unit is cm^{-1}. The reciprocal of wavenumber has units of length and it is consequently called wavelength λ. Commonly used units are angstroms, 1 Å $= 10^{-10}$ meter, and microns, $1 \mu = 10^{-6}$ meter. Table 6.3 summarizes these relations (recall Table 6.2).

The concept of the average internal energy of a molecule is perhaps the most complicated one in Table 6.3. We have separated the average total energy into two parts, the average translational kinetic energy,

$$\bar{E}_t = \tfrac{1}{2}m\bar{v^2} = \tfrac{3}{2}kT \tag{6.40}$$

and the average energy polyatomic molecules carry due to vibrations and rotations,

$$\bar{E}_{vr} = mu - \tfrac{3}{2}kT = mc_vT - \tfrac{3}{2}kT \tag{6.41}$$

Table 6.3   Momentum and Energy Relations

	Monatomic Molecule	Photon
Velocity	v	c
Momentum	$p_m = mv$	$p_m = h\nu_f/c = h\nu = h/\lambda$
Energy	$E_t = \frac{1}{2}mv^2$	$E = h\nu_f = hc\nu = hc/\lambda$
	$\bar{E}_t = \frac{1}{2}m\bar{v}^2 = \frac{3}{2}kT$	
	$\bar{E}_{vr} = m\hat{u} - \frac{3}{2}kT$	

However, a simplifying factor is that many polyatomic gases, particularly diatomic ones, are vibrationally unexcited at ordinary temperatures. If they are vibrationally unexcited but rotationally excited,

$$mc_v = (3 + N_r)\left(\tfrac{1}{2}k\right) \tag{6.42}$$

where N_r is the number of rotational degrees of freedom (zero for a monatomic gas, two for a linear molecule, and three for a nonlinear one). Whether or not this is so can be inferred from the ratio of specific heats $\gamma = c_p/c_v$. For an ideal gas

$$mc_p = mc_v + k \qquad (c_p = c_v + R, R = \mathcal{R}/M) \tag{6.43}$$

$$\gamma = 1 + \frac{k}{mc_v} = 1 + \frac{2}{3 + N_r} \tag{6.44}$$

It γ drops below this value at high temperatures, it is a sign that the vibrational modes are becoming excited.

For example, consider molecular nitrogen N_2, which is imagined to be a linear molecule and can thus rotate about two mutually perpendicular axes;

$$\gamma \doteq 1 + \tfrac{2}{5} = 1.4$$

The value holds up to approximately 1000°C. Or consider water vapor, a nonlinear molecule with three degrees of rotation:

$$\gamma \doteq 1 + \tfrac{2}{6} = 1.33$$

This value holds true for moderate pressures and temperatures up to roughly 200°C.

6.6.3   Distributions at Thermodynamic Equilibrium

Picture an enclosure at thermodynamic equilibrium filled with a monatomic gas or filled with photons (or both, since interactions will not occur when the molecules are not electronically excited). The molecules are distributed on the average uniformly over the volume with a total number per unit volume $\mathcal{N} = P/kT$. (Recall $k = 1.38054 \times 10^{-23}$ joule/°K.) They are colliding randomly and, as a result on the average moving uniformly in all directions (that is, if a small volume of them were suddenly plucked from the enclosure and placed at the center of a large sphere, and the molecules continued in motion without further collisions, the number eventually hitting the sphere per unit area would be uniform over the sphere). The number of molecules $d\mathcal{N}$ whose energies lie between E and $E + dE$ is given by Maxwell-Boltzmann statistics as

$$d\mathcal{N} = \mathcal{N}\, \frac{2}{\sqrt{\pi}} \left(\frac{E}{kT}\right)^{1/2} \exp\left(-\frac{E}{kT}\right) \frac{dE}{kT} \qquad (6.45)$$

Bose-Einstein statistics gives the number of photons per unit volume $d\mathcal{N}$ whose energies lie between E and $E + dE$

$$d\mathcal{N} = \frac{8\pi}{(hc/kT)^3} \frac{(E/kT)^2}{\exp(E/kT) - 1} \frac{dE}{kT} \qquad (6.46)$$

We take Eqs. (6.45) and (6.46) as the foundation for what follows. The student is instructed to take them on faith. There is no attempt here to derive them. A good course or text in statistical thermodynamics or statistical mechanics is recommended for those who would like to appreciate their foundation. Let us note that Eq. (6.45) can be used to derive many of the thermodynamic properties experimentally observed in gases (not too near their critical points). Equation (6.46) will be used in what follows to derive the Planck and Stefan radiation laws which are experimentally observed to be valid.

It should be apparent that the number of molecules between two energies E_1 (say 0.01 ev) and E_2 (say 0.1 ev) can be found by integrating from E_1 to E_2 with respect to E. Changing variables of integration from E to $E^* = E/kT$ gives

$$\Delta\mathcal{N} = \mathcal{N}\, \frac{2}{\sqrt{\pi}} \int_{E_1/kT}^{E_2/kT} E^{*\,1/2} \exp(-E^*)\, dE^* \qquad (6.47)$$

The student may readily carry out such an integration, for example, using a Simpson routine on a digital computer. However, one is more often interested in how many carriers crossing a surface per unit time have energies between E_1 and E_2 and how much power they transfer.

6.6.4 Surface Fluxes

Consider a surface of area dA located inside our enclosure at thermodynamic equilibrium. Particles coming up through the surface in a particular direction have a velocity normal to the surface of $v \cos \theta$ where θ is the angle between the normal and direction of interest. Figure 6.10 shows this projection.

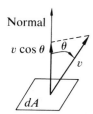

Figure 6.10 Normal velocity component.

But how many particles come in this direction? To answer such a question we imagine a sphere with its center at the centroid of our small surface, and we mark off an area dA_s on the sphere. Then we can more properly ask, how many particles come from dA in directions which cross the small area dA_s on the sphere?

The number of particles per unit volume in the energy range between E and $E + dE$ is $d\mathcal{N}$. The number per unit volume times the normal velocity component times the area is then the number crossing the surface per unit time. The fraction of these particles which would cross dA_s is dA_s/A_s where A_s is the total area of our imagined sphere $4\pi r^2$; because, as we have said, the particle trajectories are uniformly distributed in direction. Denoting the flow across the plane containing dA as $d\dot{N}$, we can write

$$d\dot{N} = d\mathcal{N} v \cos\theta \, dA \left(\frac{dA_s}{A_s}\right) \tag{6.48}$$

6.6.5   Solid Angles

In thinking about particles traveling in directions crossing a small area dA_s on a sphere we find it convenient to use the concept of a solid angle. Recall that a plane angle in radians (unitless) is the arc length on a circle divided by the radius of the circle and that there are consequently 2π radians in a complete circle. In the same manner let a solid angle $d\omega$ in steradians (unitless) be the area on a sphere divided by the radius squared,

$$d\omega = \frac{dA_s}{r^2} \tag{6.49}$$

There are then 4π steradians in a sphere. We refer to the particles crossing an area dA_s on a sphere as having directions lying within the solid angle $d\omega$ given by Eq. (6.49). The term dA_s/A_s in Eq. (6.48) can then be written

$$\frac{dA_s}{A_s} = \frac{r^2 \, d\omega}{4\pi r^2} = \frac{d\omega}{4\pi} \tag{6.50}$$

Equation (6.48) is therefore written

$$d\dot{N} = d\mathcal{N} \, v \cos\theta \, dA \frac{d\omega}{4\pi} \tag{6.51}$$

It is well to have a coordinate system with which to specify direction. Polar coordinates are convenient. Figure 6.11 shows polar coordinates adopted

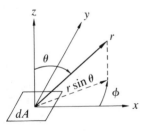

Figure 6.11   Polar coordinates.

here to be a polar angle θ measured from the z-axis, the surface normal, and an azimuthal angle ϕ measured from the x-axis to the projection of the particle trajectory in the x-y plane.

It is apparent in Figure 6.11 that, if one were to vary ϕ by $d\phi$, the lever arm in the x-y plane of length $r \sin \theta$ would strike an arc of $(r \sin \theta)\, d\phi$ perpendicular to the plane containing the surface normal and r. Varying θ by $d\theta$ would strike an arc of length $r\, d\theta$ lying in the plane. The area on the sphere of radius r struck out would be $r\, d\theta\, r \sin \theta\, d\phi$. Consequently, from Eq. (6.49) the solid angle swept out by varying θ by $d\theta$ and ϕ by $d\phi$ is

$$d\omega = \frac{r\, d\theta\, r \sin \theta\, d\phi}{r^2} = \sin \theta\, d\theta\, d\phi \tag{6.52}$$

To sweep out a sphere one would vary θ from 0 to π and ϕ from 0 to 2π. We can readily verify that

$$\int_0^{4\pi} d\omega = \int_0^{2\pi} \int_0^{\pi} \sin \theta\, d\theta\, d\phi = 4\pi \tag{6.53}$$

Similarly, to sweep out a hemisphere θ would be varied from 0 to $\pi/2$ and ϕ from 0 to 2π.

6.6.6 Radiant Intensity

To obtain the power-crossing our surface dA we simply multiply $d\dot{N}$ given by Eq. (6.51) by the energy carried by each particle. There results for photons (velocity $v = c$)

$$d\dot{Q} = E\, d\dot{N}$$

$$d\dot{Q} = (hcv)(d\mathcal{N}\, c \cos \theta\, dA\, d\omega/4\pi) \tag{6.54}$$

where $d\mathcal{N}$ is given by Eq. (6.46). Rearrangement of Eq. (6.54) yields

$$I_P = \frac{d\dot{Q}}{(dA \cos \theta)\, d\omega\, dv} = \frac{2hc^2 v^3}{\exp(hcv/kT) - 1}$$

a quantity independent of direction. It is termed the Planckian spectral radiant intensity in engineering literature (Planckian radiance in physics literature). The left-hand equation of the two relations above, when rearranged, plays the same role in radiative transport as Fourier's law does in diffusive transport,

$$\bullet \qquad d\dot{Q} \equiv I\, dA \cos \theta\, d\omega\, dv \tag{6.55}$$

The relation is written without the subscript P because the source need not be Planckian in nature for a spectral radiant intensity to be defined. The right-hand expression gives the Planckian or black body spectral radiant intensity

$$\bullet \qquad I_P(v, T) = \frac{2hc^2 v^3}{\exp(hcv/kT) - 1} \tag{6.56}$$

The term **black body** is used because thermodynamic equilibrium radiation is approached in the laboratory by having a small opening in a suitably designed large cavity with isothermal walls. When the cavity is cold, it appears black to the eye, because nearly every photon entering the cavity from the room is absorbed. (In an ideal cavity every photon entering would be absorbed.)

6.6.7   Radiant Flux

To obtain the total flow of power crossing dA in all directions to the hemisphere above it and for photons of all possible wavenumbers from zero to infinity we simply integrate Eq. (6.55) over $d\omega$ and $d\nu$. Consider first the integration with respect to solid angle over the hemisphere. Substituting Eq. (6.52) into Eq. (6.55) and integrating over a hemisphere gives the heat flow per unit area per wavenumber as

$$q^+(\nu) = \frac{dq^+}{d\nu} = \int_0^{2\pi} \int_0^{\pi/2} I^+ \cos\theta \sin\theta \, d\theta \, d\phi \tag{6.57}$$

where the $+$ denotes that the flux is leaving the surface and q is, as before, $\dot{Q}/A$. The quantity $q^+(\nu) \, d\nu$ is the flux, power per unit area, crossing dA whose photon wavenumbers lie between ν and $\nu+d\nu$. In engineering terminology $q^+(\nu)$ is termed the spectral radiosity; spectral because of the narrow band of the spectrum $d\nu$ in question, and radiosity because the photons leaving in all possible directions are included.

We know a Planckian radiator has an intensity I^+ independent of direction. Such a source is said to be diffuse. For a diffuse source I^+ may be factored out from under the integral in Eq. (6.57). The integration then yields simply π, as shown below,

$$\int_0^{2\pi} \cos\theta \, d\omega = \int_0^{2\pi} \int_0^{\pi/2} \cos\theta \sin\theta \, d\theta \, d\phi = 2\pi \int_0^{\pi/2} \cos\theta \sin\theta \, d\theta$$

Let $u = \sin^2\theta$, $du = 2\sin\theta\cos\theta \, d\theta$.

$$\int_0^{2\pi} \cos\theta \, d\omega = \pi \int_0^1 du = \pi \tag{6.58}$$

For a perfectly diffuse source Eq. (6.57) thus becomes

$$q^+(\nu) = \pi I^+ \tag{6.59}$$

To obtain the total flux, we integrate $q^+(\nu)$ over all possible photon wavenumbers from zero to infinity. Note that the probability of zero wavenumber photons is zero, from Eq. (6.56), like $2kTc\nu^2$ in the limit, and that for infinite wavenumber is also zero, like $2hc^2\nu^3 \exp(-hc\nu/kT)$. From a practical point of view we do not need to start at zero or integrate all the way to infinity. To be rigorous, however, we include all possibilities by integrating over the complete range.

$$q_P^+ = \int_0^\infty \pi I_P(\nu, T) \, d\nu$$

$$q_P{}^+ = \int_0^\infty \frac{2\pi hc^2 \nu^3}{\exp\left(hc\nu/kT\right) - 1}\, d\nu$$

Changing variables of integration from ν to $\zeta = hc\nu/kT$ yields

$$q_P{}^+ = \frac{2\pi k^4 T^4}{h^3 c^2} \int_{\zeta=0}^\infty \frac{\zeta^3\, d\zeta}{\exp\left(\zeta\right) - 1}$$

The definite integral has the value $\pi^4/15$; therefore

$$q_P{}^+ = \frac{2\pi^5 k^4}{15 h^3 c^2} T^4 \tag{6.60}$$

The coefficient multiplying T^4 is the Stefan-Boltzmann constant σ

$$\bullet \qquad \sigma = \frac{2\pi^5 k^4}{15 h^3 c^2} = 5.6697 \times 10^{-8}\ \text{w/m}^2\ {}^\circ\text{K}^4 \tag{6.61}$$

$$= 1.712 \times 10^{-9}\ \text{Btu/hr ft}^2\ {}^\circ\text{R}^4$$

6.6.8 The External Fraction

We have found that the total flux, power per unit area, leaving the opening in a black body cavity is σT^4. To obtain the total flux of radiation of photons with wavenumbers greater than a certain value ν_1 we would integrate not from zero to infinity but from ν_1 to infinity. For example, photons more energetic than 1.1 ev operate a silicon solar cell, so that we might wish to know what fraction of the radiant power is carried by photons more energetic than this value. The fraction f would be

$$f = \frac{\displaystyle\int_{\nu_1}^\infty \pi I_P\, d\nu}{\sigma T^4} = \frac{\displaystyle\int_{\nu_1}^\infty \frac{2\pi hc^2 \nu^3}{\exp\left(hc\nu/kT\right) - 1}\, d\nu}{\left(2\pi^5 k^4 / 15 h^3 c^2\right) T^4} \tag{6.62a}$$

Again, changing variables from ν to $\zeta = hc\nu/kT$ gives

$$f\!\left(\frac{hc\nu_1}{kT}\right) = \frac{15}{\pi^4} \int_{\zeta = hc\nu_1/kT}^\infty \frac{\zeta^3\, d\zeta}{\exp\left(\zeta\right) - 1} \tag{6.62b}$$

Table 6.4 shows this function. It is, in statistics terminology, the ogive to the Planck distribution.

To obtain the fraction of the power between two wavenumbers ν_1 and ν_2 ($\nu_2 > \nu_1$) we simply subtract the fraction above the higher wavenumber from that above the lower one

$$\Delta f = f\!\left(\frac{hc\nu_1}{kT}\right) - f\!\left(\frac{hc\nu_2}{kT}\right) \tag{6.63}$$

For example, 90% of black body radiant power is carried by photons with wavenumbers greater than $\nu_1 = 1.53\, kT/hc$, while 10% of the power is carried by photons with wavenumbers less than $\nu_2 = 6.55\, kT/hc$. Thus 80% of Planckian radiant power is carried by photons with wavenumbers between $\nu_1 = 1.53\, kT/hc$ and $\nu = 6.55\, kT/hc$.

Table 6.4   The External Fraction for the Planck Distribution

$\dfrac{hc\nu}{kT}$	f	$\dfrac{hc\nu}{kT}$	f
0.0	1.00	3.50	0.50
0.628	0.99	3.75	0.45
0.811	0.98	4.02	0.40
1.06	0.96	4.30	0.35
1.24	0.94	4.61	0.30
1.53	0.90	4.97	0.25
1.83	0.85	5.37	0.20
2.10	0.80	5.88	0.15
2.34	0.75	6.55	0.10
2.58	0.70	7.35	0.06
2.80	0.65	7.97	0.04
3.03	0.60	8.96	0.02
3.26	0.55	9.90	0.01

We can use the above numbers to characterize the wavenumber range of greatest interest for a given temperature. For example, consider a room temperature black body at $T = 300°K$, a *red-hot* black body at 1000°K, a black body at a temperature typical for a lamp filament of 3000°K, and a black body at a temperature of the outer layers of our sun, 5700°K. Table 6.5 shows the wavenumbers and wavelengths encompassing 80% of the radiation. The values are found from the preceding values $hc\nu/kT = 1.53$, 6.55, and the value of $hc/k \doteq 1.44$ cm °K. Note that the lower wavenumber corresponds to the longer wavelength.

Table 6.5   Wavenumbers and Wavelengths Encompassing 80% of Planckian Radiation

Source Temperature, °K	Wavenumber Range, cm^{-1}	Wavelength Range, microns
300	319–1365	7.3–31.4
1000	1062–4550	2.2–9.4
3000	3190–13,650	0.73–3.1
5700	6050–25,900	0.39–1.7

6.6.9   Molecular Flux

Equation (6.51) gives the number of molecules per unit time crossing an area dA inside solid angle $d\omega$ and with energies between E and $E + dE$. If we wish to know the total number of molecules per unit area crossing in all directions and with all possible energies, we integrate over all directions in the hemisphere and over all possible energies from zero to infinity. We have already found that

$$\int_{2\pi} \cos \theta \, d\omega = \pi$$

Hence we obtain the result after the first integration simply by replacing $\cos \theta \, d\omega$ in Eq. (6.51) with π. Denoting the flux as J^+, we have

$$dJ^+ = \tfrac{1}{4}v\frac{d\mathcal{N}}{dE}\,dE \tag{6.64}$$

Substituting Eq. (6.45), using $E = \tfrac{1}{2}mv^2$ to express v as $(2E/m)^{1/2}$, and integrating with respect to E gives

$$J^+ = \int_0^\infty \frac{1}{4}\left(\frac{2E}{m}\right)^{1/2} \mathcal{N}\, \frac{2}{\pi^{1/2}}\left(\frac{E}{kT}\right)^{1/2} \exp\left(-\frac{E}{kT}\right)\frac{dE}{kT}$$

$$= \frac{1}{4}\left(\frac{8kT}{\pi m}\right)^{1/2} \mathcal{N}\int_0^\infty xe^{-x}\,dx, \qquad \text{where } x = \frac{E}{kT}$$

$$J^+ = \frac{1}{4}\mathcal{N}\left(\frac{8kT}{\pi m}\right)^{1/2}$$

Comparison of Eq. (6.64) with this result shows that the quantity $(8kT/\pi m)^{1/2}$ is simply the mean speed. We therefore write the molecular flux in a particularly simple form

$$J^+ = \tfrac{1}{4}\mathcal{N}\bar{v}, \qquad \bar{v} = \left(\frac{8kT}{\pi m}\right)^{1/2} \tag{6.65}$$

To obtain the power transferred by a monatomic molecular flux we multiply the flux of carriers by the energy carried. From Eqs. (6.51) and (6.45)

$$d\dot{Q}^+ = E\,d\dot{N} = E d\mathcal{N}\, v \cos\theta\, dA\,\frac{d\omega}{4\pi}$$

$$dq^+ = \frac{d\dot{Q}^+}{dA} = E\mathcal{N}\,\frac{2}{\pi^{1/2}}\left(\frac{E}{kT}\right)^{1/2} \exp\left(-\frac{E}{kT}\right)\frac{dE}{kT}\,v\cos\theta\,\frac{d\omega}{4\pi}$$

Again we substitute for v the relation $(2E/m)^{1/2}$ and integrate over a hemisphere with respect to $d\omega$ and over all possible energies from zero to infinity.

$$q^+ = \frac{1}{4\pi}\int_{2\pi}\cos\theta\, d\omega\,\frac{2\mathcal{N}}{\pi^{1/2}}\int_0^\infty E\left(\frac{E}{kT}\right)^{1/2}\left(\frac{2E}{m}\right)^{1/2}\exp\left(-\frac{E}{kT}\right)\frac{dE}{kT}$$

$$= \frac{1}{4}\left(\frac{8kT}{\pi m}\right)^{1/2}kT\mathcal{N}\int_0^\infty x^2e^{-x}\,dx$$

$$q^+ = \tfrac{1}{4}\bar{v}\mathcal{N}(2kT) = KJ^+(2kT) \tag{6.66a}$$

The average energy per carrier in the volume is not $2kT$; it may be found by multiplying Eq. (6.45) by E and integrating with respect to E from zero to infinity. For a monatomic gas it is $\tfrac{3}{2}kT$. However, the more energetic molecules in the volume are faster and cross an area dA more frequently. It is for this reason that the energy flux is obtained by multiplying the molecular flux by $2kT$ rather than a lesser value.

As noted in Section 6.6.2, polyatomic molecules carry energy not only in their translational kinetic energy but also in their rotations and vibrations. These latter energies are quantized; that is, only certain discrete values are allowed, although in the case of rotation the energy levels may be so closely spaced that they may often be regarded as continuous. The average internal energy of such a molecule is mu, where $\hat{u}$ is the internal energy per unit mass. Of this total, the amount contained in the rotational and vibrational energies is on the average $mu - \frac{3}{2}kT$, since the translational energy on the average is $\frac{3}{2}kT$. If the vibrational-rotational energy is independent of the translational energy (a good approximation), then the power crossing (in one direction) a plane in a polyatomic gas is

$$q^+ \stackrel{3}{=} J^+ (m\hat{u} + \tfrac{1}{2}kT) \tag{6.66b}$$

A complicating factor is the fact that during a molecule-surface collision, the vibrational energies are usually transferred much less efficiently than are the translational and rotational energies. As might be imagined, there is a probabilistic nature to surface-molecule interactions. Some molecules which impinge upon a surface stick for a rather long time, during which period their energies are *accommodated* to the distribution characteristic of the surface temperature. Others stick not at all or only very briefly so that little energy is transferred. In the main, a high fraction, say 90%, of the incident molecules stick long enough to many types of dirty engineering surfaces to give up their translational and rotational energies and acquire those characteristic of the surface. We have already noted that many molecules are vibrationally unexcited at ordinary temperatures; thus this complication does not affect transport in those cases.

6.7 SPECTRAL VARIATIONS IN WALL CHARACTERISTICS

6.7.1 The Surface System Concept

Thus far in our *second pass* we have obtained the fluxes of photons or molecules across an *imaginary* surface inside a large enclosure at thermodynamic equilibrium. We wish to use rates under the ideal conditions of thermodynamic equilibrium as a measure of the transport rates at a wall. We restrict ourselves in this introductory treatment to a type of wall whose temperature is sufficiently uniform so that we can, if need be, measure it unambiguously; that is, we consider here only solid walls and rule out perhaps a poorly conducting porous wall such as a fiberglass filter pad. For a solid opaque wall we construct *two* imaginary surfaces, the *s*-surface in the medium just outside the wall, and the *m*-surface sufficiently deep within the wall so that no photons or molecules penetrate from the surface and so that Fourier's law of conduction will hold there. We apply the **First Law of Thermodynamics** (the energy accounting principle) to the system between the *s*- and *m*-surfaces. The volume between these boundaries constitutes a surface system. Such a system is often called (loosely) in engineering, a *surface*, which term was used in Section 6.5. Emphasis is placed on thermal radiation in what follows, but the same type of analysis can be applied for molecules.

Photon energy fluxes crossing the s-surface are the outgoing flux (spectral radiosity) $q_r^+(\nu)$ and the incoming flux, the spectral irradiation, $q_r^-(\nu)$. The net spectral radiant flux going away from the wall system is then

$$q_{r,s}(\nu) = q_r^+(\nu) - q_r^-(\nu) \tag{6.67}$$

There could also be some conduction at the s-surface $q_{c,s}$. At the m-surface there is only conduction in the absence of convection or mass transfer. The heat flux in this instance is

$$q_m = -\left(k\frac{\partial T}{\partial y}\right)_m \tag{6.68}$$

where y is measured in the direction of the outward normal to the surface. The mass between the s- and m-surfaces is ordinarily quite small, so that even during a fast transient it is not likely that the rate of energy stored in this mass is worth considering. We can therefore write, in the absence of convection,

$$-\left(k\frac{\partial T}{\partial y}\right)_m = q_{r,s} + q_{c,s} \tag{6.69}$$

This energy balance states simply that the energy reaching a surface by conduction is radiated and conducted away across the s-plane in the medium outside the wall.

6.7.2 Emittance, Absorptance, and Reflectance

An experiment can be performed as follows: Heat a wall to temperature T_w in a space with cold black walls and observe the spectral radiant intensity of the surface from an area dA_w on the surface and in directions close to θ, ϕ (Figure 6.11) within a solid angle $d\omega$ subtended by a mirror or lens focusing the power on a detector behind a filter. The filter restricts the photons received by the detector to those with wavenumbers between ν and $\nu + d\nu$ (wavelengths between $\lambda - d\lambda$ and λ, where $\lambda = 1/\nu$ and $d\lambda = d\nu/\nu^2$). The detector might be a thermistor or thermocouple in the middle of a thin rectangular or circular fin so that the radiant power heats the fin and changes the resistance or emf of the sensor. Compare the signal so obtained with that from a Planckian black body cavity at the same temperature. Since the surrounds are black and cold, all the radiation from the wall must have been emitted and not reflected. We call the ratio of the observed intensities the spectral directional emittance (also called emissivity)

$$\epsilon(\theta, \phi, \nu, T_w) = \frac{I_w(\theta, \phi, \nu)}{I_P(\nu, T_w)} \tag{6.70}$$

In the same way the total directional emittance is defined as

$$\epsilon_T(\theta, \phi, T_w) = \frac{I_{w,T}(\theta, \phi)}{\sigma T_w^4/\pi} \tag{6.71}$$

We can also perform an experiment in which we measure how much a wall, at temperature T_w, is heated by directing onto area dA_w a beam from solid angle $d\omega$ about direction θ, ϕ. A heater is situated behind the wall; its power supply is initially adjusted to balance losses so that the wall is at temperature T_w. When the external radiant source is turned on, the heater power supply is correspondingly reduced to maintain the wall temperature at T_w. The power transfer change at the m-surface is then simply equal to the decrease in power supply to the heater. We define the following:

Spectral directional absorptance

$$\alpha(\theta, \phi, \nu, T_w) = \frac{\Delta q_m(\nu) \, d\nu \, dA_w}{q^-(\nu) \, d\nu \, dA_w} \tag{6.72}$$

Total directional absorptance

$$\alpha_T(\theta, \phi, T_e, T_w) = \frac{\Delta q_m \, dA_w}{q^- \, dA_w} \tag{6.73}$$

where, if the external source is black,

$$q^-(\nu) \, d\nu \, dA_w = I_P(\nu, T_e) \cos\theta \, dA_w \, d\omega \, d\nu \tag{6.74}$$

$$q^- \, dA_w = (\sigma T_e^4/\pi) \cos\theta \, dA_w \, d\omega \tag{6.75}$$

The remaining power from the source must have been *reflected*, that is, it must have been delivered to the surrounds which see the s-surface. (Recall that our surface system is opaque, that is, it is not transmitting.) It can be arranged to detect the change in power transferred to the surrounds when the source is turned on and find that, within experimental uncertainties, the power not absorbed is indeed reflected so that energy is conserved. Note that when the wall material does not fluoresce, the reflected power has the same wavenumbers as the incident flux. We define the following:

Spectral directional reflectance

$$\rho(\theta, \phi, \nu, T_w) = \frac{\Delta q^+(\nu)}{q^-(\nu)} \tag{6.76}$$

Total directional reflectance

$$\rho_T(\theta, \phi, T_e, T_w) = \frac{\Delta q^+}{q^-} \tag{6.77}$$

where the denominators are given by Eqs. (6.74) and (6.75). (Note that ρ_T and α_T are T_e-dependent since the ν wavenumber distribution is T_e-dependent.) For an opaque surface system, conservation of energy requires that

$$\rho(\theta, \phi, \nu, T_w) + \alpha(\theta, \phi, \nu, T_w) = 1 \tag{6.78}$$

$$\rho_T(\theta, \phi, T_e, T_w) + \alpha_T(\theta, \phi, T_e, T_w) = 1 \tag{6.79}$$

These relations between reflectance and absorptance are thus a direct consequence of the First Law of Thermodynamics.

Tables 6.6 and 6.7 show measured spectral reflectances of a few sur-

Table 6.6 Reflectances of Some Paints and Coatings at an Angle of Incidence of 25° from the Normal

	Reflectance at Room Temperature					
λ, microns	3M Black Velvet	Hard Anodized Aluminum	Anodized Titanium	White Epoxy Paint	Flame Sprayed Alumina	Aluminum Paint
0.3	0.03	0.05			0.40	
0.35	0.03	0.06			0.52	
0.4	0.03	0.07		0.40	0.66	0.75
0.45	0.03	0.07		0.88	0.71	0.75
0.5	0.03	0.07	0.47	0.90	0.73	0.75
0.6	0.03	0.07	0.52	0.85	0.76	0.74
0.7	0.03	0.07	0.52	0.79	0.77	0.71
0.8	0.03	0.07	0.52	0.91	0.77	0.69
1.0	0.03	0.08	0.50	0.92	0.75	0.72
1.5	0.03	0.10	0.50	0.70	0.68	0.75
2.0	0.04	0.15	0.48	0.57	0.49	0.77
3.0	0.04	0.08	0.11	0.07	0.27	0.77
4.0	0.04	0.26	0.24	0.10	0.47	0.77
5.0	0.05	0.30	0.24	0.10	0.38	0.78
6.0	0.04	0.17	0.18	0.09	0.12	0.78
8.0	0.09	0.04	0.17	0.07	0.02	0.74
10.0	0.05	0.02	0.10	0.09	0.02	0.78
12.0	0.05	0.16	0.09	0.07	0.26	0.79
15.0	0.06	0.17	0.12	0.10	0.21	0.80
20.0	0.06	0.20	0.15	0.16	0.25	0.81
30.0	0.03	0.20		0.19		
40.0	0.03			0.20		

faces used in technology. The data were obtained with angles of incidence θ of 25° and with T_w at room temperature. Additional data are shown in Appendix B, Tables B.14 and B.15.

A useful relation between the spectral directional absorptance and spectral directional emittance may be derived as follows. The **Principle of Detailed Balancing of Statistical Thermodynamics** (from which the **Second Law of Thermodynamics**, in its classical form, results) states that two systems cannot exchange energy when their energy levels are populated with thermodynamic equilibrium populations of the same temperature. Consider two small surfaces; surface 1 is nonblack and surface 2 is black. The power emitted by 1 and absorbed by 2 is

$$\epsilon(\theta, \phi, \nu, T_1) \pi I_P(\nu, T_1) A_1 F_{1-2}$$

The power emitted by 2 and absorbed by 1 is $\alpha(\theta, \phi, \nu, T_1)\pi I_P(\nu, T_2)A_2 F_{2-1}$. When $T_1 = T_2$ these rates must be equal. Since $A_1 F_{1-2} = A_2 F_{2-1}$, we have

$$\bullet \qquad \alpha(\theta, \phi, \nu, T_1) = \epsilon(\theta, \phi, \nu, T_1) \qquad (6.80)$$

This relation equating spectral directional absorptance to spectral directional emittance is sometimes called **Kirchoff's Law.**

Table 6.7   Reflectances of Some Bright Metals at an Angle of Incidence of 25° from the Normal

	Reflectance at Room Temperature					
λ, microns	Aluminum	Chromium	Copper	Gold	Stainless Steel	Titanium
0.3	0.95	0.48		0.20		0.29
0.35	0.95	0.52		0.22	0.39	0.35
0.4	0.93	0.57		0.25	0.43	0.41
0.45	0.93	0.60		0.26	0.46	0.44
0.5	0.92	0.61	0.47	0.40	0.47	0.47
0.6	0.89	0.63	0.77	0.83	0.51	0.52
0.7	0.88	0.63	0.86	0.89	0.54	0.53
0.8	0.86	0.63	0.90	0.92	0.56	0.57
1.0	0.92	0.60	0.94	0.957	0.66	0.56
1.5	0.96	0.66	0.968	0.966	0.72	0.60
2.0	0.965	0.74	0.971	0.973	0.75	0.66
3.0	0.971	0.81	0.971	0.975	0.80	0.71
4.0	0.974	0.855	0.978	0.977	0.83	0.76
6.0	0.979	0.912	0.980	0.978	0.86	0.80
8.0	0.981	0.922	0.980	0.978	0.88	0.83
10.0	0.982	0.935	0.982	0.980	0.89	0.85
15.0	0.985	0.950	0.982	0.980	0.91	0.87
20.0	0.986	0.953	0.982	0.980	0.923	0.89
30.0	0.987	0.964			0.938	0.91
40.0	0.988	0.970			0.947	0.92

6.7.3   Total and Hemispherical Characteristics

As we have seen, we can speak of either a spectral characteristic or a total characteristic. It is obvious that the total value can be calculated from spectral values; for example, consider absorptance. From Eqs. (6.72) and (6.74)

$$\Delta q_m(\nu) = \alpha(\theta, \phi, \nu, T_w) I_P(\nu, T_e) \cos \theta \, d\omega \qquad (6.81)$$

If wavenumbers of all possible values are considered, the total Δq_m is obtained by integration

$$\Delta q_m = \int_0^\infty \Delta q_m(\nu) \, d\nu = \int_0^\infty \alpha(\theta, \phi, \nu, T_w) I_P(\nu, T_e) \, d\nu \cos \theta \, d\omega \qquad (6.82)$$

But from Eqs. (6.73) and (6.75)

$$\Delta q_m = \alpha_T(\theta, \phi, T_e, T_w) \sigma T_e^4 \cos \theta \frac{d\omega}{\pi} \qquad (6.83)$$

Comparison of Eqs. (6.83) and (6.82) gives

$$\alpha_T(\theta, \phi, T_e, T_w) = \frac{\int_0^\infty \alpha(\theta, \phi, \nu, T_w)\pi I_P(\nu, T_e)\, d\nu}{\sigma T_e^4} \qquad \textbf{(6.84a)}$$

If we introduce the quantity f defined by Eq. (6.62a), we can write

$$\bullet \quad \alpha_T(\theta, \phi, T_e, T_w) = \int_0^\infty \alpha(\theta, \phi, \nu, T_w)\left[\frac{-df(\mathpzc{h}c\nu/\mathpzc{k}T_e)}{d\nu}\right] d\nu = \int_0^1 \alpha\, df \qquad \textbf{(6.84b)}$$

A procedure for calculating $\alpha_T(\theta, \phi, T_e, T_w)$ is then as follows: Graph $\alpha(\theta, \phi, \nu, T_w)$ versus $f(\mathpzc{h}c\nu/\mathpzc{k}T_e)$ and integrate graphically over the interval 0 to 1. Of course numerical integration is to be preferred over graphical integration, but graphical integration has been described to make the procedure clear. In numerical integration a list of absorptances would be read into the computer, perhaps at preselected wavenumbers for which f had desired values. Then a Simpson or Gaussian summing would be performed.

Just as one likes to know the total emittance for all wavelengths, one likes to know the hemispherical emittance for all directions. A black body emits a radiant flux σT_w^4 of all wavelengths and in all directions in a hemisphere. Using this measuring stick, we define the total hemispherical emittance of a real body as

$$\epsilon_{TH}(T_w) = \frac{q^+}{\sigma T_w^4}, \qquad (q^- = 0) \qquad \textbf{(6.85a)}$$

We can relate the total hemispherical emittance to the total directional emittance by an integration similar to that in Eq. (6.57). First we return to Eq. (6.55),

$$dq^+\, dA_w = I_w \cos\theta\, dA_w\, d\omega$$

Then we introduce the directional emittance from Eq. (6.71),

$$dq^+ = \epsilon_T(\theta, \phi, T_w)\,(\sigma T_w^4/\pi)\, \cos\theta\, d\omega$$

Finally we integrate over the hemisphere

$$\frac{q^+}{\sigma T_w^4} = \frac{1}{\pi}\int_0^{2\pi}\int_0^{\pi/2} \epsilon_T(\theta, \phi, T_w) \cos\theta \sin\theta\, d\theta\, d\phi$$

$$\epsilon_{TH}(T_w) = \frac{1}{\pi}\int_0^{2\pi}\int_0^{\pi/2} \epsilon_T(\theta, \phi, T_w) \cos\theta \sin\theta\, d\theta\, d\phi \qquad \textbf{(6.85b)}$$

For a surface with azimuthally random roughness or pigmentation ϵ is not a function of ϕ. In such a case Eq. (6.85b) can be written

$$\epsilon_{TH}(T_w) = \int_0^1 \epsilon_T(\theta, T_w)\, d(\sin^2\theta) \qquad \textbf{(6.86)}$$

In this case the integration can be readily accomplished by regarding $\epsilon_T(\theta, T_w)$ to be a function of $\sin^2\theta$.

Hemispherical absorptance and reflectance refer to the fraction of the irradiation absorbed and reflected, respectively when the irradiation is perfectly diffuse, that is, when the incoming intensity I^- is independent of direction θ, ϕ. The same kind of averaging indicated by Eq. (6.85b) then applies.

To conclude this section we note that, for an opaque surface system, there are three system characteristics with which the engineer is often concerned: emittance, absorptance, and reflectance. The two types of averages which are of interest are the total average

$$\bullet \qquad \epsilon_T(T_w) = \frac{1}{\sigma T_w{}^4} \int_0^\infty \epsilon(\nu, T_w) \pi I_P(\nu, T_w) \, d\nu$$

$$\bullet \qquad \alpha_T(T_w, T_e) = \frac{1}{\sigma T_e{}^4} \int_0^\infty \alpha(\nu, T_w) \pi I_P(\nu, T_e) \, d\nu$$

$$\bullet \qquad \rho_T(T_w, T_e) = \frac{1}{\sigma T_e{}^4} \int_0^\infty \rho(\nu, T_w) \pi I_P(\nu, T_e) \, d\nu$$

and the hemispherical average

$$\bullet \qquad \epsilon_H = \frac{1}{\pi} \int_0^{2\pi} \int_0^{\pi/2} \epsilon(\theta, \phi) \cos\theta \sin\theta \, d\theta \, d\phi$$

$$\bullet \qquad \alpha_H = \frac{1}{\pi} \int_0^{2\pi} \int_0^{\pi/2} \alpha(\theta, \phi) \cos\theta \sin\theta \, d\theta \, d\phi$$

$$\bullet \qquad \rho_H = \frac{1}{\pi} \int_0^{2\pi} \int_0^{\pi/2} \rho(\theta, \phi) \cos\theta \sin\theta \, d\theta \, d\phi$$

The two types of averages can be taken independently or be combined. For example, the total directional emittance is the ratio of the total intensity of all wavenumbers emitted in a certain direction divided by the total Planckian intensity, and the total hemispherical emittance is the ratio of the total flux of all wavenumbers and in all directions emitted divided by the total flux for the Planckian radiator $\sigma T_w{}^4$.

For molecules the absorptance is often referred to as the *accommodation coefficient* or *sticking coefficient*, for it is a measure of the fraction of the molecules striking the wall which stick and accommodate themselves (by exchanging energy) to the wall temperature. The notion of a *spectral* quantity is the same, that is, a quantity which applies only to energies between E and $E + dE$. The concept of a *total* characteristic is the same with the averaging being done over a Maxwell-Boltzmann distribution. The concept of a *hemispherical* characteristic is entirely analogous to the case for photon transport.

6.7.4   *EXAMPLE* **The Calculation of the Total Emittance of Firebrick**

Consider the *flame sprayed alumina* in Table 6.6. Firebrick is often made of alumina and operates typically at 2500°F. Find the total emittance of firebrick at 2500°F using the data in Table 6.6.

Since the data in Table 6.6 were obtained at a specimen temperature of approximately 80°F, we cannot calculate $\epsilon(T)$ for $T = 2500°F$. However, we can calculate $\alpha(T_w, T_e)$, where $T_w = 80°F$, $T_e = 2500°F$. If $\alpha(\nu, T_w)$ is not a strong function of T_w, the value $\alpha(T_w, T_e)$ is a reasonable approximation to $\epsilon(T_e)$.

$$\alpha(T_w, T_e) = \int_0^1 \alpha(\nu, T_w) \, df\left(\frac{hc\nu}{kT_e}\right)$$

From Table 6.4 and a graph of Table 6.6 ($\alpha = 1 - \rho$, $\lambda = 1/\nu$) we construct Table 6.8, a table of $\alpha(\nu, T_w)$ versus f. Note $hc/kT_e = 1.44$ cm °K/1643°K $= (1/1141)$ cm. Integration then yields an approximate value for $\epsilon(T_e)$ of 0.55.

Table 6.8   Computation Table for Example 6.7.4

f	$hc\nu/kT$	ν	λ	α	f	$hc\nu/kT$	ν	λ	α
(1)	(2)	(3)	(4)	(5)	(1)	(2)	(3)	(4)	(5)
		(2)1141	10⁴/(3)				(2)1141	10⁴/(3)	
0.01	9.90	11,300	0.885	0.24	0.55	3.26	3720	2.69	0.64
0.02	8.96	10,220	0.979	0.24	0.60	3.03	3460	2.89	0.73
0.04	7.97	9090	1.101	0.26	0.65	2.80	3190	3.14	0.70
0.06	7.35	8390	1.192	0.27	0.70	2.58	2945	3.40	0.64
0.10	6.55	7480	1.338	0.29	0.75	2.34	2670	3.75	0.56
0.15	5.88	6710	1.490	0.32	0.80	2.10	2400	4.16	0.53
0.20	5.37	6130	1.632	0.37	0.85	1.83	2090	4.79	0.58
0.25	4.97	5670	1.763	0.41	0.90	1.53	1746	5.73	0.84
0.30	4.61	5260	1.902	0.48	0.94	1.24	1415	7.06	0.94
0.35	4.30	4910	2.035	0.51	0.96	1.06	1210	8.27	0.98
0.40	4.02	4590	2.18	0.54	0.98	0.811	925	10.81	0.80
0.45	3.75	4280	2.34	0.56	0.99	0.628	716	13.97	0.78
0.50	3.50	3990	2.51	0.60					

6.8   RADIANT TRANSPORT TO NONGRAY WALLS

6.8.1   The Radiant Energy Balance on an External Surface

Equation (6.55) is fundamental to radiant transport. It applies to both the power radiated $d\dot{Q}^+$ (to a remote object) and the power received $d\dot{Q}^-$ (from a remote object) at the s-surface on an element of area dA.   In both cases the solid angle $d\omega$ is that of the remote object as seen from dA; for $d\dot{Q}^+$, however, the outgoing intensity I^+ (the intensity of dA) is used, whereas for $d\dot{Q}^-$ the incoming intensity I^- (the intensity of the object) is used.

Consider, for example, an element of an opaque surface on the outside of a space vehicle on an interplanetary mission, far from the sun in solar radii. Suppose that the only significant radiant source "seen" by the surface is the sun.

To simplify the problem we take the sun to be a black body at $T_S = 5750°K$. In this case Eq. (6.55) yields

$$d\dot{Q}^- = I_P(\nu, T_S) \, dA \cos \theta_S \, d\omega_S \, d\nu$$

$$dq^- = d\dot{Q}^-/dA = I_P(\nu, T_S) \cos \theta_S \, d\omega_S \, d\nu$$

where $\cos \theta_S$ is the angle from a ray coming in from the sun to the normal of surface dA on the space vehicle, and $d\omega_S$ is the solid angle subtended by the sun when viewed from the vehicle.

The fraction of the irradiation absorbed is dq^- times the absorptance $\alpha(\theta, \phi, \nu, T_w)$. The area element itself radiates to all of space. The net loss of radiation by the element is then

$$q_r = \int_0^\infty \int_0^{2\pi} \int_0^{\pi/2} \epsilon(\theta, \phi, \nu, T_w) I_P(\nu, T_w) \cos \theta \sin \theta \, d\theta \, d\phi \, d\nu$$

$$- \int_0^\infty \alpha(\theta_S, \phi_S, \nu, T_w) I_P(\nu, T_S) \cos \theta_S \, d\omega_S \, d\nu$$

$$q_r = \epsilon_{TH} \sigma T_w^4 - \alpha_T(\theta_S, \phi_S, T_w, T_S)(\sigma T_S^4/\pi) \cos \theta_S \, d\omega_S$$

The quantity $\sigma T_S^4 \, d\omega_S/\pi$ is just $I_S \, d\omega_S$, so we can write

$$q_r = \epsilon_{TH} \sigma T_w^4 - \alpha_T(\theta_S, \phi_S, T_w, T_S) I_S \, d\omega_S \cos \theta_S \qquad (6.87)$$

At the earth's distance from the sun r_0 (one astronomical unit $= 93 \times 10^6$ miles), $I_S \, d\omega_S$ is approximately 435 Btu/hr ft² or 1380 w/m². Since $d\omega_S$ is the projected area of the sun divided by distance squared, at a distance r from the sun,

$$d\omega_S = d\omega_0 (r_0^2/r^2) \qquad (6.88)$$

which allows $I_S \, d\omega_S$ to be conveniently scaled to other values of r from the values given above.

If the vehicle is spherical of radius R, isothermal, uniformly finished, and at steady-state dissipating internal power $\dot{Q}_i$ (say from a radioisotope source), the integral of the flux over the area (only one-half is exposed to solar radiation) yields

$$\dot{Q}_i = \epsilon_{TH}(T_w) \sigma T_w^4 (4\pi R^2) - \alpha_{TH}(T_w, T_S)(I_S \, d\omega_s)(\pi R^2) \qquad (6.89)$$

This relation may be used to solve for T_w or, from a design point of view, to solve for suitable pairs of ϵ_{TH} and α_{TH} as a function of distance from the sun and internal power, since the designer chooses surfaces to fix ϵ_{TH} and α_{TH}.

6.8.2   EXAMPLE   Passive Temperature Control of a Spacecraft

Specify the surface finish of a spacecraft to obtain a temperature of 300°K. Assume the space vehicle is spherical and isothermal at one astronomical, unit from the sun and has negligible internal power dissipation.

First we find what Eq. (6.89) requires. For $\dot{Q}_i = 0$ it shows that

$$\frac{\alpha_{TH}}{\epsilon_{TH}} = \frac{4\sigma T_w^4}{I_S \, d\omega_S}$$

From Table 6.1 (or from a simple calculation) $\sigma T^4 = 459$ w/m², and $I_s \, d\omega_s = 1380$ w/m².

$$\frac{\alpha_{TH}}{\epsilon_{TH}} = 1.33$$

Table B.15 in Appendix B shows total emittance and total solar absorptance for a number of possible coatings. The values are directional ones, but for preliminary design they may be used to approximate hemispherical values. An aluminized silicone resin paint has very nearly the desired characteristics, $\alpha_{TH} = 0.27$ and $\epsilon_{TH} = 0.20$. An alternative approach is to use a checkered surface with a fraction F_1 of vacuum deposited aluminum, $\alpha_{TH} = 0.10$ and $\epsilon_{TH} = 0.03$, and the remainder $(1 - F_1)$ of white potassium zirconium silicate $\alpha_{TH} = 0.13$, $\epsilon_{TH} = 0.86$.

$$\frac{\alpha_{TH}}{\epsilon_{TH}} = \frac{F_1 \alpha_1 + (1 - F_1) \alpha_2}{F_1 \epsilon_1 + (1 - F_1) \epsilon_2} = 1.33$$

$$F_1 = \frac{1.33 \epsilon_2 - \alpha_2}{\alpha_1 - 1.33 \epsilon_1 - \alpha_2 + 1.33 \epsilon_2}$$

$$F_1 = \frac{1.33 (0.86) - 0.13}{0.10 - 0.04 - 0.13 + 1.33 (0.86)} = 0.943$$

Thus the designer may use a basic aluminum finish checkered or dotted on 5.7% of its surface with the white inorganic paint.

6.8.3 An "Internal" Surface

Consider an opaque surface on the inside of a space vehicle or other enclosure. We first consider the exchange between an element dA_1 and another, dA_2. Eq. (6.55) gives

$$d\dot{Q}^+_{1-2} = I_1^+ \, dA_1 \cos \theta_1 \, d\omega_{1-2} \, dv$$

$$d\dot{Q}^+_{1-2} = I_1^+ \, dA_1 \cos \theta_1 \, \frac{dA_2 \cos \theta_2}{r^2_{1-2}} \, dv$$

Similarly

$$d\dot{Q}^-_{1-2} = I_2^+ \, dA_1 \cos \theta_1 \frac{dA_2 \cos \theta_2}{r^2_{1-2}} \, dv$$

The net power transfer is then

$$d\dot{Q}_{1-2} = (I_1^+ - I_2^+) \frac{\cos \theta_1 \cos \theta_2 \, dA_1 \, dA_2}{r^2_{1-2}} \, dv$$

Multiplying and dividing by π allows us to write

$$d\dot{Q}_{1-2} = (\pi I_1^+ - \pi I_2^+) d(A_1 F_{1-2}) dv \tag{6.90}$$

where

$$d(A_1 F_{1-2}) = \frac{\cos \theta_1 \cos \theta_2 \, dA_1 \, dA_2}{\pi r^2_{1-2}} \tag{6.91}$$

The shape factor introduced in Eq. (6.15) is thus seen to arise in a perfectly natural way from Eq. (6.55).

To proceed we must find πI_1^+ and πI_2^+. If all the surfaces are perfectly diffuse, then I^+ is not a function of θ, ϕ, and we may write

$$\pi I_i^+ = \alpha_{H,i}(T_i, \nu)\pi I_P(\nu, T_i) + \rho_{H,i}(T_i, \nu)q_{r,i}^-(\nu)$$

$$q_{r,i}^- = \sum_{j=1}^{m} F_{i-j}\pi I_j^+ \tag{6.92}$$

Equation (6.92) may be seen to be in the very same form as Eq. (6.23). Therefore, all that follows Eq. (6.23) in Section 6.5 holds for *spectral values* provided the total black body radiosity σT_j^4 is replaced with the spectral value $\pi I_P(\nu, T_j)$. For example, Eq. (6.36a) becomes

$$\dot{Q}_{i-k}(\nu) = A_i F_{i-k}[\pi I_P(\nu, T_i) - \pi I_P(\nu, T_k)]\frac{\alpha_i(\nu)\alpha_k(\nu)}{\alpha_{\text{avg}}(\nu)}$$

Thus total transfer for a spherical enclosure is just

$$\dot{Q}_{i-k} = A_i F_{i-k}\int_0^\infty \frac{\alpha_i(\nu)\alpha_k(\nu)}{\alpha_{\text{avg}}(\nu)}[\pi I_P(\nu, T_i) - \pi I_P(\nu, T_k)]\,d\nu \tag{6.93}$$

To perform the integration when T_i and T_k are not close to the same temperature, Eq. (6.84b) is used,

$$\left(\frac{\alpha_i\alpha_k}{\alpha_{\text{avg}}}\right)_{T_k} = \int_0^1 \frac{\alpha_i(\nu)\alpha_k(\nu)}{\alpha_{\text{avg}}(\nu)}\left[\frac{-df(hc\nu/kT_k)}{d\nu}\right]d\nu \tag{6.94}$$

$$\dot{Q}_{i-k} = A_i F_{i-k}\left\{\left(\frac{\alpha_i\alpha_k}{\alpha_{\text{avg}}}\right)_{T_i}\sigma T_i^4 - \left(\frac{\alpha_i\alpha_k}{\alpha_{\text{avg}}}\right)_{T_k}\sigma T_k^4\right\} \tag{6.95}$$

But when T_i and T_k are nearly the same, it is best to write Eq. (6.93) in the form

$$\dot{Q}_{i-k} = A_i F_{i-k}\left\{\int_0^\infty \frac{\alpha_i(\nu)\alpha_k(\nu)}{\alpha_{\text{avg}}(\nu)}\frac{\partial}{\partial T}(\pi I_P(\nu, T)\,d\nu\right\}(T_i - T_k) \tag{6.96}$$

$$\dot{Q}_{i-k} = A_i F_{i-k}\left(\frac{\alpha_i\alpha_k}{\alpha_{\text{avg}}}\right)_{TI}4\sigma T^3(T_i - T_k) \tag{6.97}$$

where the subscript *TI* denotes an *internal* average

$$\left(\frac{\alpha_i\alpha_k}{\alpha_{\text{avg}}}\right)_{TI} = \int_0^\infty \frac{\alpha_i(\nu)\alpha_k(\nu)}{\alpha_{\text{avg}}(\nu)}\left[\frac{-df_i(hc\nu/kT)}{d\nu}\right]d\nu \tag{6.98}$$

and

$$f_i\left(\frac{hc\nu}{kT}\right) = \frac{1}{4\sigma T^3}\int_\nu^\infty \frac{\partial}{\partial T}(\pi I_P(\nu, T))\,d\nu \tag{6.99}$$

The adjective *internal* is employed, because the situation where T_i is nearly equal to T_k occurs commonly inside enclosures such as rooms or space vehicles. This internal fraction is tabulated in Table 6.9.

Table 6.9   The Internal Fraction for the
Planck Distribution

$\dfrac{hc\nu}{kT}$	f_i	$\dfrac{hc\nu}{kT}$	f_i	$\dfrac{hc\nu}{kT}$	f_i
0.0	1.00	3.39	0.70	6.11	0.25
0.934	0.99	3.66	0.65	6.56	0.20
1.19	0.98	3.93	0.60	7.11	0.15
1.52	0.96	4.19	0.55	7.85	0.10
1.76	0.94	4.46	0.50	8.72	0.06
2.13	0.90	4.75	0.45	9.39	0.04
2.50	0.85	5.05	0.40	10.45	0.02
2.82	0.80	5.37	0.35	11.48	0.01
3.11	0.75	5.72	0.30	∞	0.0

6.8.4 *EXAMPLE* Heat Transfer in an Aluminum Melting Furnace

Aluminum ingots (*sows*) are melted in a furnace very roughly spherical in shape for the purpose of alloying and recasting. The aluminum charge is covered with a layer of oxide (*dross*) which raises its absorptance and inhibits further oxidation. If the dross thickness is not very closely controlled, its upper face will heat to a temperature close to that of the furnace wall. Calculate the net radiant heat transfer in millions of Btu per hour to 400 ft² of dross surface at 2475°F from the furnace wall at 2525°F, neglecting furnace gas radiation. Both the dross and firebrick have roughly the spectral characteristics of *flame sprayed alumina* in Table 6.6.

Since the average absorptance in the enclosure is that of alumina, Eq. (6.98) reduces to

$$\alpha_{TI} = \int_0^\infty \alpha(\nu) \left[\frac{-df_i(hc\nu/kT)}{d\nu} \right] d\nu = \int_0^1 \alpha(\nu)\, df_i(hc\nu/kT)$$

As in Example 6.7.4, we construct a table (Table 6.10) of $\alpha = 1 - \rho$ versus f_i for $T = 2500°F$. Integration yields an approximate value of 0.475. Since the dross surface is fairly flat and is completely enclosed by the furnace we can take $F_{1-2} \doteq 1$. Then Eq. (6.97) yields

$$\dot{Q}_{1-2} = (400)\,(1)\,(0.475)\,(4)\,(1.712 \times 10^{-9})\,(2500 + 460)^3\,(50)$$
$$\dot{Q}_{1-2} = (190)\,(178)\,(50) = 1.7 \times 10^6 \text{ Btu/hr}$$

Note that our answer would have been only 16% too high, if we had used the total characteristic $\epsilon_{TE} = 0.55$ calculated in Section 6.7.4 in place of $\alpha_{TI} = 0.475$ computed here.

Table 6.10   Computation Table for Example 6.8.4

f_i (1)	hcv/kT (2)	ν (3) (2)1141	λ (4) 10^4/(3)	α (5)	f_i (1)	hcv/kT (2)	ν (3) (2)1141	λ (4) 10^4/(3)	α (5)
0.01	11.48	13,100	0.76	0.23	0.55	4.19	4780	2.09	0.52
0.02	10.45	11,900	0.84	0.24	0.60	3.93	4480	2.23	0.51
0.04	9.39	10,700	0.94	0.25	0.65	3.66	4170	2.40	0.52
0.06	8.72	9940	1.01	0.25	0.70	3.39	3860	2.59	0.60
0.10	7.85	8950	1.12	0.25	0.75	3.11	3550	2.82	0.72
0.15	7.11	8110	1.23	0.26	0.80	2.82	3210	3.11	0.69
0.20	6.56	7490	1.34	0.27	0.85	2.50	2850	3.51	0.58
0.25	6.11	6960	1.44	0.30	0.90	2.13	2430	4.13	0.53
0.30	5.72	6520	1.53	0.33	0.94	1.76	2005	4.99	0.62
0.35	5.37	6130	1.63	0.37	0.96	1.52	1730	5.77	0.84
0.40	5.05	5760	1.74	0.42	0.98	1.19	1360	7.27	0.93
0.45	4.75	5410	1.85	0.48	0.99	0.934	1063	9.4	0.98
0.50	4.46	5090	1.97	0.50	1.00	0	0	∞	0

6.9   SUMMARY

Chapter 6 has presented the subject of radiation transfer and some aspects of free molecule transport twice in succession. In Sections 6.2 to 6.5 the macroscopic view was taken. The concept of a *black surface* and fluxes from black surfaces were presented. The idea of a *shape factor* was introduced to account for geometrical effects. Real surfaces were approximated by assuming a *gray* behavior. Finally, in the first pass the radiosity-irradiation-network formulations of transfer were shown to be a fairly powerful tool for solving practical problems.

In the "second pass," Sections 6.6 to 6.8, we covered the ideas of spectral distributions, spectral properties, and total fluxes and properties, the latter being obtained by integration over the spectrum. Hemispherical averages were obtained by integration over all directions in a hemisphere. The concept of radiant intensity, which had been avoided in the first pass, played a central role in the treatment of the second pass. From it the shape factor was seen to arise in a natural way. The radiosity-irradiation-network formulations of the first half of the chapter were seen to apply to spectral radiation as well as to total radiation. After such formulations were used to find expressions for the spectral flux, integration over the spectrum was shown to be easily accomplished by using one or another of two fractional functions, the external fraction $f(hcv/kT)$ or the internal fraction $f_i(hcv/kT)$. The former, f in Table 6.4, is used when the surface temperatures are significantly different, and the latter, f_i in Table 6.9, is used when each surface has nearly the same absolute temperature.

EXERCISES

1. Consider two infinite black parallel walls 1°F different in temperature. What is the net thermal radiation transfer between them in Btu/hr ft² when one is 70°F and the other 71°F? What is the answer when one is 69°F and the other 71°F? (*Note:* When T_1 and T_2 are very close together, $[\sigma T_1{}^4 - \sigma T_2{}^4]$ will be directly proportional to $[T_1 - T_2]$. Find the proportionality factor by a Taylor series expansion.)

2. Consider a small object in a large enclosure. What is the net radiation transfer from the object to the surrounds in the limit when $\epsilon_2 A_2/\epsilon_1 A_1$ approaches infinity and the temperature difference $T_1 - T_2$ is very small compared to the absolute level $(T_1 + T_2)/2$? How does the answer to this question bear upon *Newton's Law of Cooling*? What is the value of the radiation contribution to a Newton's Law of Cooling heat transfer coefficient when the small object has an emittance of 0.85 and a temperature of 75°F?

3. A vacuum chamber is in the shape of a sphere 20 ft in diameter. Passing material into and out of the chamber releases 1 liter/min of air at 1 atm and 20°C into the chamber. It is desired to maintain a pressure of 10^{-5} mm Hg within the chamber by having a segment of the spherical wall act as a cryo-pump. This part of the wall is cooled by a cryogen and condenses molecules which strike it. If the cold wall condensed all molecules which struck it and emitted none, what fraction of the sphere would it have to cover? Assume the remainder of the wall is at 300°K.

4. A clear sky has been observed to have an effective emittance of approximately 0.65 (the exact value depending upon the amount of water vapor and other infrared absorbing-emitting gases present) based on the temperature of the air a few feet above the ground. Estimate the rate at which an orange hanging from a tree cools versus its temperature. Assume the following conditions:

   Air temperature = 35°F
   Convective heat transfer coefficient = 0.50 Btu/hr ft² °F
   Emittance of orange peel $\epsilon_1 = \alpha_1 = 0.90$
   Shape factor from orange to sky $F_{1-3} = 0.25$
   Shape factor from orange to grass and leaves $F_{1-2} = 0.75$
   Emittance of grass and leaves = 1.0
   Temperature of grass and leaves = equilibrium temperature for zero net heat flux when the shape factor from the grass and leaves to the sky $F_{2-3} = 0.50$. (Remember $\Sigma_j F_{2-j} = 1$.)

   Will the convective heating, when the orange is 32°F, exceed the radiative cooling? If not, will having the farmer turn on a fan to increase h_c from 0.5 to 2.0 Btu/hr ft² °F prevent freezing? (*Hint:* Find T_2 first by making an energy balance on a unit area $A_2 = 1$. Note that *grass and leaves* can be regarded as a single surface and, since $A_1 F_{1-2} = A_2 F_{2-1}$, F_{2-1} is negligible. Then assuming

T_1 and reasonable values for A_1, m_1, and $c_{p,1}$, find Q_1 and $\partial T_1/\partial t$ for each assumed value of T_1.)

5. The solar constant is the flux incident on a unit surface normal to the sun's rays at the earth's mean distance to the sun (outside the earth's atmosphere). It has a value of approximately 1.96 cal/cm² min. How does this value agree with observations that the outer atmosphere of the sun is approximately 5750°K, and that the diameter of the sun subtends 32 min of arc when viewed from earth?

6. What force would be exerted by photons on a spherical space vehicle 20 m in diameter at 1 astronomical unit (the earth's distance from the sun) when

   (a) The sphere is perfectly black and the walls are isothermal?
   (b) The sphere is perfectly specular (mirrorlike) and perfectly reflecting?

   What g-acceleration would a 1 kg balloon of such a diameter be subjected to?

7. A total radiometer output V is a linear function of the radiosity of the surface it views, when the surface fills the entire field of view. Often the instrument is calibrated to read out temperature; that is, the signal $V-V_0$ is processed as follows:

$$T_{\text{dial}} = \left\{ T_0{}^4 + (T_{\text{ref}}^4 - T_0{}^4) \frac{V-V_0}{V_{\text{ref}}-V_0} \right\}^{1/4} - 460$$

where T_0 is the instrument temperature and T_{ref} is a reference black body source temperature. Readings were made with such a device by sighting onto the interior walls of a long tunnel furnace 10 ft high and 10 ft wide which was heated indirectly by a muffle around the exterior. Averages of several readings are as follows:

Wall of Furnace	Temperature
Top	2020°F
Left side	2430°F
Right side	2380°F
Bottom	2000°F

If the walls are gray and diffuse with an emittance of 0.48, what are the true temperatures of each wall? What is the net radiant flux at each wall?

8. Assuming the sun is approximately a black body radiator at 5750°K, plot a fractional function versus wavelength using the information in Table 6.4. Calculate the solar absorptance of aluminum, chromium, and gold using the fractional function and the data given in Table 6.7.

9. Suppose the effect of the earth's atmosphere on the solar spectrum is roughly approximated by assuming

   (a) The sun is a black body radiator at 5750°K.
   (b) All photons with wavelengths shorter than 0.4 μ are absorbed.
   (c) Three-fourths of the photons with wavelengths between 0.4 and 1.8 μ are transmitted on a clear day.

(d) All photons with wavelengths longer than 1.8 μ are absorbed.

Estimate what the solar irradiation $I_S \, d\omega_S$ would be at the earth's surface. Plot a fractional function versus wavelength for terrestrial solar radiation based on these approximations. (*Hint:* The fractional function will have the same shape as that in Exercise 8 in the interval $0.4 < \lambda < 1.8\mu$ but will start at zero at $\lambda = 0.4$ and will rise to unity at $\lambda = 1.8$.)

10. Using the fractional function from Exercises 8 and 9, compute the solar absorptance of the white epoxy paint in Table 6.6 for extraterrestrial solar radiation and terrestrial solar radiation. Compare the value obtained to values for white paints in Table B.15 of Appendix B.

11. Calculate the fraction F_1 of aluminum paint and $F_2 = 1 - F_1$ of vacuum evaporated aluminum which would make a cubical space vehicle at 1 astronomical unit from the sun have a temperature of 300°K. Assume that the reflectances of the aluminum paint shown in Table 6.6 are independent of direction and that the spectral hemispherical absorptances of evaporated aluminum are roughly 1.3 times the near normal values.

12. A space vehicle is in the shape of a hollow equilateral tetrahedron; one side facing the sun is striped 40% with a flat black paint with a solar absorptance for normal incidence of 0.97 and an external total hemispherical emittance at room temperature of 0.89 (see Table 6.6). The other 60% of this surface and 100% of the other outside surfaces of the remaining three sides are painted with aluminum paint with a solar absorptance of 0.27 and an external total hemispherical emittance of 0.22 (see Table 6.6). All inside surfaces are painted black with an internal total hemispherical emittance of 0.88. Estimate the temperature of each of the four walls, neglecting heat conduction effects.

13. If it is desired to have the three shaded sides of the space vehicle described in Exercise 12 operate at 70°F and the sunny side as close to 70°F as possible, recommend a new set of external surface finishes using the data given in Table B.15 to function at

   (a) the earth's distance from the sun, 1 astronomical unit, and

   (b) half the earth's distance from the sun, 0.5 astronomical unit.

14. Two infinite parallel plates, one at 2100°F and the other at 840°F, face each other. They are made of platinum and have a spectral absorptance given approximately by

$$\alpha(\theta, \phi, \lambda, T_w) \doteq \alpha(\lambda, T_w) [\tfrac{1}{2} \cos \theta + \tfrac{1}{2} \sec \theta] \leqslant 1$$

where $\alpha(\lambda, T_w)$ is a function of wavelength λ and four parameters $A(T_w)$, $B(T_w)$, C, and $\lambda_{12}(T_w)$ as shown at the top of Table B.14. The parameters are as follows:

$$A \doteq 0.034(T/T_0)^{0.94}, \ T_0 = 550°R$$
$$B \doteq 0.577 - 2.89 \, Af(\lambda_x/\lambda_{12}), \ \lambda_x = 1.7 \ \mu$$
$$C \doteq 0$$
$$\lambda_{12} = 7(T/T_0)^{-0.94} \ \mu$$

$$f(\lambda_x/\lambda_{12}) = \left\{ \frac{(1 + \lambda_x{}^2/\lambda_{12}{}^2)^{1/2} - 1}{\lambda_x{}^2/2\lambda_{12}{}^2} \right\}^{1/2}$$

Find the radiation heat transfer between the plates accounting for spectral selectivity but assuming diffuse surfaces at hemispherically averaged values.

REFERENCES

Radiation Transfer Texts

There are a number of texts on radiation transfer; in alphabetical order they are:

Hottel, H. C., and A. F. Sarofim, *Radiative Transfer*. New York: McGraw-Hill, Inc., 1967.

This text covers the entire subject and is particularly good in its treatment of scatter from particles such as dust or soot.

Kreith, F., *Radiation Heat Transfer*. Scranton, Pa.: International Textbook Co., 1962.

This text emphasizes spacecraft temperature control.

Love, T. J., *Radiative Heat Transfer*. Columbus, Ohio: Merrill Publishing Co., 1968.

This text contains tables for Gaussian and Laguerre integration, has a section on experimental techniques, and gives some property values.

Siegel, R., and J. R. Howell, *Thermal Radiation Heat Transfer*. New York: McGraw-Hill, Inc., 1972.

This text contains a description of the Monte Carlo technique.

Sparrow, E. M., and R. D. Cess, *Radiation Heat Transfer*. Belmont, Calif.: Brooks Cole Publishing, 1966.

This text contains a good treatment of simultaneous radiation and convection.

Wiebelt, J. A., *Engineering Radiation Heat Transfer*. New York: Holt, Rinehart and Winston, Inc., 1966.

This text has a good treatment of surface-to-surface radiation transfer.

Molecular Theory

The definitive work in the field is:

Hirschfelder, J. O., C. F. Curtiss, and R. B. Bird, *Molecular Theory of Gases and Liquids*. New York: John Wiley, 1954.

Chapter 2 contains some good descriptive material for the beginning student.

A development of emissions from surfaces is given by:

> Knuth, E. L., *Introduction to Statistical Thermodynamics*. New York: McGraw-Hill, Inc. 1966 (see Appendix I).

Molecular drag and heating rates on high altitude spacecraft can be computed using the simple theory presented in:

> Oppenheim, A. K., "Generalized Theory of Convective Heat Transfer in Free-Molecule Flow," *Journal of the Aeronautical Sciences*, Vol. 20, pp. 49–58 (1953).
> Rohsenow, W. M., and H. Choi, *Heat, Mass and Momentum Transfer*. Englewood Cliffs, N. J.: Prentice-Hall, 1961, (see Chapter 11.)

Radiation Tables

The internal and external fractional functions are given in:

> Czerny, M., and A. Walther, *Tables of the Fractional Functions for the Planck Distribution Law*. Berlin: Springer Verlag, 1961.

An older work which is quite readable is:

> Dunkle, R. V., "Thermal Radiation Tables and Applications," *Transactions of the American Society of Mechanical Engineers*, Vol. 76, pp. 549–552 (1954).

Solar Radiation

A recent work which gives a good list of references is:

> Thekaekara, M. P., "Solar Irradiance Curves and Absorptance of Satellite Coatings," *Solar Energy*, Vol. 12, pp. 205–215 (1968).

A book which explains how to find the solar flux on the earth's surface is:

> Threlkeld, J. L., *Thermal Environmental Engineering*. Englewood Cliffs, N. J.: Prentice-Hall, 1962.

Shape Factors

A nice compilation can be found in:

> Hamilton, D. C., and W. R. Morgan, *Radiant Interchange Configuration Factors*, NACA TN 2836. 1952.

Recently a number of computer programs have been developed, for example:

> Toups, K. A., CONFAC II, Technical Documentary Report FDL-TDR-64-43, North American Aviation, Inc., Downey, California,

prepared for the Air Force Flight Dynamics Laboratory, Wright-Patterson Air Force Base, Ohio.

Radiation Properties

A survey of properties of surfaces and molecular gases is given by:

Edwards, D. K., "Radiative Transfer Characteristics of Materials," *Journal of Heat Transfer*, Vol. 91, pp. 1–15 (1969).

Gas radiation properties are also surveyed by:

Tien, C. L., "Thermal Radiation Properties of Gases," *Advances in Heat Transfer*, Vol. 5, pp. 254–324. New York: Academic, 1968.

CHAPTER 7

TRANSPORT PROPERTIES FROM MEAN FREE PATH CONSIDERATIONS

7.1 INTRODUCTION

On the one hand we have phenomenological descriptions of diffusive transport processes: Fourier's and Fick's laws and Newton's law of viscosity. On the other hand we have the picture of transport used in the previous chapter, that of a carrier, such as a molecule, conveying mass, momentum, and energy as it moves from one place to another. In this chapter we attempt to reconcile the two views by using a simple model of molecular carriers, the rigid billiard ball model and some heuristic kinetic theory arguments. In this manner we are able to establish that the viscosity, thermal conductivity, and density-mass-diffusivity product of a dilute gas are independent of pressure and increase roughly with the square root of temperature. Formulas for the transport properties, derived from a more exact kinetic theory of gases, are also presented. Values of the viscosity, thermal conductivity, and mass diffusivity may be readily calculated from these formulas. Neutron diffusion is briefly discussed from mean free path considerations. The chapter closes with an analysis of the photon contribution to thermal conductivity in *super-insulation* or other porous materials.

7.2   THE MEAN FREE PATH

7.2.1   The Collision Mean Free Path

How far, on the average, does a molecule move before being involved in a collision? The answer to this question is a length known as the *mean free path*. We imagine molecules to be rigid spheres of diameter d, so that the center of a given molecule lies at a distance of two molecular radii $(d_1/2 + d_2/2)$ from that of another molecule at the instant of collision. If we regard the center of molecule 2 as a point projectile streaking toward a target molecule 1 of the same diameter, the silhouette of the target is a circle of radius $d_1/2 + d_2/2 = d$. The *collision cross-section*, the area of that circle, is thus

$$\sigma_c = \pi d^2 \tag{7.1}$$

The target is a moving target. The projectile closes on the target not with velocity $\mathbf{v}_2$ but relative velocity $\Delta \mathbf{v} = \mathbf{v}_2 - \mathbf{v}_1$. However, this detail does not affect the order of magnitude of the velocity. The order of magnitude remains the order of the mean molecular speed, which from Eq. (6.65) is

$$\bar{v} = \left(\frac{8kT}{\pi m}\right)^{1/2} \tag{7.2}$$

In fact, it may be shown without much difficulty from Eq. (6.45) that the average magnitude of the relative velocity between two molecules which collide is

$$\overline{\Delta v} = \sqrt{2}\,\bar{v} \tag{7.3}$$

We may just as well regard the point center of molecule 2 as fixed and molecule 1, having a collision cross section of πd^2, as moving at velocity $\overline{\Delta v}$. Although a given molecule follows a tortuous path, during time t it moves a distance relative to a target of $\overline{\Delta v}t$. The probability of a collision during that time is the probability of finding a center of a molecule within the volume swept out, the volume being $\overline{\Delta v}t\pi d^2$. Since the number of molecules per unit volume, on the average, is $\mathcal{N} = P/kT$, the average number of collisions which our average molecule would experience would be $\overline{\Delta v}t\pi d^2\mathcal{N}$. The average time between collisions is then

$$t_c = \frac{1}{\overline{\Delta v}\pi d^2\mathcal{N}} = \frac{1}{\sqrt{2}\,\bar{v}\pi d^2\mathcal{N}} \tag{7.4}$$

and during this time our average molecule moves a distance

$$l = vt_c = \frac{1}{\sqrt{2}\,\pi d^2\mathcal{N}} \tag{7.5}$$

This distance is the mean free path.

The rigid sphere model of a molecule is quite naive. A more realistic model embodies a weak *attractive* force between two molecules at distances several times the distance d at which strong repulsive forces start to come into action. The weak forces will cause a noticeable effect on two relatively slowly moving molecules but will have a negligible effect on two relatively fast moving ones. The shorter range repulsive forces remain effective for the fast pairs. The potential energy tied up in a pair of molecules obeys a relationship of the form shown in Figure 7.1. The potential energy ϕ is just the integral of the interaction force F: thus $F = -d\phi/dr$, where r is the separation distance. To the left of the minimum in the potential energy curve the molecules repel each other, while to the right there is the attractive force. A detailed discussion of this more realistic model of molecular interaction will be given later in the chapter. For the present we stress only one consequence of the model, namely, that slow moving, low temperature molecules appear larger to one another than do fast moving, high temperature ones.

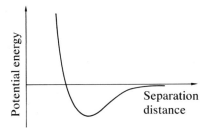

Figure 7.1   An attractive-repulsive intermolecular potential.

7.2.2   The Transport Mean Free Path

For our purposes it is sufficient to accept that the transport mean free path is on the same order of magnitude as the collision mean path. This section is essentially a digression to argue the plausibility of the transport mean free path being somewhat longer than the collision mean free path. On a first reading the student may go directly to Section 7.3.

If a *collision* resulted in no essential change in the trajectory of a molecule, we would not consider that collision in transport calculations. We could retain the view taken in Chapter 6 that the carrier moves undeviatingly along a straight line path. Detailed consideration of the flux of particles per unit solid angle, and the changes in the flux due to collisions, shows that the effective collision cross-section affecting transport rates is

$$\bullet \qquad \sigma_t = (1 - \overline{\cos \theta}) \sigma_c \qquad (7.6)$$

where $\overline{\cos \theta}$ is the average angle that a particle is deflected. If there were no deflection, $\theta = 0$; then we would not count the collision. If a collision resulted in perfect backscattering with $\theta = \pi$, then such a collision would count twice as much as one which merely turned the particle through $\pi/2$.

The angle θ is the angle of deflection measured from the original trajectory when we regard the target molecule as stationary. Figure 7.2a shows the situation envisioned. The projectile closes on the target at velocity Δv and after collision is deflected angle θ. Figure 7.2b shows the situation viewed from the

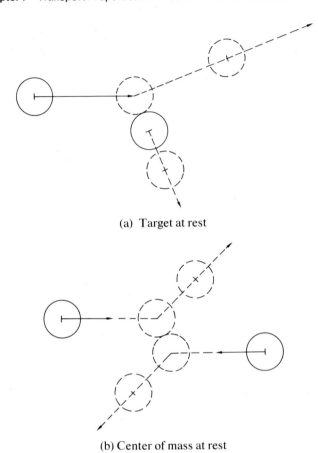

(a) Target at rest

(b) Center of mass at rest

Figure 7.2   Particle trajectories before and after an elastic collision.

center of mass coordinate system. In this case both particles (with the same mass for simplicity) are moving closer to the center of mass before the collision and recoil away symmetrically afterwards. Conservation of the components of momentum and conservation of energy serve to fix the trajectories after collision.

In the center of mass coordinates the scattering is isotropic for rigid spheres, that is, the number of particles per unit solid angle after the collision will be independent of θ_s. The probability of a collision causing a turning into solid angle $d\omega = 2\pi \sin \theta_s \, d\theta_s$ is the projected area of that region of the sphere between $\phi = \theta_s/2$ and $\phi = (\theta_s + d\theta_s)/2$ (see Figure 7.3) divided by the total projected area. The probability of scattering into $d\omega = 2\pi \sin \theta_s \, d\theta_s$ is therefore

$$dP_{\theta_s} = \frac{(2\pi R \sin \phi \, R \, d\phi) \cos \phi}{\pi R^2} = 2 \sin \phi \cos \phi \, d\phi$$

and

$$d\omega = 2\pi \sin \theta_s \, d\theta_s$$

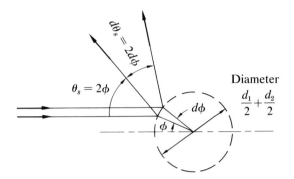

Figure 7.3 Scattering in the center of mass system.

Thus

$$\frac{dP_{\theta_s}}{d\omega} = \frac{2 \sin \phi \cos \phi \, d\phi}{2\pi \sin \theta_s \, d\theta_s} = \frac{(\sin 2\phi) \, d\phi}{2\pi \sin \theta_s \, d\theta_s} = \frac{1}{4\pi}$$

The average $\cos \theta$ will then be weighted by this uniform probability,

$$\overline{\cos \theta} = \int_{\omega=0}^{4\pi} \cos \theta \, \frac{dP_{\theta_s}}{d\omega} \, d\omega = \frac{1}{4\pi} \int_{\omega=0}^{4\pi} \cos \theta \, d\omega$$

$$\overline{\cos \theta} = \frac{1}{4\pi} \int_0^\pi \cos \theta \, 2\pi \sin \theta_s \, d\theta_s = \frac{1}{2} \int_0^\pi \cos \theta \sin \theta_s \, d\theta_s \qquad (7.7)$$

To relate θ to θ_s we return to Figure 7.2. In Figure 7.2b velocity components after collision relative to the center of mass are

$$v_y = \frac{\Delta v}{2} \sin \theta_s; \qquad v_x = -\frac{\Delta v}{2} \cos \theta_s$$

The center of mass in Figure 7.2a is moving in the x-direction with velocity $\Delta v/2$. Therefore, in the coordinates of Figure 7.2a, after collision the velocity components are

$$v_y = \frac{\Delta v}{2} \sin \theta_s, \qquad v_x = \frac{\Delta v}{2} (1 - \cos \theta_s)$$

so that

$$\tan \theta = \frac{v_y}{v_x} = \frac{\sin \theta_s}{1 - \cos \theta_s}$$

$$\theta = \tan^{-1} \left(\frac{\sin \theta_s}{1 - \cos \theta_s} \right) \qquad (7.8)$$

With Eq. (7.8) substituted into Eq. (7.7), the integration may be performed after some algebraic-trigonometric manipulations. The result is

$$\overline{\cos \theta} = \tfrac{2}{3} \qquad (7.9a)$$

If the molecules were unequal in mass, for example, if a light molecule of mass m_1 scattered from a heavy molecule of mass m_2, we could repeat the above derivation to obtain a more general result, of particular interest in neutron transport,

$$\overline{\cos\theta} = \frac{2m_1}{3m_2}$$ (7.9b)

When m_2 is very large compared to m_1, the velocity of the center of mass is negligible, and the scattering is isotropic so that the mean $\overline{\cos\theta}$ is nearly zero.

7.3   NET TRANSPORT ACROSS A PLANE

7.3.1   Mass Diffusivity of Isotopic Species

Consider a gas with a slight gradient in composition. Let the z-axis be in the direction of the gradient. To keep the physical situation simple let the molecules have essentially the same mass m and size d. The change in composition can be due to one isotope diffusing into another, both being very nearly the same molecular weight, say fluorides of uranium isotopes. We wish to find the net flux of species A crossing a plane normal to the z-axis. We already know the one-way flux of molecules crossing a plane in a uniform gas (our gas is very nearly uniform); from Eq. (6.5)

$$J^+ = \tfrac{1}{4}\mathcal{N}\bar{v}$$ (7.10)

Now we argue that those molecules, on the average, came from below the plane from a distance on the order of a *transport* mean free path, l_t.   Therefore the flux of molecules of species A is

$$J_A^+ = J^+x_A|_{z-blt}$$ (7.11)

where x_A is the mole fraction of species A and b is a number of order of magnitude unity.

We digress for a moment to establish a value for b, even though we already know its order of magnitude and thus are not vitally concerned about a precise value. Only those molecules which crossed normal to the plane came from distance $z-l_t$. Those that crossed at slant angle θ came on the average from $z-l_t\cos\theta$. The average $\cos\theta$ weighted by $dJ^+/d\omega$ and integrated over a hemisphere is $\tfrac{2}{3}$, since $dJ^+/d\omega$ varies as $\cos\theta$; see Eq. (6.48). Recall that the differential solid angle $d\omega$ is $\sin\theta\,d\theta\,d\phi$ in polar coordinates; then

$$b = \frac{\left(\dfrac{dJ^+}{\cos\theta\,d\omega}\right)\displaystyle\int_0^{2\pi}\int_0^{\pi/2}\cos\theta\cos\theta\sin\theta\,d\theta\,d\phi}{\left(\dfrac{dJ^+}{\cos\theta\,d\omega}\right)\displaystyle\int_0^{2\pi}\int_0^{\pi/2}\cos\theta\sin\theta\,d\theta\,d\phi} = \frac{2}{3}$$ (7.12)

Those molecules of species A crossing in the negative z-direction came on the average from $\tfrac{2}{3}l_t$ above the plane,

$$J_A^- = J^-x_A|_{z+(2/3)l_t}$$ (7.13)

We can suppose that the total number of molecules crossing the plane from below is equal to the number crossing from above, that is, the number average z-velocity is zero. Alternatively we could state that the plane is moving with the local number average velocity (molar average velocity). Therefore

$$J^+ = J^- \tag{7.14}$$

Now we obtain the net flux of molecules of species A crossing the plane,

$$J_A = J_A^+ - J_A^- = \tfrac{1}{4} \mathcal{N} \bar{v} [x_A|_{z-(2/3)l_t} - x_A|_{z+(2/3)l_t}] \tag{7.15}$$

Expanding the two terms in two Taylor's series we have

$$x_A|_{z-(2/3)l_t} = x_A|_z + \frac{dx_A}{dz}\left(-\frac{2}{3}l_t\right) + \cdots$$

$$x_A|_{z+(2/3)l} = x_A|_z + \frac{dx_A}{dz}\left(+\frac{2}{3}l_t\right) + \cdots$$

Therefore

$$J_A = \frac{1}{4} \mathcal{N} \bar{v} \left[-\frac{4}{3} l_t \frac{dx_A}{dz} \right]$$

$$J_A = -\frac{1}{3} \mathcal{N} \bar{v} l_t \frac{dx_A}{dz} \tag{7.16}$$

If Eq. (7.16) is divided by Avogadro's number N_{Av}, there results, upon setting $\mathcal{N}/N_{Av} = c$, the total molar concentration,

$$J_A = -c \left(\frac{1}{3} \bar{v} l_t\right) \frac{dx_A}{dz} \text{ moles/area-time} \tag{7.17}$$

Equation (7.17) is in the form of Fick's first law,

$$J_A = -c \mathcal{D}_{AB} \frac{dx_A}{dz}$$

Thus the mass diffusivity is simply

$$\mathcal{D}_{AB} = \tfrac{1}{3} v l_t \tag{7.18}$$

A more appropriate measure of mass transport than $\mathcal{D}_{AB}$ alone is the product $c\mathcal{D}_{AB}$ or $\rho\mathcal{D}_{AB}$; the latter becomes

$$\bullet \qquad \rho \mathcal{D}_{AB} = \frac{1}{3(1-\cos\theta)} \left(\frac{8kT}{\pi m}\right)^{1/2} \left(\frac{Pm}{kT}\right) \left(\frac{kT}{\sqrt{2}\,\pi d^2 P}\right) \tag{7.19}$$

Equation (7.19) shows us that the density-diffusivity product is independent of pressure. As the pressure increases, the number of carriers (proportional to the density) goes up, but the path which they can travel goes down, with the net result that there is no pressure dependence (unless at high pressures there are departures from an ideal gas equation of state or unless at low pressures the mean free path becomes appreciable compared to the size of the system). The temperature effect on the number of carriers and the distance they can travel likewise cancels, but since the carriers are moving faster at higher temperatures, the density-diffusivity product does increase as the square root of temperature, if d is constant. But, as we have already remarked, the effective size d varies with temperature, decreasing with increasing temperature until the temperature is sufficiently high for only the repulsive part of the potential curve of Figure 7.1 to be important. Only at these higher temperatures does $\rho \mathcal{D}$ vary like $T^{1/2}$. At lower temperatures the dependence is stronger than $T^{1/2}$, since d^2 in the denominator is decreasing with T. Empirical fits using $T^{0.8}$ or so are typically employed near room temperature. In Section 7.4 there will be presented a formula for $\mathcal{D}_{AB}$ which gives this more correct temperature dependence.

7.3.2   Viscosity

First consider the force on a freight train being loaded with ore on the fly. If the train moves with velocity v_0 and the ore is dumped with velocity component v_1 in the direction of the tracks, collisions of the lumps of ore with one another and the walls of the cars must supply sufficient impulse to accelerate them rapidly to the velocity v_0. The velocity of the ore relative to the moving train is thus $v_1 - v_0$. Let the average mass of a lump of ore be m and the number of lumps per unit area of the open hopper cars be J^+. The impulse received by the ore in time t per unit area of the car top would be

$$\int \left(\frac{F}{A}\right) dt = \int d\,(Mv) = J^+ t m\,(v_0 - v_1) \tag{7.20}$$

The average force per unit area on the train (equal and opposite) is then

$$\tau = \frac{\bar{F}}{A} = \frac{1}{t} \int F dt = J^+ m\,(v_1 - v_0) \tag{7.21}$$

Notice that when the ore velocity is less than that of the train, the sign convention indicates a drag on the train.

In the kinetic theory model we consider momentum transport in a gas to be the same as occurs when lumps of ore are loaded into a train. Consider a gas in simple shear flow. Let z be in the direction of the positive velocity gradient. Molecules coming from below a plane at z have, on the average, the mass average velocity v_m at location $z - \frac{2}{3}l_t$. The momentum flux upward relative to the velocity at the plane is

$$J^+ m \left[\left(v_m(z) - \frac{dv_m}{dz} \frac{2}{3} l_t + \cdots \right) - v_m(z) \right]$$

The momentum flux downward is carried by molecules with velocities appropriate for the location two-thirds of a transport mean free path above the plane,

$$J^- m \left[\left(v_m(z) + \frac{dv_m}{dz} \frac{2}{3} l_t + \cdots \right) - v_m(z) \right]$$

When there is no net flow in the z-direction, $J^+ = J^- = \frac{1}{4}\mathcal{N}\bar{v}$, and

$$\tau = \frac{1}{4}\mathcal{N}\bar{v}m\left[-\frac{4}{3}l_t\frac{dv_m}{dz}\right] = -\left(\frac{1}{3}\bar{v}\mathcal{N}ml_t\right)\frac{dv_m}{dz} \tag{7.22}$$

The negative sign shows that the shear force is a drag, that is, it operates in the direction opposed to the velocity of the faster moving fluid above the plane.

Comparison of Eq. (7.22) with the shear stress of a Newtonian fluid shows that kinetic theory does indeed indicate that a gas should be Newtonian with a viscosity of

$$\mu = \frac{1}{3}\bar{v}\rho l_t \qquad (\rho = \mathcal{N}m) \tag{7.23}$$

Note that this relation is exactly equal to $\rho\mathcal{D}_{AB}$ when A and B have nearly the same mass and size. We can designate this latter situation by writing $\mathcal{D}_{AA}$. The Schmidt number for such a gas is then unity,

$$Sc = \frac{\mu}{\rho\mathcal{D}_{AA}} = 1 \tag{7.24}$$

It also follows that the viscosity has a pressure and temperature dependence identical to that of the density-diffusivity product; the discussion at the end of Section 7.3.1 applies to viscosity as well.

7.3.3   Thermal Conductivity

Now consider a vibrationally unexcited gas with an imposed temperature gradient. Again let the z-axis be in the direction of the positive gradient. As before we may also assume that there is no net flow in the z-direction so that $J^+ = J^- = \frac{1}{4}\mathcal{N}v$. Consistent with our simple kinetic theory model, we can, following Section 6.3.2, estimate the energy transfer by evaluating the energy at $z - \frac{2}{3}l_t$ for the molecules moving up and at $z + \frac{2}{3}l_t$ for those moving down.

$$q_{\text{net}} = \frac{1}{4}\mathcal{N}\bar{v}(mc_vT + \frac{1}{2}kT)_{z-(2/3)l_t} - \frac{1}{4}\mathcal{N}\bar{v}(mc_vT + \frac{1}{2}kT)_{z+(2/3)l_t}$$

$$q = -\frac{1}{3}\bar{v}\mathcal{N}l_t(mc_v + \frac{1}{2}k)\frac{dT}{dz} \tag{7.25}$$

Equation (7.25) is in the form of Fourier's law of heat conduction. The pressure and temperature dependence of the thermal conductivity are seen to be identical to those for viscosity and the density-diffusivity product.

The rigid sphere model fails very badly in the case of thermal conductivity. The fast moving molecules in the tail of the Maxwell-Boltzmann distribution carry much more energy than the slow ones and, because the other molecules look smaller to fast molecules, they go much farther. We could gain some insight into this effect by postulating, say, that d^2 varies inversely with molecular velocity v, and then integrate the spectral flux over the Maxwell-Boltzmann distribution. But suffice it to say that the ratio of conductivity to viscosity is not just $c_v + k/2m$, but is appreciably larger when the translational kinetic energy is an important part of the energy carried by the molecule. The ratio of thermal conductivity to viscosity proves to be approximately

$$\frac{k}{\mu} = c_p \left[\frac{1.77\gamma - 0.45}{\gamma} \right] \qquad \left(\gamma = \frac{c_p}{c_v} \right) \tag{7.26}$$

Equation (7.26) can be rewritten in a more interesting form as

$$Pr = \frac{\mu c_p}{k} = \frac{\gamma}{1.77\gamma - 0.45} \tag{7.27}$$

and the Prandtl number is seen to be a constant. For a monatomic gas $\gamma = \frac{5}{3}$ [see Eq. (6.44)], and Eq. (7.27) predicts $Pr = 0.667$. For a vibrationally unexcited diatomic gas like air, $\gamma = \frac{7}{5}$ and $Pr = 0.69$. Both values are in good agreement with experiment.

Rearrangement of Eq. (7.27) and substitution for μ and c_p lead to

$$k = \frac{1}{3(1 - \cos\theta)} \left(\frac{8kT}{\pi m} \right)^{1/2} \left(\frac{Pm}{kT} \right) \left(\frac{kT}{\sqrt{2}\,\pi d^2 P} \right) \frac{(5 + N_r)k}{2m} \frac{1}{Pr}$$

$$k = \frac{(5 + N_r)k^{3/2}T^{1/2}}{3\pi^{3/2}(1 - \cos\theta)d^2 m^{1/2}Pr} \tag{7.28}$$

where, as discussed in Chapter 6, $N_r = 0$ for a monatomic molecule, 2 for a linear molecule, and 3 for a nonlinear molecule, Notice that a small light molecule will have a higher value of k than a large heavy molecule. For example, at 100°F the value of k for H_2 is 0.11 Btu/hr ft °F, while for air it is only 0.016 Btu/hr ft °F.

7.4   ESTIMATION OF GAS TRANSPORT PROPERTIES

7.4.1   The Chapman-Enskog Kinetic Theory of Gases

In Sections 7.2 and 7.3 a simple kinetic theory of gases was presented in order to provide a model, on a microscopic scale, for physical reasoning about transport mechanisms. The Chapman-Enskog kinetic theory of gases is based both on a more realistic physical model and a rigorous mathematical development. Some features of the theory to give an indication of the range of validity of the results are as follows:

1.   The density of mixture must be low enough for three body collisions to occur with negligible frequency. But, except at very low temperatures, pressures less than 100 atm are low enough to ensure that this condition is met.
2.   The model assumes monatomic molecules, but little error is introduced by applying the results to polyatomic gases. The viscosity and mass diffusivity are not appreciably affected by the internal degrees of freedom. The thermal conductivity, as we have seen,

depends on both the translational energy and the energy of the internal degrees of freedom; the so-called *Eucken correction* has been introduced to account for the additional contribution.

3. The theory neglects higher than first order spatial derivatives of temperature, pressure, concentration, and so on. Thus the results are inapplicable when gradients change abruptly, for example, within a shock wave.

The forces acting between a pair of molecules during a collision is characterized by the potential energy of interaction ϕ. This concept was introduced in Section 7.2.1; the functional form of ϕ was illustrated in Figure 7.1. An empirical representation of the potential energy function, which has proven fairly successful, is the Lennard-Jones 6-12 potential model

$$\phi(r) = 4\epsilon\left[\left(\frac{\sigma}{r}\right)^{12} - \left(\frac{\sigma}{r}\right)^{6}\right] \tag{7.29}$$

where σ, the *collision diameter*, is the value of r for which $\phi(r) = 0$ and ϵ is the maximum energy of attraction between a pair of molecules. The model exhibits weak attraction, due to London dispersion forces, at large separations (like r^{-6}) and strong repulsion, due to electron cloud overlapping, at small separations (nearly like r^{-12}). Table 7.1 lists values of σ and ϵ for a number of chemical species.

The Lennard-Jones model describes a spherically symmetrical force field and hence is intended for use with nonpolar, nearly symmetrical molecules (for example, O_2, He, CO). Indeed, the Chapman-Enskog theory is, strictly speaking, only valid for molecules with spherically symmetrical force fields. Molecules with appreciable dipole moments (for example, H_2O, NH_3) or which are highly elongated (for example, C_3H_6, $n-C_6H_{14}$) interact with potentials which are angle dependent. For polar molecules the Stockmayer potential model, which adds an angle dependent factor to the Lennard-Jones expression, has been successfully used. However, for many practical purposes it has been found adequate to use the Lennard-Jones potential even for polar and elongated molecules. The usual practice is to *determine* the parameters σ and ϵ by matching theoretical viscosity predictions with experimental data. In this way experimental viscosity data are extrapolated outside the original temperature range; also the same values of σ and ϵ usually prove to be the best available for the estimation of thermal conductivity and mass diffusivity.

7.4.2 Formulas Based on the Lennard-Jones Potential

Use of the Lennard-Jones potential in the Chapman-Enskog kinetic theory of gases yields for the viscosity of a pure monatomic gas

$$\bullet \qquad \mu = 4.158 \times 10^{-8} \frac{\sqrt{MT}}{\sigma^2 \Omega_\mu} \text{ lb}_f \text{ sec/ft}^2 \tag{7.30}$$

where T is in °R and σ is in angstrom units ($1\text{Å} = 10^{-8}$ cm). The quantity Ω_μ is the

Table 7.1   Force Constants for the Lennard-Jones Potential Model[a]

Species	σ, Å	$\frac{\epsilon}{k}$, °K	Species	σ, Å	$\frac{\epsilon}{k}$, °K	Species	σ, Å	$\frac{\epsilon}{k}$, °K
Al	2.655	2750	CH_3CCH	4.761	252	Li_2O	3.561	1827
AlO	3.204	542	C_3H_8	5.118	237	Mg	2.926	1614
Al_2	2.940	2750	$n\text{-}C_3H_7OH$	4.549	577	N	3.298	71
Air	3.711	79	$n\text{-}C_4H_{10}$	4.687	531	NH_3	2.900	558
Ar	3.542	93	$iso\text{-}C_4H_{10}$	5.278	330	NO	3.492	117
C	3.385	31	$n\text{-}C_5H_{12}$	5.784	341	N_2	3.798	71
CCl_2	4.692	213	C_6H_{12}	6.182	297	N_2O	3.828	232
CCl_2F_2	5.25	253	$n\text{-}C_6H_{14}$	5.949	399	Na	3.567	1375
CCl_4	5.947	323	Cl	3.613	131	NaCl	4.186	1989
CH	3.370	69	Cl_2	4.217	316	NaOH	3.804	1962
$CHCl_3$	5.389	340	H	2.708	37	Na_2	4.156	1375
CH_3OH	3.626	482	HCN	3.630	569	Ne	2.820	33
CH_4	3.758	149	HCl	3.339	345	O	3.050	107
CN	3.856	75	H_2	2.827	60	OH	3.147	80
CO	3.690	92	H_2O	3.737	32	O_2	3.467	107
CO_2	3.941	195	H_2O_2	4.196	289	S	3.839	847
CS_2	4.483	467	H_2S	3.623	301	SO	3.993	301
C_2	3.913	79	He	2.551	10	SO_2	4.112	335
C_2H_2	4.033	232	Hg	2.969	750	Si	2.910	3036
C_2H_4	4.163	225	I_2	5.160	474	SiO	3.374	569
C_2H_6	4.443	216	Kr	3.655	179	SiO_2	3.706	2954
C_2H_5OH	4.530	363	Li	2.850	1899	UF_6	5.967	237
C_2N_2	4.361	349	LiO	3.334	450	Xe	4.047	231
$C_2H_2CHCH_3$	4.678	299	Li_2	3.200	1899	Zn	2.284	1393

[a]   Taken largely from R. A. Svehla, NASA TR R-132, 1962.

collision integral and is tabulated in Table 7.2; it is a weak function of temperature, becoming very nearly constant at high temperatures.

The Chapman-Enskog kinetic theory shows that, for a monatomic gas, the relation between thermal conductivity and viscosity is

$$k = \frac{5}{2} c_v \mu, \qquad \left(c_v = \frac{3}{2} \frac{\mathcal{R}}{M} \right) \qquad (7.31)$$

Therefore

$$k_{\text{monatomic}} = 3.583 \times 10^{-2} \frac{\sqrt{T/M}}{\sigma^2 \Omega_k} \text{ Btu/hr ft °F} \qquad (7.32)$$

where again T is in °R and σ is in Å. The collision integral for thermal conductivity is identical to that for viscosity, $\Omega_k = \Omega_\mu$. For polyatomic gases we can simply

Table 7.2 Collision Integrals for the Lennard-Jones Potential Model[a]

$\dfrac{kT}{\epsilon}$	$\Omega_\mu = \Omega_k$	$\Omega_{\mathscr{D}}$	$\dfrac{kT}{\epsilon}$	$\Omega_\mu = \Omega_k$	$\Omega_{\mathscr{D}}$	$\dfrac{kT}{\epsilon}$	$\Omega_\mu = \Omega_k$	$\Omega_{\mathscr{D}}$
0.30	2.785	2.662	1.60	1.279	1.167	3.80	0.9811	0.8942
0.35	2.628	2.476	1.65	1.264	1.153	3.90	0.9755	0.8888
0.40	2.492	2.318	1.70	1.248	1.140	4.00	0.9700	0.8836
0.45	2.368	2.184	1.75	1.234	1.128	4.10	0.9649	0.8788
0.50	2.257	2.066	1.80	1.221	1.116	4.20	0.9600	0.8740
0.55	2.156	1.966	1.85	1.209	1.105	4.30	0.9553	0.8694
0.60	2.065	1.877	1.90	1.197	1.094	4.40	0.9507	0.8652
0.65	1.982	1.798	1.95	1.186	1.084	4.50	0.9464	0.8610
0.70	1.908	1.729	2.00	1.175	1.075	4.60	0.9422	0.8568
0.75	1.841	1.667	2.10	1.156	1.057	4.70	0.9382	0.8530
0.80	1.780	1.612	2.20	1.138	1.041	4.80	0.9343	0.8492
0.85	1.725	1.562	2.30	1.122	1.026	4.90	0.9305	0.8456
0.90	1.675	1.517	2.40	1.107	1.012	5.0	0.9269	0.8422
0.95	1.629	1.476	2.50	1.093	0.9996	6.0	0.8963	0.8124
1.00	1.587	1.439	2.60	1.081	0.9878	7.0	0.8727	0.7896
1.05	1.549	1.406	2.70	1.069	0.9770	8.0	0.8538	0.7712
1.10	1.514	1.375	2.80	1.058	0.9672	9.0	0.8379	0.7556
1.15	1.482	1.346	2.90	1.048	0.9576	10.0	0.8242	0.7424
1.20	1.452	1.320	3.00	1.039	0.9490	20.0	0.7432	0.6640
1.25	1.424	1.296	3.10	1.030	0.9406	30.0	0.7005	0.6232
1.30	1.399	1.273	3.20	1.022	0.9328	40.0	0.6718	0.5960
1.35	1.375	1.253	3.30	1.014	0.9256	50.0	0.6504	0.5756
1.40	1.353	1.233	3.40	1.007	0.9186	60.0	0.6335	0.5596
1.45	1.333	1.215	3.50	0.9999	0.9120	70.0	0.6194	0.5464
1.50	1.314	1.198	3.60	0.9932	0.9058	80.0	0.6076	0.5352
1.55	1.296	1.182	3.70	0.9870	0.8998	90.0	0.5973	0.5256
						100.0	0.5882	0.5170

[a] Taken from J. O. Hirschfelder, R. B. Bird, and E. L. Spotz, *Chem. Revs.*, vol. 44, p. 205 (1949).

replace Eq. (7.31) by Eq. (7.26), which in a more convenient form is

$$k = k_{\text{monatomic}} + \rho\mathscr{D}_{ii}\left(c_p - \frac{5}{2}\frac{\mathscr{R}}{M}\right) \tag{7.33}$$

where $\rho\mathscr{D}_{ii} = \mu/Sc_{ii}$, and the self-diffusion Schmidt number is $Sc_{ii} \doteq 0.76$. Equation (7.33) becomes

$$k = k_{\text{monatomic}} + 1.32\left(c_p - \frac{5}{2}\frac{\mathscr{R}}{M}\right)\mu \tag{7.34}$$

where the second term is called the **modified Eucken correction**. We see that a table of specific heats is required for the estimation of thermal conductivity in polyatomic gases.

The mass diffusivity, in the form of the binary diffusion coefficient, is given by

$$\bullet \qquad \mathscr{D}_{12} = 8.283 \times 10^{-7} \frac{\sqrt{T^3 \left(\frac{1}{M_1} + \frac{1}{M_2} \right)}}{\sigma_{12}^2 \Omega_{\mathscr{D}} P} \text{ ft}^2/\text{sec} \qquad (7.35)$$

where T is in °R, P is in atm, and σ_{12} is in Å. The intermolecular potential field for a pair of unlike molecules, species 1 and 2, is approximated as

$$\phi_{12}(r) = 4\epsilon_{12} \left[\left(\frac{\sigma_{12}}{r} \right)^{12} - \left(\frac{\sigma_{12}}{r} \right)^6 \right] \qquad (7.36)$$

The collision integral for mass diffusion $\Omega_{\mathscr{D}}$ differs from Ω_μ; values for $\Omega_{\mathscr{D}}$ are tabulated in Table 7.2. The Lennard-Jones parameters σ_{12} and ϵ_{12} must be obtained from the empirical relations

$$\sigma_{12} = \tfrac{1}{2}(\sigma_1 + \sigma_2) \qquad (7.37a)$$

$$\epsilon_{12} = \sqrt{\epsilon_1 \epsilon_2} \qquad (7.37b)$$

where, as was mentioned before, the values of σ and ϵ for the individual species have usually been estimated from viscosity data.

7.4.3   Mixture Rules

It remains to prescribe how the viscosity and thermal conductivity of a gas mixture can be estimated from the values for the pure species. C. R. Wilke simplified the rigorous kinetic theory prediction by introducing elements of the rigid sphere model into the final result. The formulas are simple and have proven to be quite adequate.

$$\bullet \qquad \mu_{\text{mix}} = \sum_{i=1}^{n} \frac{x_i \mu_i}{\sum_{j=1}^{n} x_j \Phi_{ij}} \qquad (7.38)$$

$$\bullet \qquad k_{\text{mix}} = \sum_{i=1}^{n} \frac{x_i k_i}{\sum_{j=1}^{n} x_j \Phi_{ij}} \qquad (7.39)$$

where

$$\Phi_{ij} = \frac{\left[1 + \left(\frac{\mu_i}{\mu_j} \right)^{1/2} \left(\frac{M_j}{M_i} \right)^{1/4} \right]^2}{\sqrt{8} [1 + (M_i/M_j)]^{1/2}}$$

The important feature of these formulas is that the weighting is essentially with mole (number) fraction, as we would expect from simple kinetic theory.

In Chapters 3 and 5 and later in Chapter 10, the mass diffusivity of a dilute species 1 in a mixture of n components, $\mathscr{D}_{1m}$, is used. A simple kinetic theory argument along the lines of that in Section 7.3 yields

$$\mathscr{D}_{1m} \doteq \frac{(1-x_1)}{\sum\limits_{i=2}^{n} (x_i/\mathscr{D}_{1i})}, \qquad x_1 \ll 1 \qquad (7.40)$$

The important feature here is that the addition of an amount of light gas (say species 3) small enough to leave m_1 essentially unaffected does not have a great effect on $\rho\mathscr{D}_{1m}$ of the mixture even though $\mathscr{D}_{13}$ is much larger than the other $\mathscr{D}_{1i}$ values.

7.4.4 EXAMPLE Estimation of the Viscosity of a Gaseous Fuel

A gaseous fuel has a composition of 70% by volume methane (CH_4) and 30% propane (C_3H_8). Calculate the viscosity of the mixture at 80°F and at 2 atm pressure.

First we calculate the viscosity of the pure species CH_4 and C_3H_8 at $80°F = 540°R = 300°K$. Equation (7.30) reads

$$\mu = 4.158 \times 10^{-8} \frac{\sqrt{MT}}{\sigma^2 \Omega_\mu} \text{ lb}_f \text{ sec/ft}^2$$

Using Tables 7.1 and 7.2:

Species	M	$\sigma(\text{Å})$	$\frac{\epsilon}{k}$ (°K)	$\frac{kT}{\epsilon}$	Ω_μ
1. CH_4	16	3.758	149	2.013	1.173
2. C_3H_8	44	5.118	237	1.266	1.416

$$\mu_{CH_4} = 4.158 \times 10^{-8} \frac{\sqrt{(16)(540)}}{(3.758)^2(1.173)} = 2.33 \times 10^{-7} \text{ lb}_f \text{ sec/ft}^2$$

$$\mu_{C_3H_8} = 4.158 \times 10^{-8} \frac{\sqrt{(44)(540)}}{(5.118)^2(1.416)} = 1.73 \times 10^{-7} \text{ lb}_f \text{ sec/ft}^2$$

For the mixture of mole fraction x_1 of methane equal to 0.7 and x_2 of propane 0.3, we use Eq. (7.38),

$$\mu_{\text{mix}} = \sum_{i=1}^{2} \frac{x_i \mu_i}{\sum\limits_{j=1}^{2} x_j \Phi_{ij}}$$

$$= \frac{x_1 \mu_1}{x_1 \Phi_{11} + x_2 \Phi_{12}} + \frac{x_2 \mu_2}{x_1 \Phi_{21} + x_2 \Phi_{22}}$$

where

$$\Phi_{ij} = \frac{\left[1 + \left(\frac{\mu_i}{\mu_j}\right)^{1/2}\left(\frac{M_j}{M_i}\right)^{1/4}\right]^2}{\sqrt{8}[1 + (M_i/M_j)]^{1/2}}$$

$$\Phi_{11} = \Phi_{22} = 1$$

$$\Phi_{12} = \frac{\left[1 + \left(\frac{2.33}{1.73}\right)^{1/2}\left(\frac{44}{16}\right)^{1/4}\right]^2}{\sqrt{8}\,(1 + 16/44)^{1/2}} = 1.88$$

$$\Phi_{21} = \frac{\left[1 + \left(\frac{1.73}{2.33}\right)^{1/2}\left(\frac{16}{44}\right)^{1/4}\right]^2}{\sqrt{8}\,(1 + 44/16)^{1/2}} = 0.508$$

$$\mu_{mix} = \frac{(0.7)\,(2.33 \times 10^{-7})}{(0.7)\,(1) + (0.3)\,(1.88)} + \frac{(0.3)\,(1.73 \times 10^{-7})}{(0.7)\,(0.508) + (0.3)\,(1)}$$
$$= 2.08 \ \text{lb}_f \ \text{sec/ft}^2$$

7.4.5   *EXAMPLE*   Estimation of the Diffusivity of SO_2 in Air

Let us calculate the binary diffusion coefficient of SO_2 in air at 800°F and 1 atm.

Equation (7.35) reads

$$\mathscr{D}_{12} = 8.283 \times 10^{-7} \frac{\sqrt{T^3\left(\frac{1}{M_1} + \frac{1}{M_2}\right)}}{\sigma_{12}^2 \Omega_{\mathscr{D}} P} \ \text{ft}^2/\text{sec}$$

Using Table 7.1:

Species	M	$\sigma(\text{Å})$	$\frac{\epsilon}{k}$ (°K)
1. SO_2	64	4.112	335
2. Air	29	3.711	79

$$\sigma_{12} = \tfrac{1}{2}(\sigma_1 + \sigma_2) = \tfrac{1}{2}(4.112 + 3.711) = 3.911$$

$$\frac{\epsilon_{12}}{k} = \sqrt{\left(\frac{\epsilon_1}{k}\right)\left(\frac{\epsilon_2}{k}\right)} = \sqrt{(335)\,(79)} = 163$$

$$T = 800°\text{F} = 1260°\text{R} = 700°\text{K}$$

$$\frac{kT}{\epsilon_{12}} = \frac{700}{163} = 4.30$$

and from Table 7.2, $\Omega_{\mathscr{D}} = 0.8694$.

$$\mathscr{D}_{12} = \frac{8.283 \times 10^{-7}\sqrt{(1260)^3\left(\frac{1}{64} + \frac{1}{29}\right)}}{(1)\,(3.911)^2(0.8694)}$$
$$= 6.24 \times 10^{-4} \ \text{ft}^2/\text{sec}$$

or

$$\mathscr{D}_{12} = 2.24 \ \text{ft}^2/\text{hr}$$

7.4.6   *EXAMPLE* **Estimation of the Thermal Conductivity of Methane**

We wish to calculate the thermal conductivity of methane CH_4 at 170°F and 1 atm pressure.

As a first step we calculate the thermal conductivity of CH_4, as if it were monatomic, and its viscosity.

$$T = 170°F = 630°R = 350°K$$
$$M = 16$$

From Table 7.1, $\sigma = 3.758$ Å, $\epsilon/k = 149°K$, $kT/\epsilon = 350/149 = 2.35$; from Table 7.2 $\Omega_\mu = \Omega_k = 1.114$. Using Eq. (7.32) we find

$$k_{\text{monatomic}} = 3.583 \times 10^{-2} \frac{\sqrt{(630)(1/16)}}{(3.758)^2(1.114)} = 1.429 \times 10^{-2} \text{ Btu/hr ft °F}$$

Using Eq. (7.30) we get

$$\mu = 4.158 \times 10^{-8} \frac{\sqrt{(630)(16)}}{(3.758)^2(1.114)} = 2.650 \times 10^{-7} \text{ lb}_f \text{ sec/ft}^2$$

The actual thermal conductivity of CH_4 is then obtained from Eq. (7.34),

$$k = k_{\text{monatomic}} + 1.32\left(c_p - \frac{5}{2}\frac{\mathcal{R}}{M}\right)\mu$$

To proceed we need a value for c_p. Since the temperature is low, we can illustrate the remaining steps using the simple formulas presented in Chapter 6, which neglect vibrational excitation. In general, an accurate value from experiment or statistical thermodynamics should be used. From Eqs. (6.42) and (6.43)

$$c_p = (5 + N_r)\left(\frac{1}{2}\frac{\mathcal{R}}{M}\right), \qquad N_r = 3 \text{ for } CH_4$$

$$\left(c_p - \frac{5}{2}\frac{\mathcal{R}}{M}\right) = \frac{3}{2}\frac{\mathcal{R}}{M}$$

$$= \frac{(3)(1545)}{(2)(778)(16)} = 0.186 \text{ Btu/lb °R}$$

We require consistent units.

$$\mu = 2.65 \times 10^{-7} \text{ lb}_f \text{ sec/ft}^2$$
$$= 3.07 \times 10^{-2} \text{ lb/hr ft}$$

Thus there results

$$k = 1.429 \times 10^{-2} + 1.32(0.186)(3.07 \times 10^{-2})$$
$$= 2.183 \times 10^{-2} \text{ Btu/hr ft °F}$$

7.5   NEUTRON TRANSPORT

7.5.1   Introduction

Neutron transport is a transfer process of fundamental importance to nuclear engineers. Similar techniques to those used for molecular and photon transport are used. Atomic nuclei can be regarded as having *cross sections* for scattering of neutrons, parasitic absorption of neutrons, and absorption of neutrons leading to fission. Because the fission cross section (that is, the effective $\pi \overline{d^2}$ of a uranium nucleus for a *collision* leading to fission) is larger for slower moving neutrons, a much smaller amount of fuel is needed to achieve criticality (and a slower more easily controlled process). For this reason one may "thermalize" or "slow down" the highly energetic neutrons expelled during fission. This slowing down process is accomplished by allowing the fast neutrons to collide with hydrogen or other light atoms in the *moderator* until the neutrons approach a Maxwell-Boltzmann distribution corresponding to the moderator temperature. Neutrons are also a precious commodity. They can be used on the one hand for triggering fission or breeding fuel capable of fissioning or on the other hand they can be uselessly absorbed to prevent them from causing unwanted reactions in, say, a human body. For this reason a *reflector* is placed around the core of a nuclear reactor in much the same way a blanket of thermal insulation may be placed around a water heater.

7.5.2   Neutron Diffusion

Many interesting problems which can be viewed as neutron diffusion arise in the design of a thermal reactor core. Thermal neutrons appear in the moderator from the slowing down process. The number of neutrons per unit volume builds up in the moderator away from the fuel rods and control rods. In the vicinity of the fuel and control rods neutrons are depleted by absorption processes. The neutrons thereupon diffuse from the regions rich in neutrons to those less rich in much the same way that heat diffuses from hot to cold regions. A Fick's law of diffusion will relate the net flux J to the negative of the gradient in number density.

Rather than use the number density workers in reactor core design use the number density-velocity product $\mathcal{N}\bar{v}$ and refer to it as flux ϕ. This flux is more convenient, because it governs the rates of absorption and fission. Since v is absorbed in ϕ, the units of *diffusivity* D are the same as l_t. The diffusion relation equivalent to Fick's first law becomes

$$J = -\frac{1}{3\mathcal{N}_s \sigma_s (1 - \overline{\cos \theta})} \frac{d\phi}{dz} = -\mathrm{D}\frac{d\phi}{dz} \qquad (7.41)$$

where $\mathcal{N}_s$ is the number density of scatterers and σ_s is the cross section per scatterer. Table 7.3 shows some typical values.

Table 7.3   Thermal Neutron Diffusivities

Material	Density	$\mathcal{N}_s$	$\sigma_s(1 - \overline{\cos \theta})$	l_t	D
	gm/cm³	Molecules/cm³	cm²/molecule	cm	cm
H_2O	1.00	0.0334×10^{24}	62.9×10^{-24}	0.48	0.16
D_2O	1.10	0.0331×10^{24}	12.6×10^{-24}	2.40	0.80
C	1.60	0.0803×10^{24}	4.54×10^{-24}	2.74	0.91

7.6   PHOTON CONTRIBUTIONS TO THERMAL CONDUCTIVITY

7.6.1   Radiation Conductivity

As a final example of how long mean free path transport relations may collapse to a diffusion relation, consider the transport through thermal radiation shielding. For example, many layers of aluminized mylar are used to form *super-insulation*. Suppose that there is gas present between the layers so that both molecular conduction and *radiation conduction* occur.

Following the procedure spelled out in the preceding chapter one may use a radiation network to analyze the transport between two adjacent shields. For emissivity independent of wavelength there results

$$q = \frac{\epsilon}{2 - \epsilon} (\sigma T_i{}^4 - \sigma T_{i+1}^4) \qquad (7.42)$$

If the temperature difference is small, Eq. (7.42) may be linearized

$$q_r = \frac{4\sigma T^3 \epsilon}{2 - \epsilon} (T_i - T_{i+1}) \qquad (7.43)$$

where T is the mean temperature. The conduction through the gas is

$$q_c = \frac{k_{\text{gas}}}{\delta} (T_i - T_{i+1}) \qquad (7.44)$$

Comparing Eqs. (7.43) and (7.44) shows we can include the radiation contribution as a form of *conduction*.

$$k_{\text{total}} = k_{\text{gas}} + \frac{4\delta\epsilon\sigma T^3}{2 - \epsilon} \qquad (7.45)$$

so that

$$q_{\text{total}} = \frac{k_{\text{total}}}{\delta} (T_i - T_{i+1}) \qquad (7.46)$$

The photon contribution to radiation shielding is thus

$$k_{\text{rad}} = \frac{4\delta\epsilon\sigma T^3}{2 - \epsilon} \qquad (7.47)$$

In this case the shielding spacing δ is seen to play the role of a transport mean free path.

7.6.2 Accounting for Variable Conductivity

In Eq. (7.45) we see that the effective conductivity varies with temperature significantly. Temperature-dependent conductivity can be accounted for neatly by introducing

$$\bullet \qquad \Phi = \int_{T_0}^{T} k(T)\, dT \qquad (7.48)$$

For example, in conduction through a pipe wall the governing equation was shown in Chapter 2 to be

$$0 = \frac{1}{r}\frac{d}{dr}\left(rk\frac{dT}{dr}\right) \qquad (7.49)$$

We observe that

$$\frac{d\Phi}{dr} = \frac{d\Phi}{dT}\frac{dT}{dr} = k(T)\frac{dT}{dr} \qquad (7.50)$$

so that Eq. (7.49) becomes

$$0 = \frac{1}{r}\frac{d}{dr}\left(r\frac{d\Phi}{dr}\right) \qquad (7.51)$$

This is the equation previously solved for constant conductivity. Integrating twice and imposing the boundary conditions $\Phi = \Phi_1$ at $r = R_1$ and $\Phi = \Phi_2$ at $r = R_2$ gives

$$\frac{\Phi - \Phi_1}{\Phi_2 - \Phi_1} = \frac{\ln\dfrac{r}{R_1}}{\ln(R_2/R_1)} \qquad (7.52)$$

The heat flux is

$$\dot{Q} = -k\frac{dT}{dr}\bigg|_{r=R_1} 2\pi R_1 L$$

$$\dot{Q} = -\frac{d\Phi}{dr}\bigg|_{r=R_1} 2\pi R_1 L \qquad (7.53)$$

Differentiating (7.52) and substituting into (7.53) gives

$$\dot{Q} = \frac{2\pi L}{\ln\dfrac{R_2}{R_1}}(\Phi_1 - \Phi_2) \qquad (7.54)$$

where from Eqs. (7.48) and (7.45)

$$\Phi_1 - \Phi_2 = \int_{T_2}^{T_1} k\, dT = \int_{T_2}^{T_1}\left\{k_{gas}(T_0)\cdot\left(\frac{T}{T_0}\right)^{0.8} + \frac{4\delta\epsilon\sigma T^3}{2-\epsilon}\right\} dT$$

$$\Phi_1 - \Phi_2 = \frac{1}{1.8}k_{gas}(T_0)T_0^{-0.8}(T_1^{1.8} - T_2^{1.8}) + \frac{\delta\epsilon}{2-\epsilon}(\sigma T_1^4 - \sigma T_2^4) \qquad (7.55)$$

The conductivity of the gas was imagined to vary with $T^{0.8}$. It may be noted that $\dot{Q}$ is no longer linear in $T_1 - T_2$ unless T_1 and T_2 are sufficiently close to one another to justify linearization.

7.7   SUMMARY

Simplified kinetic theory was used to derive the temperature and pressure dependencies for the mass diffusivity, viscosity, and thermal conductivity of dilute gases. It was shown that the density-diffusivity product, the viscosity, and the thermal conductivity of a rigid molecule gas is pressure independent and varies with temperature as $T^{1/2}$. The Prandtl and Schmidt numbers were found to be constants.

Heuristic arguments for the behavior of attractive-repulsive, nonrigid molecules were given, and Lennard-Jones parameters for use in the Chapman-Enskog relations were tabulated. Very important were the rules for estimating the transport properties of gas mixtures.

Neutron transport and photon contributions to thermal conductivity were discussed briefly.

EXERCISES

1. (a) What is the transport mean free path of a molecule with a diameter of 3.7 Å and a molecular weight of 28 at 1 atm and 300°K?
    (b) What is the self-diffusion coefficient?
    (c) What is the viscosity?
    (d) What is the thermal conductivity ($\gamma = 1.4$, $c_p = 0.25$ Btu/lb °R)?
    Use the simple rigid sphere kinetic theory results.
2. Estimate the Prandtl number for water vapor at room temperature merely from the fact that it is a triatomic nonlinear molecule not vibrationally excited at that temperature.
3. Calculate the viscosity of sodium vapor Na_2 at 1500°F, 0.5 atm pressure.
4. Calculate the thermal conductivity of a mixture of combustion products CO_2, H_2O, and N_2 with mass fractions 0.19, 0.12, and 0.69, respectively, at 1100°F and 22 psia. Take the values of k for the individual species from Appendix B.
5. Determine the binary diffusion coefficients in air for (a) helium, (b) methane, and (c) carbon tetrachloride at 80°F and 1 atm pressure.
6. Suppose a boiling water reactor contains 90% (by volume) liquid water and 10% saturated steam in small bubbles at 500°C. What is the effective neutron diffusivity of the steam-water mixture for thermal neutrons?
7. There are 20 sheets of aluminized (both sides) mylar insulating a 2 in. outside diameter pipeline contained within a 4 in. inside diameter vacuum

jacket. What is the photon contribution to the conductivity at a temperature of $-50°C$? Assume

(a) The aluminum emittance is independent of temperature at 0.03.
(b) The aluminum emittance varies with temperature according to 0.03 $(T/300)^{1/2}$ where T is in $°K$.

8. How would Eq. (7.47) be modified for mylar aluminized on just one side so that the emittance was 0.03 on the aluminized side and 0.3 on the mylar side?
9. How would Eq. (7.45) be modified if the gas transport mean free path were much longer than the plate spacing δ?
10. Imagine that a polyurethane foam consists of bubbles of gas trapped by resin. Propose a model with which to estimate the effective thermal conductivity of the foam. Assume that the gas has a short mean free path compared to the bubble diameters; that the resin has a short mean free path for photons and a high emittance; that the bubbles are of a uniform size of diameter δ; and that the void fraction ϵ_v is known from the densities of the foam, the resin, and the gas.

REFERENCES

Molecular Theory of Gases

The standard reference in this field is:

Hirschfelder, J. O., C. F. Curtiss, and R. B. Bird, *Molecular Theory of Gases and Liquids*. New York: John Wiley, 1954.

The beginning student can benefit from pages 8-34, in particular.

Two short works which are quite readable are:

Guggenheim, E. A., *Elements of the Kinetic Theory of Gases*. Oxford: Pergamon, 1960.
Westenberg, A. A., "A Critical Survey of the Major Methods for Measuring and Calculating Dilute Gas Transport Properties," *Advances in Heat Transfer*, Vol. 3, pp. 253–302. New York: Academic, 1966.

The applied worker interested in calculating numbers should become familiar with:

Svehla, R. A., *Estimated Viscosities and Thermal Conductivities of Gases at High Temperatures*, NACA TR R-132, 1962.
Reid, R. C., and T. K. Sherwood, *The Properties of Gases and Liquids, Their Estimation and Correlation*, 2nd Ed. New York: McGraw-Hill, 1966.

Neutron Diffusion

An introductory description is given in:

Glasstone, S., and M. D. Edlund, *The Elements of Nuclear Reactor Theory*. New York: Van Nostrand-Reinhold, 1952. (See Chapters 3, 5, and 14.)

CHAPTER 8

TURBULENCE

8.1 INTRODUCTION

Turbulence is a fact of life in most practical fluid-flow problems with which an engineer must deal. Even though a pump may be running smoothly or an aircraft may be in steady flight through apparently calm air, careful measurements will disclose that the flow velocities are fluctuating rapidly with time inside a pipe connected to the pump or near the wing of the aircraft. Only in rather exceptional circumstances, flow of a very viscous fluid at low speeds in small passages or around small objects or flow of a very low density fluid, will the fluid flow steadily. Flows which are steady on even a small scale are said to be laminar, and those which are unsteady on a small scale are said to be turbulent.

Everyday experience shows turbulence increases the transfer of heat, mass, and momentum. The smoke from an ember or cigarette rises at first in a fairly steady laminar fashion with little noticeable mass or heat transfer laterally into the air. But after a few inches the smoke may be seen to fluctuate in the flow direction and break up. There is then rapid transfer leading to dilution of the smoke.

From a "pure" analyst's point of view turbulence presents awesome difficulties. The governing differential equations presented in Appendix A are already nearly intractable for steady laminar flow, if the geometry is complex. But should a solution to the steady flow equations be obtained, say numerically using a finite difference approximation, it may have no relevance to nature; for the time-dependent equations (which we suppose are relevant) permit other solutions. We can picture, mathematically, a steady laminar flow in a pipe at a Reynolds number of $10^5 (\gg 2300)$. But suppose we disturb the flow by, say, vibrating the pipe wall a little. The time-dependent equations then show that the small disturbance would grow with the passage of time. One might compare the inherently unstable nature of this flow with the inherently unstable nature of a flag in a (subsonic) breeze.

From an engineer's point of view, turbulence is a fact of life, as remarked previously, but not an altogether unpleasant one. The fact that it does increase transfer rates means that a small unit can be made to transfer heat or mass at a high rate, provided that the price of a high momentum transfer rate can be paid. For the applied scientist and engineer analysis of turbulence can be made, to the extent of extracting a number of useful quantitative expressions for predicting transfer rates, by combining logical reasoning with empiricism.

8.2 FLUCTUATIONS AND AVERAGES

The primary instrument for studying turbulence is the thin wire probe. A hot wire heated electrically cools to a somewhat lower temperature when the velocity of an isothermal fluid increases. A cold wire experiences a change in temperature when the medium in which it is immersed changes in temperature. The change in temperature affects the electrical resistance of the wire which unbalances a bridge circuit and results in a signal which is recorded. The wire is typically platinum sheathed in silver and drawn to a very fine size. Two small prongs may hold (by soldered connections to the silver) a wire 1 mm long or somewhat less. A drop of nitric applied to the wire on the tip of a negative electrode, connected to a 1.5 volt dry cell with a 10,000 ohm resistor in series with the prongs, etches away the silver to expose the platinum. The exposed diameter is too small to be seen without a microscope; it is a fraction of a micron (10^{-6} meter). Multiple wire configurations on one probe may allow affects of directional selectivity or temperature fluctuations to cancel out. Small wires of this sort are capable of detecting fluctuations in velocity and temperature with a time resolution in microseconds (with the help of negative feedback). Figure 8.1 shows a sketch of a measurement made with such a probe.

Consider a quantity A (say a velocity component) which varies rapidly with time. The mean value is

$$\bar{A} = \frac{1}{\mathscr{T}} \int_0^{\mathscr{T}} A(t)\, dt \qquad (8.1)$$

Typical turbulent fluctuations occur so rapidly that a period $\mathscr{T}$ of a few seconds or less may suffice to fix a good mean value. The difference A' between $A(t)$ and $\bar{A}$ is the perturbation or fluctuation in A ($A = \bar{A} + A'$). Of course, the average value

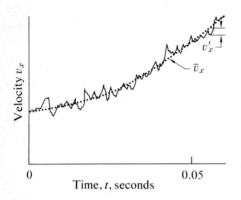

Figure 8.1   Oscillation of a velocity component.

of A' is then zero, because the integral, Eq. (8.1), of a sum is the sum of the integrals,

$$\bar{A} = \overline{\bar{A} + A'} = \bar{\bar{A}} + \bar{A}' = \bar{A} + \bar{A}', \qquad \bar{A}' = 0 \tag{8.2}$$

A mean flow $\bar{v}_x$ may be steady, but the instantaneous velocity v_x is composed of $\bar{v}_x$ and a fluctuation v_x'.

Consider two quantities $A(t)$ and $B(t)$ (for example, a velocity component and a temperature or two velocity components). Each has an average value $\bar{A}$ and $\bar{B}$, respectively. The average of the product AB is

$$\overline{AB} = \overline{(\bar{A} + A')(\bar{B} + B')} = \bar{A}\bar{B} + \overline{\bar{A}B'} + \overline{A'\bar{B}} + \overline{A'B'}$$

But the average of a constant times a fluctuation whose average is zero is also zero.

$$\overline{\bar{A}B'} = \bar{A}\bar{B}' = 0$$

$$\overline{A'\bar{B}} = \bar{A}'\bar{B} = 0$$

$$\overline{AB} = \bar{A}\bar{B} + \overline{A'B'} \tag{8.3}$$

If A' and B' were uncorrelated, then $\overline{A'B'}$ would be zero. However, if A' tends to be positive when B' is positive or negative when B' is negative, the average is clearly nonzero.

Now consider flow of a cold fluid over a hot wall. We have seen in Chapters 3, 4, and 5 that convection in a particular direction results from multiplying the mass flow rate per unit area by the quantity transferred contained in a unit mass of the fluid. In the case of heat transported in, say, the y-direction

$$q_{\text{conv},y} = \rho v_y c_p T \tag{8.4}$$

The engineer is, of course, interested only in the average q, $\bar{q}$

$$\bar{q}_{\text{conv},y} = \overline{\rho v_y c_p T} \tag{8.5}$$

Consider for simplicity a constant property fluid for which, of the terms in Eq. (8.5), only the velocity and temperature fluctuate. From Eq. (8.3)

$$\bar{q}_{\text{conv},y} = \rho c_p \overline{v_y T} = \rho c_p \overline{(\bar{v}_y + v_y')(\bar{T} + T')}$$

$$\bar{q}_{\text{conv},y} = \rho c_p \bar{v}_y \bar{T} + \rho c_p \overline{v_y' T'} \tag{8.6}$$

The net effect of the fluctuations is seen in Eq. (8.6) to increase the convective flux over and above that which would occur if there were no fluctuations, if $\overline{v_y' T'}$ is finite and positive.

Similar considerations hold for mass and momentum transfer:

$$\bar{n}_{\text{conv},y} = \rho \bar{v}_y \bar{m} + \rho \overline{v_y' m'} \tag{8.7}$$

$$\tau_{\text{conv},yx} = \rho \bar{v}_y \bar{v}_x + \rho \overline{v_y' v_x'} \tag{8.8}$$

The quantity $\overline{v_y' v_x'}$ is called a Reynolds stress after the famous fluid mechanician, Osborne Reynolds. The subscript on mass fraction m is omitted to simplify notation.

Hot wire measurements show that the rather abrupt changes observed in the behavior of Nu, Nu_m, and c_f versus Re are accompanied by the onset of fluctuations. Table 5.1 lists correlations of Nu and c_f versus Re in two categories, one labeled laminar flow for Re less than some critical value and another labeled turbulent flow for Re greater than the critical value. For a pipe this *transition Reynolds number* is about 2300 (more for coiled pipes); for a flat plate it is in the range 10^5 to 5×10^5. For boundary layer flows, a transition Reynolds number based upon laminar boundary layer thickness is perhaps a better criterion for transition. Anything which tends to thicken a boundary layer (heating of an expansible gas, a pressure which climbs in the direction of flow, blowing through a porous wall into the boundary layer) tends to produce transition at a lower value of Re_x, while a favorable pressure gradient (negative $\partial P/\partial x$), cooling, or suction tends to delay transition to a larger value of Re_x.

The more rapid growth of Nu with Re occurring after transition is thus known to be due to contributions of the turbulent terms in Eqs. (8.6) to (8.8).

8.3   THE PRANDTL MIXING LENGTH

8.3.1   Eddies as Carriers of Heat, Mass, and Momentum

Equations (8.6) to (8.8) alone are useless for practical engineering. All they do is give us some understanding of how turbulence might affect transfer rates. Just from the elementary ideas used to derive them, we gain no assurance that v_x' and v_y' are correlated or, if they are, we do not know whether $\overline{v_y' v_x'}$ would be negative or positive. A very simple model proposed by Prandtl can serve to convince us that $\overline{v_y' v_x'}$ is negative when $\partial \bar{v}_x/\partial y$ is positive and can lead us to some useful results.

Prandtl hypothesized that heat, mass, and momentum were carried by *eddies*, little whirlwinds of fluid generated in the main flow, which fly or swim about for a short time before becoming unrecognizable. He supposed that, at a plane of constant y parallel to and a short distance away from a wall, a number of eddies per unit area per unit time J^+ must be passing upward in the positive y-direction, and a number J^- passing downward. Let the mass of an eddy be on the average m. Let the mass average velocity normal to the wall be zero; then

$$mJ^+ = mJ^- = mJ \qquad (8.9)$$

An eddy is imagined to travel, on the average, a mean free path l before it loses its identity by mixing with the rest of the fluid. The eddy mean free path l is called the **mixing length**.

8.3.2 Eddy Diffusivity

With this simple kinetic-theory-like picture of turbulence we can formulate the transfer rates of heat, mass, and momentum as was done in Chapter 7. At the point of origin of the eddy the quantity T, m, or v_x is imagined to have its average value. From

$$q_{\text{turb}, y} = mJc_p\bar{T}\big|_{y-\frac{2}{3}l} - mJc_p\bar{T}\big|_{y+\frac{2}{3}l}$$

we get

$$q_{\text{turb}, y} = -\frac{4}{3}lmJc_p\frac{\partial\bar{T}}{\partial y}. \qquad (8.10a)$$

From

$$j_{\text{turb}, y} = mJ\bar{m}\big|_{y-\frac{2}{3}l} - mJ\bar{m}\big|_{y+\frac{2}{3}l}$$

we get

$$j_{\text{turb}, y} = -\frac{4}{3}lmJ\frac{\partial\bar{m}}{\partial y} \qquad (8.11a)$$

And from

$$\tau_{\text{turb}, yx} = mJ\bar{v}_x\big|_{y-\frac{2}{3}l} - mJ\bar{v}_x\big|_{y+\frac{2}{3}l}$$

we find

$$\tau_{\text{turb}, yx} = -\frac{4}{3}lmJ\frac{\partial\bar{v}_x}{\partial y} \qquad (8.12a)$$

Recall Fourier's law of heat conduction, Fick's first law of diffusion, and Newton's law of viscosity,

$$q_y = -k\frac{\partial T}{\partial y} = -\rho c_p\frac{k}{\rho c_p}\frac{\partial T}{\partial y} = -\rho c_p\alpha\frac{\partial T}{\partial y}$$

$$j_y = -\rho\mathcal{D}\frac{\partial m}{\partial y}$$

$$\tau_{yx} = -\mu\frac{\partial v_x}{\partial y} = -\rho\frac{\mu}{\rho}\frac{\partial v_x}{\partial y} = -\rho\nu\frac{\partial v_x}{\partial y} \qquad \left(\frac{\partial v_y}{\partial x} = 0\right)$$

Equations (8.10a) to (8.12a) are of this form with an eddy diffusivity,

$$\epsilon = \frac{4}{3}\frac{lmJ}{\rho}$$

We are thus led to write

$$q_{\text{turb},y} = -\rho c_p \epsilon \frac{\partial \bar{T}}{\partial y} \tag{8.10b}$$

$$j_{\text{turb},y} = -\rho \epsilon \frac{\partial \bar{m}}{\partial y} \tag{8.11b}$$

$$\tau_{\text{turb},y} = -\rho \epsilon \frac{\partial \bar{v}_x}{\partial y} \tag{8.12b}$$

where according to the simple mixing length theory all the eddy diffusivities are the same.

Adding the flux resulting from the ordinary molecular encounters to that contributed by turbulence results in

$$q_y = -\rho c_p (\epsilon + \alpha) \frac{\partial \bar{T}}{\partial y} \tag{8.10c}$$

$$j_y = -\rho (\epsilon + \mathcal{D}) \frac{\partial \bar{m}}{\partial y} \tag{8.11c}$$

$$\tau_{yx} = -\rho (\epsilon + \nu) \frac{\partial \bar{v}_x}{\partial y} \tag{8.12c}$$

8.3.3   Reynolds Analogy

At first glance Eqs. (8.10c) to (8.12c) appear no more useful than Eqs. (8.6) to (8.8), for we do not know the eddy diffusivity except at a rigid wall where it is clearly zero (from $v_y' = 0$ or $l = 0$). But the fact that we believe that all the ϵ's are the same (at least on an order of magnitude basis) can lead us to a powerful relation known as Reynolds analogy. Consider boundary layer flow on a flat plate as shown in Figures A.4 and A.5 of Appendix A. The governing equations are the following:

Conservation of mass

$$\frac{\partial \bar{v}_x}{\partial x} + \frac{\partial \bar{v}_y}{\partial y} = 0$$

Conservation of momentum

$$\rho \bar{v}_x \frac{\partial \bar{v}_x}{\partial x} + \rho \bar{v}_y \frac{\partial \bar{v}_x}{\partial y} = \frac{\partial}{\partial y} \left[\rho (\epsilon + \nu) \frac{\partial \bar{v}_x}{\partial y} \right]$$

Conservation of energy

$$\rho c_p \bar{v}_x \frac{\partial \bar{T}}{\partial x} + \rho c_p \bar{v}_y \frac{\partial \bar{T}}{\partial y} = \frac{\partial}{\partial y} \left[\rho c_p (\epsilon + \alpha) \frac{\partial \bar{T}}{\partial y} \right]$$

Conservation of species

$$\rho \bar{v}_x \frac{\partial \bar{m}}{\partial x} + \rho \bar{v}_y \frac{\partial \bar{m}}{\partial y} = \frac{\partial}{\partial y} \left[\rho (\epsilon + \mathcal{D}) \frac{\partial \bar{m}}{\partial y} \right]$$

For a gas ν, α, and $\mathscr{D}$ are all of the same order of magnitude, that is, $Pr \simeq 1$, $Sc \simeq 1$. If they are all equal, then it is clear that the momentum equation, energy equation, and species equation are of identical form. If the boundary conditions are the same, then the solutions for all three quantities

$$\frac{\bar{v}_x}{v_e}, \quad \frac{\bar{T}-T_s}{T_e-T_s}, \quad \frac{\bar{m}-m_s}{m_e-m_s}$$

will be identical.

As a result of the sameness of the dimensionless velocity, temperature, and mass fraction profiles in the boundary layer, Eqs. (8.10c) to (8.12c) indicate

$$\frac{q}{\rho c_p(T_s-T_e)} = \frac{j}{\rho(m_s-m_e)} = \frac{\tau}{\rho v_e}$$

Dividing by v_e results in dimensionless quantities encountered in Chapters 4 and 5

$$St = St_m = c_f/2 \qquad (Sc = Pr = 1) \tag{8.13a}$$

Equation (8.13a) is clearly a result of great practical importance. For example, a designer can estimate the mass transfer coefficient for flow through a duct of a particular geometry simply from knowledge of the heat transfer for that geometry. While the near equality of St and St_m holds (for Pr and Sc close to unity) in similar flow situations, their near equality to $c_f/2$ breaks down when strong pressure gradients act on submerged bodies or protrusions and when the flow separates. The friction coefficient is affected much more than the heat and mass transfer by wall roughness and the presence of pipe fittings such as elbows, tees, or valves.

When Pr and Sc are not unity, a rough account of the departure from unity can be taken by using for pipe flow

$$\bullet \qquad\qquad St\,Pr^{2/3} \doteq St_m Sc^{2/3} \doteq c_f/2 \tag{8.13b}$$

8.3.4   The Log Law from Prandtl's Mixing Length

Let us return to Eq. (8.8) which gives the shear stress due to turbulence as

$$\tau_{\text{turb}} = \rho\overline{v_y'v_x'} \tag{8.14}$$

An eddy moving upward has a positive v_y' ($+y$ is up) and comes from the vicinity of the wall where velocities are low. Such an eddy finds itself surrounded by fluid whose $\bar{v}_x$ velocity is greater. The difference between the eddy's velocity v_x and $\bar{v}_x$ is then a negative value of v_x'.

$$v_x' \simeq -\frac{2}{3}l\frac{\partial \bar{v}_x}{\partial y}, \qquad v_y' > 0 \tag{8.15}$$

Similarly those eddies moving downward have a negative v_y' and a positive v_x' similar to Eq. (8.15). Consistent with our view that the eddies are moving randomly in all directions with average paths of l we estimate the magnitudes of the fluctuations to be equal,

$$|v_x'| = |v_y'| = |v_z'|$$

so that when v_x' is negative, with the value given by Eq. (8.15), v_y' is positive with the same order of magnitude

$$v_y' \simeq \frac{2}{3} l \frac{\partial \bar{v}_x}{\partial y} \tag{8.16}$$

Equation (8.14) thus becomes

$$\tau_{\text{turb}} = -\rho \frac{4}{9} l^2 \left(\frac{\partial \bar{v}_x}{\partial y}\right)^2 \tag{8.17}$$

And from Eq. (8.12b) we see that

$$\epsilon = \frac{4}{9} l^2 \left|\frac{\partial \bar{v}_x}{\partial y}\right| \tag{8.18}$$

The negative sign in Eq. (8.17) means that the fluid of lesser y experiences a shear in the direction of the flow, while that of greater y experiences a drag which tends to retard it. With turbulent flow there is a region adjacent to the wall for which the convective terms of the momentum equation may be neglected, compared to the shear stresses. In this region

$$0 = \frac{\partial}{\partial y}\left[\rho(\epsilon + \nu) \frac{\partial \bar{v}_x}{\partial y}\right]$$

Integrating gives

$$\rho(\epsilon + \nu)\frac{\partial \bar{v}_x}{\partial y} = -\tau = \text{constant} = \tau_s$$

This relation shows that the shear stress is nearly constant at the wall value, a result in agreement with experiment. Equation (8.12c) may then be written (allowing the sign to be absorbed in τ_s) as

$$\tau_s = \rho \frac{4}{9} l^2 \left(\frac{\partial \bar{v}_x}{\partial y}\right)^2$$

$$\frac{\tau_s}{\rho} = \frac{4}{9} l^2 \left(\frac{\partial \bar{v}_x}{\partial y}\right)^2 \tag{8.19}$$

This relation holds only when $\epsilon \gg \nu$, because we have neglected the kinematic viscosity in Eq. (8.12c). From inspection of Eq. (8.19) the units of τ_s/ρ are of a velocity squared; it is the custom to call this velocity the *friction velocity* denoted by v_τ.

$$v_\tau^2 = \frac{\tau_s}{\rho} = \frac{4}{9} l^2 \left(\frac{\partial \bar{v}_x}{\partial y}\right)^2 \tag{8.20}$$

But, by definition, $\tau_s = (c_f/2)\rho v_e^2 = (c_f/2)\rho V^2$, so

$$v_\tau = \sqrt{\frac{c_f}{2}} v_e \tag{8.21}$$

(For pipe flow c_f is based upon $V = v_{\text{avg}}$, so $v_\tau = \sqrt{c_f/2}\, v_{\text{avg}}$.)

We can argue that the mixing length can be at most the distance to the wall y, since a downward moving eddy can go no further. If we guess that l might be some fraction of this upper limit, we try

$$\tfrac{2}{3}l = \mathcal{K}y \tag{8.22}$$

Then, substituting in the square root of Eq. (8.20) and integrating will give

$$v_\tau \int \frac{dy}{y} = \int \mathcal{K} d\bar{v}_x$$

$$v_\tau \ln y = \mathcal{K}\bar{v}_x + \text{constant}$$

$$v^+ = \frac{\bar{v}_x}{v_\tau} = \frac{1}{\mathcal{K}} \ln y + \text{constant} \tag{8.23}$$

Equation (8.23) leads us to believe that if we were to measure $\bar{v}_x$ experimentally and divide it by $\sqrt{c_f/2}\, v_e$, we would obtain a logarithmic velocity distribution whose slope on a semilog plot would be a **universal** constant somewhat greater than one (since $\mathcal{K}$ is imagined to be somewhat less than one). Experiment does indicate such a log law with a universal *von Karman constant* $\mathcal{K}$ on the order of $0.4 (1/\mathcal{K} = 2.5)$.

8.3.5   Universal Velocity Distributions

Very near the wall the motion of eddies is damped and the eddy diffusivity ϵ goes to zero; Eq. (8.12c) then gives

$$\tau_s = \rho \nu \frac{\partial \bar{v}_x}{\partial y}, \qquad v_\tau^{\,2} = \nu \frac{\partial \bar{v}_x}{\partial y}$$

Integrating gives

$$\frac{\bar{v}_x}{v_\tau} = \frac{v_\tau y}{\nu}$$

Introducing the dimensionless quantities $v^+ = \bar{v}_x/v_\tau$ and $y^+ = v_\tau y/\nu$ yields

$$v^+ = y^+ \tag{8.24}$$

This expression together with Eq. (8.23) suggests that a plot of v^+ versus y^+ would be **universal**. Such a plot is shown in Figure 8.2. The experimental data do support this idea.

Various workers have chosen to fit this curve in various ways. A fit of v^+ versus y^+ is known as a **law of the wall** and enables analytical-numerical calculations to be made for turbulent transport rates. Entirely equivalent to a fit of v^+ versus y^+ under conditions of constant shear stress is a fit of ϵ^+ versus y^+, where $\epsilon^+ = (\epsilon + \nu)/\nu = dy^+/dv^+$, as may be seen by making Eq. (8.12c) dimensionless. Such a fit is also called a law of the wall. One which has been patched together to agree with experiment is that given by Van Driest

$$\epsilon^+ = \frac{\epsilon + \nu}{\nu} = \tfrac{1}{2} + \tfrac{1}{2}\{1.0 + 4\mathcal{K}^2 y^{+2}[1.0 - \exp{(-y^+/y_t^+)}]^2\}^{1/2} \tag{8.25}$$

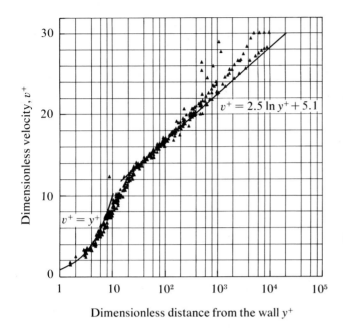

Figure 8.2 Universal velocity distribution for turbulent flow near a wall.

where

$$\mathscr{K} = 0.4, \qquad y_t^+ = 26$$

When y^+ goes to zero, ϵ^+ goes to one, and v^+ is equal to y^+. When y^+ becomes somewhat larger than 26, ϵ^+ goes like y^+, and the log law results. In between a smooth transition is made.

If the law of the wall is used at too large distances from the wall, it yields, of course, erroneous results, but often these errors do not have serious consequences, as most of the resistance to transfer is near the wall. For example, one may use the law of the wall clear to the center of a pipe and not arrive at predictions of transfer significantly in error. There have been proposals of universal law for such regions as the core of a pipe of radius R. Reichardt's law of the pipe, for example, gives

$$\frac{\epsilon}{\nu} = \mathscr{K} y^+ \frac{(1+r/R)(1+2r^2/R^2)}{6} \tag{8.26}$$

where r is the radius from the center of the pipe, R is the radius to the pipe wall, $\mathscr{K}$ is the von Karman constant 0.4, and y^+ is the dimensionless distance from the wall

$$y^+ = \frac{(R-r)v_\tau}{\nu}$$

This relation may be patched onto the law of the wall at the value of $y = R - r$ where the curves of ϵ versus y cross.

8.3.6   *EXAMPLE*   A Sample Calculation of Eddy Diffusivity for Water Flow in a Pipe

Let us use Eq. (8.25) to calculate ϵ and y at $y^+ = 26$ for 140°F water flowing in a 1.0 in. inside diameter pipe at a bulk velocity of 5 ft/sec. The location $y^+ = 26$ is interesting, because it is where ϵ is increasing rapidly with y^+. For water at 140°F, $\nu = 5.14 \times 10^{-6}$ ft²/sec.

To find y we require v_τ. From Eq. (8.21)

$$v_\tau = \sqrt{\frac{c_f}{2}} v_{\text{avg}}$$

In order to obtain c_f from Figure 8.3 we require the Reynolds number

$$Re_D = \frac{(5.0)(1.0/12)}{5.14 \times 10^{-6}} = 81{,}000$$

$$f = 0.019, \qquad \frac{c_f}{2} = \frac{f}{8} = 0.00238$$

$$v_\tau = (0.0488)(5) = 0.244 \text{ ft/sec}$$

At the location $y^+ = v_\tau y / \nu = 26$, the value of y is

$$y = \frac{\nu y^+}{v_\tau} = \frac{(5.14 \times 10^{-6})(26)}{0.244} = 5.48 \times 10^{-4} \text{ ft} = 0.0066 \text{ in.}$$

Thus the distance from the pipe wall to where ϵ is already growing fast is a little more than the thickness of a piece of wrapping paper.

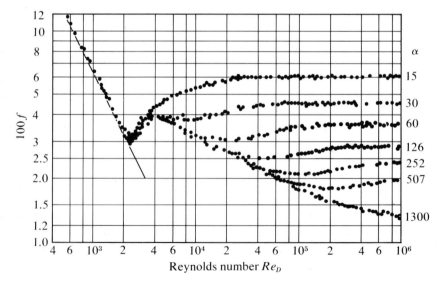

Figure 8.3   Friction factor for pipe flow. The parameter α is the ratio of pipe radius to height of roughness; for $\alpha = 1300$ consider the pipe smooth.

At $y^+ = 26$, Eq. (8.25) gives

$$\epsilon^+ = \frac{\epsilon + \nu}{\nu} = \tfrac{1}{2} + \tfrac{1}{2}\{1.0 + 4(0.4 \times 26)^2[1.0 - \exp(-1)]^2\}^{1/2}$$
$$= \tfrac{1}{2} + \tfrac{1}{2}\{1.0 + 4(10.4)^2(0.632)^2\}^{1/2}$$
$$= 7.10$$

At this location of only six-thousandths of an inch from the pipe wall we find that the turbulent transport is already over six times as large as the molecular transport!

8.3.7 *EXAMPLE* An Analysis of Heat Transfer for Pipe Flow

In Section 4.5 we have previously considered established, constant-property, low-speed flow in a pipe heated with a constant wall heat flux when the flow was laminar. With an expression for eddy diffusivity one can repeat the analysis. The governing momentum equation, Eq. (4.22), for the pressure-shear force balance becomes (dropping overscores to simplify notation)

$$0 = -\frac{1}{\rho}\frac{dP}{dz} + \frac{1}{r}\frac{d}{dr}\left[r(\epsilon + \nu)\frac{dv_z}{dr}\right] \tag{8.27}$$

while that for the heat balance, Eq. (4.29), becomes

$$v_z\frac{dT}{dz} = \frac{1}{r}\frac{d}{dr}\left[r(\epsilon + \alpha)\frac{dT}{dr}\right] \tag{8.28}$$

An overall force balance shows

$$-\frac{dP}{dz}\,dz\,\pi R^2 = 2\pi R\,dz\,\tau_s = 2\pi R\,dz\,(\rho v_\tau^2)$$
$$-\frac{1}{\rho}\frac{dP}{dz} = \frac{2v_\tau^2}{R} \tag{8.29}$$

Likewise an overall heat balance shows (recall that $dT/dz = dT_b/dz$)

$$\rho c_p v_{z,\mathrm{avg}}\pi R^2\frac{dT}{dz}\,dz = 2\pi R\,dz\,q_s$$
$$\frac{dT}{dz} = \frac{2}{Rv_{z,\mathrm{avg}}}\frac{q_s}{\rho c_p} = \frac{2v_\tau T_\tau}{Rv_{z,\mathrm{avg}}} \tag{8.30}$$

where

$$T_\tau \equiv \frac{q_s}{\rho c_p v_\tau} \tag{8.31}$$

If we like, we can regard $dP/dz = \Delta P/L$ known and q_s known, and v_{avg} and $T_s - T_b$ unknown. In this case we would know both v_τ and T_τ from Eqs. (8.29) and (8.31). Then Eqs. (8.27) and (8.28) become

$$0 = \frac{2v_\tau^2}{R} + \frac{1}{r}\frac{d}{dr}\left[r(\epsilon + \nu)\frac{dv_z}{dr}\right] \tag{8.32}$$

$$\frac{2v_\tau T_\tau}{Rv_{z,\mathrm{avg}}} v_z = \frac{1}{r}\frac{d}{dr}\left[r(\epsilon+\alpha)\frac{dT}{dr}\right] \tag{8.33}$$

Equation (8.32) must be solved first for v_z, subject to two boundary conditions:

$$r=0:\quad \frac{dv_z}{dr}=0 \quad \text{(It follows from } v_z(0) \text{ being bounded.)}$$

$$r=R:\ v_z=0$$

Integrating Eq. (8.32) twice, and using the two boundary conditions gives

$$v_z = \frac{v_\tau^2}{R}\int_r^R \frac{r\,dr}{(\epsilon+\nu)}$$

In order to carry out the integration numerically, we introduce Eq. (8.25), or better, Eqs. (8.25) and (8.26), and we define

$$R^+ = \frac{v_\tau R}{\nu},\quad v^+ = \frac{v_z}{v_\tau},\quad y^+ = \frac{v_\tau(R-r)}{\nu}$$

Then we find v^+ by numerical integration,

$$v^+ = \frac{1}{R^+}\int_0^{y^+} \frac{(R^+-y^+)\,dy^+}{\epsilon^+(y^+)} \tag{8.34}$$

To find $v_{z,\mathrm{avg}}$ we integrate numerically again,

$$v_{z,\mathrm{avg}} = \frac{1}{\pi R^2}\int_0^R v_z\, 2\pi r\, dr$$

Dividing both sides by v_τ and rearranging gives

$$v_{\mathrm{avg}}^+ = \frac{2}{R^{+2}}\int_0^{R^+} v^+(R^+-y^+)\,dy^+ \tag{8.35}$$

Should we desire $c_f = f/4$, we can find it from

$$\frac{c_f}{2}\rho v_{z,\mathrm{avg}}^2 = \tau_s = \rho v_\tau^2$$

$$\frac{c_f}{2} = \frac{v_\tau^2}{v_{z,\mathrm{avg}}^2} = \frac{1}{(v_{\mathrm{avg}}^+)^2} \tag{8.36}$$

In order to verify that we have proceeded correctly we may plot c_f or f versus

$$Re = \frac{v_{z,\mathrm{avg}}(2R)}{\nu} = 2v_{\mathrm{avg}}^+ R^+$$

since we know both quantities on the right side, and compare the results with data.

Equation (8.33) may now be solved subject to the two boundary conditions:

$$r=0:\quad T\neq\infty$$

$$r = R: \quad -k\frac{dT}{dy} = +q_s = +\rho c_p v_\tau T_\tau$$

If, as we have said, q_s and v_τ are known, the quantity T_τ is also known. Integrating once and making use of the first boundary condition yields

$$-\frac{2v_\tau T_\tau}{Rv_{z,\,\mathrm{avg}}} \int_y^R v_z(R-y)\,dy = (R-y)(\epsilon+\alpha)\frac{dT}{dy}$$

Integrating a second time yields

$$T = T_s - \frac{2v_\tau T_\tau}{Rv_{z,\,\mathrm{avg}}} \int_0^y \frac{1}{(R-y')(\epsilon+\alpha)} \int_{y'}^R v_z(R-y'')\,dy''\,dy' \qquad (8.37)$$

Defining T_b as in Eq. (4.31) then gives

$$T_b = T_s - \frac{4v_\tau T_\tau}{R^3 v_{z,\,\mathrm{avg}}^2} \int_0^R (R-y)v_z \int_0^y \frac{1}{(R-y')(\epsilon+\alpha)} \int_{y'}^R v_z(R-y'')\,dy''\,dy'\,dy$$

$$\frac{T_s-T_b}{T_\tau} = \frac{4}{R^{+3}v_{\mathrm{avg}}^{+2}} \int_0^{R^+} (R^+-y^+)v^+ \int_0^{y^+} \frac{1}{R^+-y^{+'}}\left(\epsilon^+ + \frac{1}{Pr}-1\right)^{-1}$$

$$\int_{y^{+'}}^{R^+} v^+(R^+-y^{+''})\,dy^{+''}\,dy^{+'}\,dy^+ \qquad (8.38)$$

The Stanton number is related to this temperature ratio through the second boundary condition.

$$St = \frac{q_s}{\rho c_p v_{z,\,\mathrm{avg}}(T_s-T_b)} = \frac{v_\tau T_\tau}{v_{z,\,\mathrm{avg}}(T_s-T_b)} = \frac{T_\tau}{v_{\mathrm{avg}}^+(T_s-T_b)} \qquad (8.39)$$

The reader can see that the integrations indicated in Eq. (8.38) may be carried out by successive numerical integration (after, say, dividing the outer integral into two, one from 0 to $R^+ - \Delta y^+$ and the other from $R^+ - \Delta y^+$ to R^+, and evaluating the second integral analytically). Then Stanton number may be found from Eq. (8.39) and plotted versus Reynolds number with Pr a parameter.

8.4 SIMPLE CALCULATIONS OF TURBULENT TRANSFER COEFFICIENTS

8.4.1 Working Relations

We have already used the relations of Table 5.1 to estimate turbulent convective transfer coefficients. We can now add the **Reynolds-Colburn analogy** in the form of Eq. (8.13b), to be used in conjunction with the friction factor data of Figure 8.3 and Table 8.1. These semi-empirical results enable the engineer to make reasonably accurate estimations of turbulent transfer rates.

Table 8.1　Some Useful Turbulent Flow Relations for Pipe Flow

1. $c_f/2 = f/8$

2. Friction factor of curved pipe

$$\frac{f_{\text{curved}}}{f_{\text{straight}}} = \left(Re_D \frac{D^2}{2R_{\text{bend}}}\right)^{1/20} \qquad Re_D > 20,000 \left(\frac{D}{2R_{\text{bend}}}\right)^{0.32}$$

3. Equivalent (hydraulic) diameter of noncircular duct cross sections

$$D_{\text{equiv}} = \frac{4A_c}{\mathscr{P}} \qquad \begin{matrix} A_c \text{ flow area} \\ \mathscr{P} \text{ wetted perimeter} \end{matrix}$$

4. Pressure drop in commercial pipe fittings, $\Delta P = \Sigma(K_{\text{loss}} \frac{1}{2}\rho V^2)$

Fitting	K_{loss}
90° Elbow	0.90
Tee	1.8
Tee	2.7
180° return bend	2.2
Gate valve, open	0.19
Angle valve, open	5.0
Globe valve, open	10.0

5. Effect of entrance region on average heat transfer: the ratio Nu (average)/Nu (fully developed). For gas flow or Pr near unity.

Entrance configuration	Pipe Length in Diameters									
	2	4	6	8	10	20	40	80	160	320
Long calming section	1.49	1.34	1.26	1.21	1.17	1.10	1.06	1.03	1.01	1.01
Open end, 90° edge	2.36	1.95	1.73	1.60	1.54	1.32	1.18	1.09	1.05	1.02
90° elbow	2.15	1.86	1.68	1.57	1.49	1.32	1.18	1.09	1.05	1.02
Tee (confluence)	1.77	1.56	1.44	1.36	1.31	1.19	1.10	1.06	1.03	1.01
90° round bend	1.63	1.44	1.34	1.28	1.24	1.16	1.10	1.05	1.03	1.01
180° return bend	1.54	1.37	1.28	1.23	1.19	1.12	1.08	1.04	1.02	1.01

8.4.2　*EXAMPLE*　Heat Supply for a Chemical Dip Bath

Let us estimate the heat transfer coefficient and pressure drop in twelve 20 ft lengths of 1 in. pipe connected in series. These pipes are to be used to heat a 6 ft deep, 6 ft wide, 21 ft long chemical dip bath used for surface finishing aluminum extrusions. The bath is maintained at 135°F by 10,000 lb/hr of water at a mean temperature of 160°F, circulated through the pipes by a pump. Water is used in preference to atmospheric pressure steam, because too high a temperature leads to unwanted deposits forming too rapidly on the heater pipes.

From the mass flow rate we find the mean velocity. The inside diameter of standard 1 in. pipe is 1.049 in. From this value we find

$$V = \frac{\dot{m}}{\rho A} = \frac{10,000[\text{lb/hr}]}{61[\text{lb/ft}^3]0.006[\text{ft}^2]3600[\text{sec/hr}]}$$

$$V = 7.59 \text{ ft/sec}$$

Then we find the Reynolds number,

$$Re_D = \frac{(7.59)(1.049/12)}{4.5 \times 10^{-6}} = 147,400$$

The flow is clearly turbulent. The dynamic pressure is

$$\frac{1}{2}\rho V^2 = \frac{1}{2}\left(\frac{61}{32.2}\right)(7.59)^2 = 54.6 \text{ lb}_f/\text{ft}^2$$

The pressure drop is now estimated. There is a length of 240 ft of pipe, not counting that connecting to the pump. Figure 8.3 yields

$$f = 0.017, \qquad f\frac{L}{D} = (0.017)\frac{(240)(12)}{1.049} = 46.7$$

In addition, there are at least eleven 180° return bends. From Table 8.1,

$$\sum K_{\text{loss}} = 11 \times 2.2 = 24.2$$

The total pressure drop in the horizontal piping is consequently

$$\Delta P = \left(\sum K_{\text{loss}} + f\frac{L}{D}\right)\frac{1}{2}\rho V^2 = (70.9)(54.6) = 3870 \text{ lb}_f/\text{ft}^2$$

or about 27 psi.

To estimate the heat transfer coefficient inside the pipe, we employ item 5 of Table 8.1 together with Table 5.1. After each 180° bend there is a run of $240/1.049 = 229$ pipe diameters. For this long length the entrance effect is small.

$$Nu_{avg}/Nu_D = 1.016$$
$$Nu_D = 0.023\, Re^{0.8}Pr^{0.33}$$
$$Nu_D = 0.023\,(147,400)^{0.8}(2.6)^{0.33} = 430$$
$$h_c = (k/D)\,Nu_{avg}$$
$$h_c = (0.384)(12/1.049)(1.016)(430)$$
$$h_c = 1920 \text{ Btu/hr ft}^2 \text{ °F}$$

This value, it must be emphasized, is not the overall heat transfer coefficient but merely the one pertaining to the inside of the tube. As shown in Chapter 2, the thermal resistance of scale in and on the pipe, the resistance of the pipe itself, and that of the free convection boundary layer outside the pipe must be added to the resistance of the inside film. Chapter 9 will take up the question of treating a complete system such as that of the pipe and bath.

8.5 SUMMARY

The way in which turbulent fluctuations increase transfer rates for heat, mass, and momentum has been briefly described. The concepts of eddy diffusivity, turbulent mixing length, and a law of the wall have been discussed. The Van Driest form of the law of the wall was used for illustrative purposes. A sketch of how such a law can be used to expand the laminar flow analysis of Section 4.5 to include turbulent contributions to the transfer processes was given in Section 8.3.7.

For many ordinary engineering purposes, however, no direct use of the law of the wall need be made. Figure 8.3, the relations in Table 5.1, and empirical data of the sort in Table 8.1 enable an engineer to make many useful estimates of convective transfer coefficients.

EXERCISES

1. Calculate the value of y and ϵ/α for flow in a 1 in. diameter tube at $y^+ = 26$ for a flow at 30 ft/sec of

    (a) water at 70°F
    (b) air at 80°F
    (c) mercury at 200°F

    Note that the ratio desired is ϵ/α, not ϵ/ν ($Pr = \nu/\alpha$). Discuss the significance of your findings. Repeat your calculations for $y^+ = R^+$.

2. Find the heat transfer coefficient and pressure drop for exhaust gases flowing through 10 ft of 1.5 in. inside diameter tubing. The gases (use the properties of air at 1000°F) come from a 285 in.[3] four cycle automotive engine running at 3000 rpm. Assume a volumetric efficiency of 85%.

3. For liquid metal flows it is very reasonable to assume $v_z = v_{z,\text{avg}}$, a constant. Show how the calculation of heat transfer in a pipe is simplified by this assumption.

4. For mass transfer into liquid solvents, Table B.12c shows that the Schmidt numbers are moderately large. Show that the ratio $(m_s - m_c)/(m_s - m_b)$ is consequently very nearly one, where m_s is the mass fraction of the solute at the surface, m_c that at the center of the pipe, and m_b the bulk value. Show how this ratio being unity simplifies the calculation of mass transfer.

5. Using the expressions given in Table 5.1 show how the heat transfer coefficient varies with velocity and pipe size.

6. Consider flow through a number N_t of tubes of length L and diameter D. For a fixed mass flow rate $\dot{m}$ the mean velocity V is, of course, given by

$$\dot{m} = \rho\left(\frac{\pi}{4}D^2\right)N_t V$$

Suppose it is desired to maintain hA constant for a fixed $\dot{m}$, where $A = \pi DLN_t$. Show a table of D, L, A, volume $(\pi/4)D^2N_tL$, and the pumping power required $\dot{m}\,\Delta P/\rho$ versus Reynolds number for $10^2 \leqslant Re_D \leqslant 10^5$. Consider an air flow at 80°F of 50 lb$_m$/sec with a desired hA of 50,000 Btu/hr °F. Show D and L in inches, A in square feet, volume in cubic feet, and pumping power in watts. *Neglect entrance effects.* Discuss what the table might mean to an engineer designing an air conditioning unit.

7. Consider the following table:

Geometry	Porosity	$4St/c_f$ for $Pr = 0.7$
Parallel plates	—	1.94
Circular pipe	—	1.54
Randomly stacked wire mesh	0.832	0.76
Randomly packed spheres	0.38	0.256
Crushed rock	0.47	0.052

The equivalent diameter of the flow passage is that indicated in Table 8.1, $4A_c/\mathscr{P} = 4$ volume/surface. Repeat Exercise 6 for a bed of randomly packed spheres and discuss the significance of the new table.

8. Discuss how roughening the wall of a pipe increases heat transfer but also increases pumping power (see Figure 8.3). Give a quantitative example based upon Exercise 6 at $Re_D = 10^5$.

9. Repeat the analysis of Section 4.7 for diffusion into a falling film, but now consider turbulent flow with the eddy diffusivity given by a simple power law with distance from the free surface, $\epsilon = ay^n$ where a is expected to be a function of film Reynolds number.

(a) Show that for small penetration distances, the mass transfer coefficient is given by

$$K = \rho\frac{n}{\pi}a^{1/n}\mathscr{D}_{12}^{1-1/n}\sin(\pi/n)$$

(b) Recently Lamourelle and Sandall reported data for absorption of CO_2, O_2, H_2, and He into a turbulently falling water film at 25°C and correlated their data as

$$\frac{K}{\rho} = 0.339\,Re^{0.839}\mathscr{D}_{12}^{0.5}\text{ ft/hr}, \qquad Re = \frac{4\Gamma}{\mu} > 1800$$

What are the corresponding values of n and a?

(c) If (following V. G. Levich) it is postulated that the damping of eddy diffusivity near the liquid surface is due to the action of surface tension, develop suitable dimensionless forms of ϵ and K.

REFERENCES

Turbulence

Three authoritative references are:

Tennekes, H., and J. L. Lumley, *A First Course in Turbulence*, Cambridge, Mass.: The MIT Press, 1972.

Hinze, J. O., *Turbulence, An Introduction to Its Mechanism and Theory*. New York: McGraw-Hill, 1959.

Schlichting, H., *Boundary-Layer Theory*. New York: McGraw-Hill 1968. (See Parts C and D.)

A good review of "laws of the wall" is given by:

Kestin, J., and P. D. Richardson, "Heat Transfer Across Turbulent, Incompressible Boundary Layers," *International Journal of Heat and Mass Transfer*, Vol. 6, pp. 147–190, 1963.

For turbulent flow in ducts, sources of useful information are:

Kays, W. M., *Convective Heat and Mass Transfer*. New York: McGraw-Hill, 1966. (See Chapters 6 and 9.)

Rohsenow, W. M., and H. Y. Choi, *Heat, Mass, and Momentum Transfer*. Englewood Cliffs, N.J.: Prentice-Hall, 1961. (See Chapters 4, 8, and 16.)

CHAPTER **9**

HEAT EXCHANGERS

AND REGENERATORS

9.1 INTRODUCTION

In this chapter we consider heat transfer between two flowing streams. Devices which provide for such transfer are called heat exchangers. Chemical and food processing, refining, power production, refrigeration, heating and air conditioning, and the operation of vehicles of all types depend upon heat exchangers in one form or another. To provide cold air for passenger comfort in aircraft or for thermal control of electronic equipment, warm, high pressure air is cooled and then expanded. In the cooling process cool fuel flows on one side of a wall while the warm, high pressure air flows on the other side. Heat flows from the warm fluid through the wall into the cold fluid. In the evaporator section of many air conditioners a stream of cold boiling freon flows in the interior of tubes while warm air flows over the finned exterior of the tubes. In the condenser section of such refrigerative air conditioners a stream of condensing hot freon transfers heat to a coolant — air, water, or jet fuel. A stream of sodium from a nuclear reactor is used to heat a flow of steam and water so that the steam can be used in a turbine to generate electrical power.

All the principles needed to analyze such situations have been developed and used before in this text. We now know how to determine convective transfer coefficients, and we know how to isolate an element of one of the flowing streams and make accountings of the thermal transfer to and from it. First, the element is taken to be the system as a whole. Second, a small element is taken in the fluid of a single-stream exchanger. After an overall heat transfer coefficient is developed, balances are made on small elements of each of two flowing streams, and governing differential equations are developed. The solutions to the equations are then manipulated to put them in the best possible form for analysis and conceptual design of systems.

9.2   HEAT BALANCES

9.2.1   An Overall Balance on a Heat Exchanger Stream

Consider two streams, shown in Figure 9.1, which flow into and out of a unit and do not mix within it. If there is net transfer between the streams, one will be colder than the other. Denote the cold one with a subscript C and the hot one with a subscript H. Let the mass flow rates be $\dot{m}_C$ and $\dot{m}_H$, respectively. From conservation of mass, these flow rates must be the same coming out as going in. The first law of thermodynamics for a steady open flow system then gives for the cold stream

$$\dot{Q}_C = \dot{m}_C \hat{h}_{C,o} - \dot{m}_C \hat{h}_{C,i} \qquad (9.1)$$

and for the hot stream

$$\dot{Q}_H = \dot{m}_H \hat{h}_{H,o} - \dot{m}_H \hat{h}_{H,i} \qquad (9.2)$$

where $\hat{h}$ is enthalpy and o and i denote out and in, respectively. If there is negligible heat transfer with the environment

$$\dot{Q}_H + \dot{Q}_C = 0 \qquad (9.3)$$

$$m_C \hat{h}_{C,o} - \dot{m}_C \hat{h}_{C,i} = \dot{m}_H \hat{h}_{H,i} - \dot{m}_H \hat{h}_{H,o} \qquad (9.4)$$

For constant specific heats and no change of phase, Eqs. (9.1) and (9.2) may be rewritten in terms of $\dot{m}c_p$, the thermal capacity rate, as

$$\dot{Q}_C = (\dot{m}c_p)_C (T_{C,o} - T_{C,i}) \qquad (9.5)$$

$$\dot{Q}_H = (\dot{m}c_p)_H (T_{H,o} - T_{H,i}) \qquad (9.6)$$

- $\qquad (\dot{m}c_p)_C (T_{C,o} - T_{C,i}) = (\dot{m}c_p)_H (T_{H,i} - T_{H,o}) \qquad (9.7)$

9.2.2   *EXAMPLE*   An Air Preheater for an Afterburner

Consider an automotive afterburner from which flows 360 lb/hr of exhaust gases at 1000°F. Cold air at 80°F is preheated to 800°F by flowing in a heat exchanger heated by the exhaust gases. The cold air, needed to supply oxygen

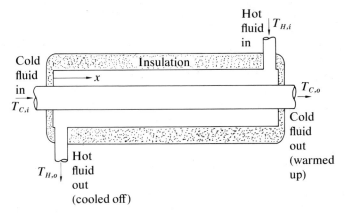

Counter-Current

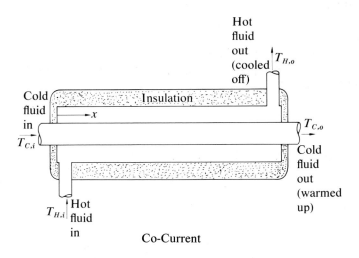

Co-Current

Figure 9.1 Counter- and co-current heat exchangers.

for the afterburner, flows at a rate of 60 lb/hr. What is the exhaust gas temperature when it leaves the preheater?

Assuming a constant specific heat of 0.25 Btu/lb °F for the air and 0.28 for the exhaust gases (since these gases contain a mass fraction of H_2O of about 0.1), we find the exit temperature very readily:

$$(60)(0.25)(800-80) = (360)(0.28)(1000-T_{H,o})$$

$$T_{H,o} = 1000 - \frac{720(0.25)}{6(0{\cdot}28)} = 893°F$$

9.3   SINGLE-STREAM HEAT EXCHANGERS

9.3.1   The Heat Balance on a Differential Element

When boiling or condensation occurs on one side of a heat exchanger surface, the temperature on that side is constant along the exchanger, equal to the saturation temperature corresponding to the pressure at which the change of phase is taking place. Only the temperature of the other stream varies. While this case is a special case of the more general situation to be analyzed in Section 9.5, some of the concepts basic to understanding exchangers can be introduced by first treating the single stream whose temperature varies.

Consider the cold stream in Figure 9.2 of Section 9.4. It might be cooling water in a condenser. The hot stream of condensing steam is at temperature T_H, the hot saturation temperature. The heat transfer from the hot stream to the cold stream is, for a length dx with heat transfer area $\mathscr{P}dx$,

$$d\dot{Q} = U\mathscr{P}dx(T_H - T_C) \tag{9.8}$$

where U is the overall heat transfer coefficient or *unit conductance* defined in Chapter 2, Section 2.7.3. Recall that for a thermal circuit having several resistances in series, U is the reciprocal of the sum of those resistances. In our problem the significant resistance is usually in the cold stream, but the resistance of the pipe wall, scale or the film of condensate on the hot side may also have to be included, as discussed in Section 9.4.

The heat flow into the element of the exchanger at x is

$$(\dot{m}_C c_p T_C)_x$$

and that out of the element at $x + dx$ is

$$(\dot{m}_C c_p T_C)_{x+dx} = (\dot{m}_C c_p T_C)_x + \dot{m}_C c_p \frac{dT_C}{dx} dx$$

where c_p has been assumed constant. The net heat carried away by the cold stream comes from the hot stream across the side wall, when the flow is steady.

$$\dot{m}_C c_p \frac{dT_C}{dx} dx = d\dot{Q} = U\mathscr{P}dx(T_H - T_C)$$

We make these equations dimensionless by defining

$$x^* = \frac{x}{L}, \qquad x = Lx^* \tag{9.9}$$

$$T^* = \frac{T_H - T_C}{T_H - T_{C,i}}, \qquad T_C = T_H - (T_H - T_{C,i})T^* \tag{9.10}$$

Substituting gives

$$-\dot{m}_C c_p (T_H - T_{C,i}) \frac{1}{L} \frac{dT^*}{dx^*} = U\mathscr{P}(T_H - T_{C,i}) T^*$$

$$\frac{dT^*}{dx^*} + \frac{U\mathscr{P}L}{\dot{m}_C c_p} T^* = 0 \qquad (9.11)$$

This first order ordinary differential equation requires one boundary condition. We may take it to be the inlet temperature at $x = 0$. Transformed into dimensionless coordinates we write the boundary condition

$$T^* = 1 \quad \text{at } x^* = 0 \qquad (9.12)$$

9.3.2   Effectiveness and Number of Transfer Units

The quantity $U\mathscr{P}L/(\dot{m}_C c_p)$ in Eq. (9.11) is dimensionless. For reasons to be more fully explained later, it is called the number of transfer units, abbreviated NTU, and symbolized by N_{tu}.

$$N_{tu} = \frac{U\mathscr{P}L}{\dot{m}_C c_p} \qquad (9.13)$$

For a given $U/\dot{m}_C c_p$, the longer the heat exchanger or the larger its perimeter the greater the number of transfer units it may be said to possess.

Equation (9.11) is solved very readily. The eigenvalue is the negative of the N_{tu},

$$T^* = \mathscr{C} e^{-N_{tu} x^*}$$

Imposition of the boundary condition, Eq. (9.12), shows that the constant is unity.

$$T^* = e^{-N_{tu} x^*} \qquad (9.14)$$

The outlet temperature is that at $x = L$ or $x^* = 1$. Replacing T^* by its definition in Eq. (9.10) gives for $x^* = 1$

$$\frac{T_H - T_{C,o}}{T_H - T_{C,i}} = e^{-N_{tu}} \qquad (9.15)$$

It is apparent from Eq. (9.15) that if the exchanger were made infinite in side area $\mathscr{P}L$, the outlet temperature of the cold stream would approach the temperature of the hot stream. Such an infinite exchanger represents the maximum attainable ideal and is said to be 100% effective. The effectiveness is thus defined as the heat actually transferred divided by the maximum attainable transfer

$$\epsilon = \frac{\dot{m}_C c_p (T_{C,o} - T_{C,i})}{\dot{m}_C c_p (T_H - T_{C,i})} \qquad (9.16)$$

Adding and subtracting $T_{C,i}$ to the numerator of the left-hand side of Eq. (9.15) and rearranging show that for our simple one-stream exchanger

$$\epsilon = 1 - e^{-N_{tu}}, \qquad N_{tu} = \ln \frac{1}{1 - \epsilon} \qquad (9.17)$$

9.3.3   *EXAMPLE* **A Condenser**

A laboratory unit for producing distilled water must condense 3 lb/hr of steam at 212°F. Cooling water is available at 70°F, and to avoid unnecessary wastage of cooling water it is desired to have a 90% effective unit. If the unit conductance U is 100 Btu/hr ft² °F, what area is necessary and, if a jacketed 0.5 in. diameter tube is used, how long should it be? How much cooling water will be required?

First an overall heat balance is made to find $\dot{m}_C$. We write

$$\dot{Q} = \dot{m}_C c_p (T_{C,o} - T_{C,i}), \qquad \dot{m}_C c_p = \frac{\dot{Q}}{T_{C,o} - T_{C,i}}$$

The required heat transfer $\dot{Q}$ is in turn given by

$$\dot{Q} = \dot{m}_H (\hat{h}_{H,i} - \hat{h}_{H,o}) = \dot{m}_H \hat{h}_{fg} = (3)(970) = 2910 \text{ Btu/hr}$$

From the definition of effectiveness, Eq. (9.16), we find $T_{C,o}$:

$$\epsilon = \frac{T_{C,o} - T_{C,i}}{T_H - T_{C,i}}, \qquad T_{C,o} = T_{C,i} + \epsilon(T_H - T_{C,i})$$

$$T_{C,o} = 70 + (0.90)(212 - 70) = 198°F$$

The product $\dot{m}_C c_p$ is then

$$\dot{m}_C c_p = \frac{2910}{198 - 70} = 22.7 \text{ Btu/hr °F}$$

Taking c_p to be approximately 1 Btu/lb °F then gives $\dot{m}_C = 22.7$ lb/hr, a little less than 3 gallons per hour.

Equation (9.17) shows that for an effectiveness of 0.90 an N_{tu} of 2.3 is needed. From Eq. (9.13) the area needed is

$$\mathscr{P}L = \frac{\dot{m}_C c_p N_{tu}}{U} = \frac{(22.7)(2.3)}{100} = 0.52 \text{ ft}^2$$

If the tube has a perimeter of $0.5\pi/12$ ft, the length necessary is

$$L = \frac{\mathscr{P}L}{\mathscr{P}} = \frac{0.52}{0.131} = 4.0 \text{ ft}$$

The length is such that it should be coiled to make the condenser compact.

9.4   OVERALL HEAT TRANSFER COEFFICIENT

Now we turn our attention to exchangers in which both streams have varying temperatures. We will consider only two types of exchangers: (1) countercurrent (counter flow) units and (2) co-current (parallel flow) ones. Note that a given physical unit can be piped either way. Figure 9.1 showed these types of arrangements. These arrangements are one-dimensional in nature; the bulk temperature of each stream varies with only one dimension along the length, say x.

There are, of course, arrangements other than counter or parallel flow, such as cross flow and mixed cross and parallel or counter flow. But these arrangements are inherently two-dimensional in nature and are beyond the scope of this introduction.

Consider the transfer of heat in a section of length dx, as shown in Figure 9.2. The heated perimeter for the cold stream is, say, $\mathscr{P}_C$. For example, if the cold fluid flows within a circular tube of inside diameter D_i, the perimeter is πD_i. The heat transfer area for length dx is $\mathscr{P}_C\, dx$. The heat transfer coefficient (film coefficient) is h_C. There is consequently a thermal resistance of the cold fluid

$$R_{F,C} = \frac{1}{\mathscr{P}_C h_C\, dx} \tag{9.18}$$

due to the surface film. There may be also on the cold side some scale, scum, or crud with resistance $R_{S,C}$. For example, if the perimeter $\mathscr{P}_C$ is coated with a thin layer $\delta_{S,C}$ thick, and the scale has conductivity $k_{S,C}$, the resistance is

$$R_{S,C} = \frac{1}{\mathscr{P}_C k_{S,C}\, dx/\delta_{S,C}} \tag{9.19}$$

On the hot side there are similarly resistances $R_{F,H}$ and $R_{S,H}$

$$R_{F,H} = \frac{1}{\mathscr{P}_H h_H\, dx} \tag{9.20}$$

$$R_{S,H} = \frac{1}{\mathscr{P}_H k_{S,H}\, dx/\delta_{S,H}} \tag{9.21}$$

There is, in addition, the thermal resistance of the wall itself, for metal walls often negligible. For example, if the wall is a tube (see Section 2.8.2)

$$R_W = \frac{\ln\,(D_o/D_i)}{2\pi k_W\, dx} \tag{9.22}$$

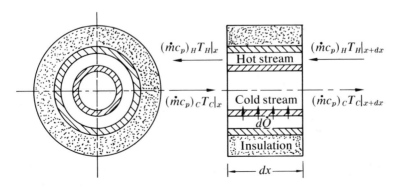

Figure 9.2   Heat balances for an element of a counterflow heat exchanger.

The total, or overall, thermal resistance is then the sum of these

$$R_{oa} = R_{F,C} + R_{S,C} + R_W + R_{S,H} + R_{F,H} \qquad (9.23)$$

It is more usual to use the reciprocal, a thermal conductance,

$$\frac{1}{R_{oa}} = (U\mathscr{P})\,dx = \frac{1}{R_{F,C} + R_{S,C} + R_W + R_{S,H} + R_{F,H}} \qquad (9.24)$$

Substitution yields

$$\bullet \quad U\mathscr{P} = \frac{1}{\dfrac{1}{\mathscr{P}_C h_C} + \dfrac{\delta_{S,C}}{\mathscr{P}_C k_{S,C}} + \dfrac{\ln\,(D_o/D_i)}{2\pi k_W} + \dfrac{\delta_{S,H}}{\mathscr{P}_H k_{S,H}} + \dfrac{1}{\mathscr{P}_H h_H}} \qquad (9.25)$$

When the wall is thin so that $D_o \doteq D_i$, the quantity $\mathscr{P}$ can be identified with $\mathscr{P}_C$ or $\mathscr{P}_H$, which are nearly equal. The quantity U is then called the overall heat transfer coefficient. In any event, Eq. (9.25) gives the $U\mathscr{P}$ product, which is all we need to describe the conductance between streams per unit length of exchanger.

9.5   COUNTER-CURRENT HEAT EXCHANGERS

9.5.1   The Governing Differential Equations

Return to the element of length dx in a counter-current heat exchanger pictured in Figure 9.2. Two balances must be made, one for the cold stream and one for the hot stream. For the cold stream

$$(mc_p)_C T_C + d\dot{Q} = (\dot{m}c_p)_C \left(T_C + \frac{dT_C}{dx}\,dx \right)$$

$$d\dot{Q} = (mc_p)_C \frac{dT_C}{dx}\,dx \qquad (9.26)$$

For the hot stream

$$(\dot{m}c_p)_H \left(T_H + \frac{dT_H}{dx}\,dx \right) = d\dot{Q} + (\dot{m}c_p)_H T_H$$

$$d\dot{Q} = (\dot{m}c_p)_H \frac{dT_H}{dx}\,dx \qquad (9.27)$$

Returning to the considerations leading up to Eqs. (9.23) and (9.25) we write

$$d\dot{Q} = \frac{T_H - T_C}{R_{oa}} = (U\mathscr{P}\,dx)(T_H - T_C) \qquad (9.28)$$

Substituting Eq. (9.28) into Eqs. (9.26) and (9.27) and canceling dx yields

$$U\mathscr{P}(T_H - T_C) = (\dot{m}c_p)_C \frac{dT_C}{dx} \qquad (9.29a)$$

$$U\mathscr{P}(T_H - T_C) = (\dot{m}c_p)_H \frac{dT_H}{dx} \qquad (9.30a)$$

9.5.2   Boundary Conditions and Solutions

For the counterflow unit, the boundary conditions are as follows (see Figure 9.1):

$$x = 0: T_H(0) = T_{H,o}, \qquad T_C(0) = T_{C,i}$$
$$x = L: T_H(L) = T_{H,i}, \qquad T_C(L) = T_{C,o} \tag{9.31}$$

Note that these conditions must be compatible with Eq. (9.7). We have two simultaneous ordinary linear differential equations subject to these conditions. Let

$$C_C = (\dot{m}c_p)_C, \qquad C_H = (\dot{m}c_p)_H \tag{9.32}$$

Equations (9.29a) and (9.30a) then can be written, respectively,

$$\frac{dT_C}{dx} = \frac{U\mathscr{P}}{C_C}(T_H - T_C) \tag{9.29b}$$

$$\frac{dT_H}{dx} = \frac{U\mathscr{P}}{C_H}(T_H - T_C) \tag{9.30b}$$

Subtracting Eq. (9.29b) from Eq. (9.30b) gives

$$\frac{d}{dx}(T_H - T_C) = U\mathscr{P}\left(\frac{1}{C_H} - \frac{1}{C_C}\right)(T_H - T_C) \tag{9.33}$$

This equation is readily solved when $U\mathscr{P}$ is taken constant,

$$T_H - T_C = \mathscr{C}_1 \exp\left[U\mathscr{P}\left(\frac{1}{C_H} - \frac{1}{C_C}\right)x\right] \tag{9.34}$$

The original governing equations can now be solved. Equation (9.30b) becomes

$$\frac{dT_H}{dx} = \frac{U\mathscr{P}}{C_H}\mathscr{C}_1 \exp\left[U\mathscr{P}\left(\frac{1}{C_H} - \frac{1}{C_C}\right)x\right]$$

If $C_H = C_C$, T_H is found to vary linearly with x. For $C_H \neq C_C$, there is obtained

$$T_H = \mathscr{C}_2 + \frac{\mathscr{C}_1}{1 - (C_H/C_C)} \exp\left[U\mathscr{P}\left(\frac{1}{C_H} - \frac{1}{C_C}\right)x\right] \tag{9.35}$$

Similarly Eq. (9.29b) yields

$$T_C = \mathscr{C}_3 - \frac{\mathscr{C}_1}{1 - (C_C/C_H)} \exp\left[U\mathscr{P}\left(\frac{1}{C_H} - \frac{1}{C_C}\right)x\right] \tag{9.36}$$

All three constants are not independent, for Eq. (9.34) must be satisfied. For this equation to be satisfied $\mathscr{C}_2 = \mathscr{C}_3$.

Two cases may now be distinguished: (1) $C_C < C_H$ and (2) $C_H < C_C$.

9.5.3   Small Cold Flow Capacity

When $C_C < C_H$, the exponents in Eqs. (9.35 to 9.36) are negative. Substituting the boundary conditions for $x = 0$, Eq. (9.31), into Eq. (9.34) yields

$$T_{H,o} - T_{C,i} = \mathscr{C}_1$$

so that Eq. (9.34) becomes

$$T_H(x) - T_C(x) = (T_{H,o} - T_{C,i}) \exp\left[-U\mathscr{P}\left(\frac{1}{C_C} - \frac{1}{C_H}\right)x\right] \qquad (9.37a)$$

For $x = L$

$$T_{H,i} - T_{C,o} = (T_{H,o} - T_{C,i}) \exp\left[-U\mathscr{P}\left(\frac{1}{C_C} - \frac{1}{C_H}\right)L\right] \qquad (9.38a)$$

Since the exponent is negative, it is clear that $T_{C,o}$ can be made to approach $T_{H,i}$ by making the exchanger long. Given any three temperatures and the flow rates, Eq. (9.7) can be used to find the fourth one, and Eq. (9.38a) can be used to solve for the length required.

To complete our solution for the temperature profiles, let us choose as given $T_{H,i}$, $T_{H,o}$, and $T_{C,i}$ where, clearly, $T_{H,i} > T_{H,o} > T_{C,i}$. We have already found $\mathscr{C}_1$ to be $T_{H,o} - T_{C,i}$. Equation (9.36) evaluated at $x = 0$ where $T_C = T_{C,i}$ gives (since $\mathscr{C}_3 = \mathscr{C}_2$)

$$T_{C,i} = \mathscr{C}_2 - \frac{T_{H,o} - T_{C,i}}{1 - (C_C/C_H)}$$

$$\mathscr{C}_2 = (T_{H,o} - R_C T_{C,i})/(1 - R_C), \quad \text{where } R_C = \frac{C_C}{C_H}$$

Equations (9.35) and (9.36) then become

$$T_H = \frac{1}{1 - R_C}\left\{T_{H,o} - R_C T_{C,i} - R_C(T_{H,o} - T_{C,i}) \exp\left[-\left(\frac{1}{C_C} - \frac{1}{C_H}\right)U\mathscr{P}x\right]\right\} \qquad (9.39)$$

$$T_C = \frac{1}{1 - R_C}\left\{T_{H,o} - R_C T_{C,i} - (T_{H,o} - T_{C,i}) \exp\left[-\left(\frac{1}{C_C} - \frac{1}{C_H}\right)U\mathscr{P}x\right]\right\} \qquad (9.40)$$

9.5.4   Small Hot Flow Capacity

With $C_H < C_C$ it is more instructive to apply the boundary conditions at $x = L$ from Eq. (9.31) to Eq. (9.34) to obtain $\mathscr{C}_1$

$$T_{H,i} - T_{C,o} = \mathscr{C}_1 \exp\left[U\mathscr{P}L\left(\frac{1}{C_H} - \frac{1}{C_C}\right)\right]$$

$$\mathscr{C}_1 = (T_{H,i} - T_{C,o}) \exp\left[-U\mathscr{P}L\left(\frac{1}{C_H} - \frac{1}{C_C}\right)\right]$$

In this case Eq. (9.34) can be written

$$T_H(x) - T_C(x) = (T_{H,i} - T_{C,o}) \exp\left[-U\mathscr{P}(L-x)\left(\frac{1}{C_H} - \frac{1}{C_C}\right)\right] \qquad (9.37b)$$

and for $x = 0$

$$T_{H,o} - T_{C,i} = (T_{H,i} - T_{C,o}) \exp\left[-U\mathscr{P}L\left(\frac{1}{C_H} - \frac{1}{C_C}\right)\right] \qquad (9.38b)$$

It is clear that $T_{H,o}$ can be made to approach $T_{C,i}$ as closely as desired by making L sufficiently long. As before, given any three of the four temperatures and the flow rates, Eq. (9.7) can be used to find the fourth temperature, and Eq. (9.38b) can be used to find the required $U\mathscr{P}L$.

9.5.5   *EXAMPLE* An Air Preheater for an Afterburner

Return to the example of Section 9.2.2. Suppose we want to find the area necessary for our preheater which must heat 60 lb/hr of cold air from 80°F to 800°F using 360 lb/hr of exhaust gas at 1000°F. In Section 9.2.2 we found using Eq. (9.7) that $T_{H,o} = 893°F$. Substituting now into Eq. (9.38a), rearranging, and taking the logarithm of the reciprocal of both sides gives

$$U\mathscr{P}L\left(\frac{1}{C_C} - \frac{1}{C_H}\right) = \ln\frac{T_{H,o} - T_{C,i}}{T_{H,i} - T_{C,o}}$$

$$U\mathscr{P}L\left(\frac{1}{(0.25)(60)} - \frac{1}{(0.28)(360)}\right) = \ln\frac{893 - 80}{1000 - 800}$$

$$U\mathscr{P}L = \frac{1.403}{0.0667 - 0.0099} = 24.7 \text{ Btu/hr °F}$$

The next step would be to calculate h_H and h_C using the techniques of Chapters 5 and 8. In order to do so we have to decide what flow passage geometry will be used. If the hot exhaust gases flow in a 1.5 in. diameter tube (and assuming physical properties for pure air),

$$Re = \frac{VD}{\nu} = \frac{\rho V(\pi D^2/4)}{\rho(\pi D/4)\nu} = \frac{\dot{m}}{\rho(\pi D/4)\nu}$$

$$Re \doteq \frac{360/3600}{(0.028)(1.5\pi/48)(87 \times 10^{-5})} = 41,700$$

$$Nu = \frac{hD}{k} = 0.023\, Re^{0.8}Pr^{0.33} \doteq 0.023(5000)(0.88) \doteq 100$$

$$h_H = \frac{0.033}{1.5/12}(100) = 26.4 \text{ Btu/ft}^2 \text{ hr °F}$$

Suppose the air-side passage, say an annulus, is sized to make h_C about the same. The surfaces of the tube might have scale deposits with resistances of 0.01 ft² hr °F/Btu for a square foot of area. The tube itself might be 0.040 in. of steel so that for 1 ft²

$$R_W = \frac{0.040/12}{30} \doteq 0.0001 \text{ ft}^2 \text{ hr °F/Btu}$$

For $\mathscr{P}dx = 1$ ft² Eq. (9.23) then gives

$$R_{oa} = \frac{1}{26.4} + 0.01 + 0.0001 + 0.01 + \frac{1}{26.4} \doteq 0.095 \text{ ft}^2 \text{ hr °F/Btu}$$

Equation (9.24) gives

$$U = \frac{1}{0.095} = 10.5 \text{ Btu/ft}^2 \text{ hr °F} = \frac{U\mathscr{P}L}{\mathscr{P}L}$$

We find that $\mathscr{P}L$ must be $24.7/10.5 = 2.35$ ft². If $\mathscr{P}$ is $1.5\pi/12 = 0.4$ ft, then L would have to be 5.9 ft. This length could be reduced by using a more compact design, for example, by using a number of triangular flow passages made by stacking V-corrugated sheets. The reader interested in such arrangements should see *Compact Heat Exchangers* by Kays and London.

9.6   CO-CURRENT HEAT EXCHANGERS

The analysis given in Section 9.5 may be repeated for parallel flow. In Figure 9.2 the direction of the flow $(\dot{m}c_p)_H$ is reversed, and Eqs. (9.27) and (9.30a) have minus signs. Equations (9.29a) and (9.30a) thus become

$$U\mathscr{P}(T_H - T_C) = (\dot{m}c_p)_C \frac{dT_C}{dx} \tag{9.41}$$

$$U\mathscr{P}(T_H - T_C) = -(\dot{m}c_p)_H \frac{dT_H}{dx} \tag{9.42}$$

The boundary conditions are now

$$x = 0: \ T_H = T_{H,i} \qquad T_C = T_{C,i}$$
$$x = L: \ T_H = T_{H,o} \qquad T_C = T_{C,o} \tag{9.43}$$

After dividing by C_C and C_H, respectively, Eqs. (9.41) and (9.42) may be added to obtain

$$U\mathscr{P}\left(\frac{1}{C_C} + \frac{1}{C_H}\right)(T_H - T_C) = -\frac{d}{dx}(T_H - T_C) \tag{9.44}$$

which yields

$$T_H - T_C = (T_{H,i} - T_{C,i})\exp\left[-U\mathscr{P}\left(\frac{1}{C_C} + \frac{1}{C_H}\right)x\right] \tag{9.45}$$

$$T_{H,o} - T_{C,o} = (T_{H,i} - T_{C,i})\exp\left[-U\mathscr{P}\left(\frac{1}{C_C} + \frac{1}{C_H}\right)L\right] \tag{9.46}$$

In this case $T_{H,o}$ and $T_{C,o}$ may be made to approach one another by making L very long, but $T_{C,o}$ cannot be made to approach $T_{H,i}$ unless C_H is much greater than C_C. For this reason parallel flow exchangers are not as effective as counterflow ones and are not used unless convenience in piping dictates the arrangement. When C_C and C_H differ an order of magnitude, the penalty associated with using parallel flow instead of counterflow is but slight.

9.7   THE LOG MEAN TEMPERATURE DIFFERENCE

9.7.1   The Area Average Temperature Difference

One way to use Eqs. (9.38a) and (9.46) is to define a mean temperature such that

$$\dot{Q} = U\mathscr{P}L(\Delta T)_{\mathrm{mean}} \tag{9.47a}$$

where $(\Delta T)_{\text{mean}}$ is clearly, from Eq. (9.28), the area average temperature difference in the exchanger. To define a proper mean ΔT we use Eq. (9.38a) or (9.46) to find $\dot{Q}/U\mathscr{P}L$ and Eqs. (9.5) and (9.6) to eliminate C_C and C_H. Equation (9.47a) can be rearranged for this purpose in the form

$$(\Delta T)_{\text{mean}} = \frac{\dot{Q}}{U\mathscr{P}L} \tag{9.47b}$$

9.7.2 Counter-Current Flow

Introducing Eq. (9.38a) we write

$$(\Delta T)_{\text{mean}} = \frac{\dot{Q}\left(\dfrac{1}{C_C} - \dfrac{1}{C_H}\right)}{\ln\dfrac{T_{H,o} - T_{C,i}}{T_{H,i} - T_{C,o}}}$$

Equations (9.5) and (9.6) give $\dot{Q}/C_C$ and $\dot{Q}/C_H$, respectively. We find

$$(\Delta T)_{\text{mean}} = \frac{(T_{C,o} - T_{C,i}) - (T_{H,i} - T_{H,o})}{\ln\dfrac{T_{H,o} - T_{C,i}}{T_{H,i} - T_{C,o}}}$$

$$(\Delta T)_{\text{mean}} = \frac{(T_{H,o} - T_{C,i}) - (T_{H,i} - T_{C,o})}{\ln\dfrac{T_{H,o} - T_{C,i}}{T_{H,i} - T_{C,o}}} \tag{9.48}$$

Equation (9.48) shows the area average to be a logarithmic mean temperature difference for the two ends of the exchanger, for we see that

$$T_{H,o} - T_{C,i} = T_H\big|_{x=0} - T_C\big|_{x=0}$$

and

$$T_{H,i} - T_{C,o} = T_H\big|_{x=L} - T_C\big|_{x=L}$$

We write Eq. (9.48) in this form

$$(\Delta T)_{\text{mean}} = \frac{[T_H(0) - T_C(0)] - [T_H(L) - T_C(L)]}{\ln\dfrac{[T_H(0) - T_C(0)]}{[T_H(L) - T_C(L)]}} \tag{9.49}$$

9.7.3 Co-Current Flow

If the procedure is repeated using Eq. (9.46) in place of (9.38a), we obtain

$$(\Delta T)_{\text{mean}} = \frac{\dot{Q}\left(\dfrac{1}{C_C} + \dfrac{1}{C_H}\right)}{\ln\dfrac{T_{H,i} - T_{C,i}}{T_{H,o} - T_{C,o}}}$$

Again introducing Eqs. (9.5) and (9.6) gives

$$(\Delta T)_{mean} = \frac{(T_{C,o} - T_{C,i}) + (T_{H,i} - T_{H,o})}{\ln \dfrac{T_{H,i} - T_{C,i}}{T_{H,o} - T_{C,o}}}$$

$$(\Delta T)_{mean} = \frac{(T_{H,i} - T_{C,i}) - (T_{H,o} - T_{C,o})}{\ln \dfrac{T_{H,i} - T_{C,i}}{T_{H,o} - T_{C,o}}} \tag{9.50}$$

But Eq. (9.50) is exactly the same as Eq. (9.49) when the inlet and outlet temperatures are written in terms of their spatial locations.

$$(\Delta T)_{mean} = \frac{[T_H(0) - T_C(0)] - [T_H(L) - T_C(L)]}{\ln \dfrac{T_H(0) - T_C(0)}{T_H(L) - T_C(L)}} \tag{9.51}$$

It should be noted that, in addition to being applicable to two different types of exchangers (parallel and counterflow units), Eqs. (9.47a) and (9.49 and 9.51) are readily applicable to the case $C_C = C_H$ not treated in Section 9.5. In this case the inlet and outlet temperature differences are the same, and the log-mean and arithmetic mean are identical. When more than one of the four inlet and outlet temperatures are unknown, the effectiveness and NTU concepts developed below are more convenient to use.

9.7.4   *EXAMPLE* **An Air Preheater**

Consider our catalytic reactor preheater described in Sections 9.2.2 and 9.5.5. What is the log-mean temperature difference?

$$(\Delta T)_{mean} = \frac{(893 - 80) - (1000 - 800)}{\ln \dfrac{893 - 80}{1000 - 800}} = \frac{613}{\ln \dfrac{813}{200}} = 437°F$$

Since, from Eq. (9.5),

$$\dot{Q} = (60)(0.25)(800 - 80) = 10,800 \text{ Btu/hr}$$

we can find $U\mathscr{P}L$ readily from Eq. (9.47a)

$$U\mathscr{P}L = \frac{10,800}{437} = 24.7$$

For $U = 10.5$, we again find $\mathscr{P}L = 2.35 \text{ ft}^2$.

9.8   EFFECTIVENESS AND NTU CONCEPTS

9.8.1   Maximum Possible Transfer

We have found that a counterflow exchanger can transfer more heat than a parallel flow exchanger, because $T_{C,o}$ can be made to approach $T_{H,i}$ (for the case $C_C < C_H$) in the former but not in the latter. Consider the maximum possible transfer with a counterflow unit. For the case $C_C < C_H$ and $L \to \infty$ we found $T_{C,o} \to T_{H,i}$, so that Eq. (9.5) yields

$$\dot{Q}_{max} = C_C(T_{H,i} - T_{C,i})$$

For the case $C_H < C_C$, we found $T_{H,o} \to T_{C,i}$ as $L \to \infty$. Equation (9.6) yields then

$$\dot{Q}_{max} = C_H(T_{H,i} - T_{C,i})$$

Both cases can be covered by writing

$$\dot{Q}_{max} = C_{min}(T_{H,i} - T_{C,i}) \tag{9.52}$$

where C_{min} is the smaller of the pair of values G_G and G_H.

9.8.2   Effectiveness

The effectiveness of an exchanger is simply the actual transfer divided by the maximum possible (that which would occur in an infinite counterflow unit),

$$\epsilon = \dot{Q}/\dot{Q}_{max} \tag{9.53}$$

Either Eq. (9.5) or (9.6) can be used:

$$\epsilon = \frac{C_C(T_{C,o} - T_{C,i})}{C_{min}(T_{H,i} - T_{C,i})} = \frac{C_H(T_{H,i} - T_{H,o})}{C_{min}(T_{H,i} - T_{C,i})} \tag{9.54}$$

9.8.3   Number of Transfer Units

Consider for simplicity equal capacities $C_C = C_H = C$. Suppose we had an infinite parallel flow heat exchanger so that the outlet streams come to thermal equilibrium. From Eq. (9.5) or (9.6),

$$T_{H,o} = T_{C,o} = \tfrac{1}{2}(T_{H,i} + T_{C,i})$$

The actual heat transfer in such a case is therefore

$$\dot{Q} = C\left(\frac{T_{H,i} + T_{C,i}}{2} - T_{C,i}\right) = C\left(\frac{T_{H,i} - T_{C,i}}{2}\right)$$

The effectiveness of such a unit is, from Eqs. (9.52) and (9.53), one half.

Now consider how long a counterflow exchanger would have to be to achieve the same effectiveness. We must return to Eqs. (9.29b, 9.30b, and 9.33) since $C_C = C_H$. In this case

$$T_H = \frac{U\mathscr{P}}{C}(T_{H,o} - T_{C,i})x + T_{H,o}$$

$$T_{H,i} = \frac{U\mathscr{P}L}{C}(T_{H,o} - T_{C,i}) + T_{H,o}$$

Setting

$$T_{H,o} = \tfrac{1}{2}(T_{H,i} + T_{C,i})$$

gives the same heat transfer as the infinite parallel flow exchanger, and

$$\frac{U\mathscr{P}L}{C} = \frac{T_{H,i} - \frac{1}{2}(T_{H,i} + T_{C,i})}{\frac{1}{2}(T_{H,i} + T_{C,i}) - T_{C,i}} = 1$$

We see that the quantity $U\mathscr{P}L/C$ is dimensionless, and that when $U\mathscr{P}L/C = 1$, a small counterflow exchanger can do the work of one infinite parallel flow unit. For this reason the dimensionless quantity $U\mathscr{P}L/C$ is called **the number of transfer units**, N_{tu}. When C_C is not equal to C_H, we use

$$N_{tu} = \frac{U\mathscr{P}L}{C_{min}} \tag{9.55}$$

It is possible to express effectiveness ϵ in terms of N_{tu} and the ratio

$$R_C = \frac{C_{min}}{C_{max}} \tag{9.56}$$

It may be shown that for the counterflow exchanger

$$\bullet \qquad \epsilon_{cf} = \frac{1 - e^{-N_{tu}(1-R_C)}}{1 - R_C e^{-N_{tu}(1-R_C)}}$$

Note that when R_C goes to zero, Eq. (9.57) reduces to Eq. (9.17), which was derived for infinite C_{max}.

For parallel flow

$$\bullet \qquad \epsilon_{pf} = \frac{1 - e^{-N_{tu}(1+R_C)}}{1 + R_C} \tag{9.58}$$

9.8.4   *EXAMPLE* Recap of the Air Preheater

Return to our catalytic afterburner preheater first described in Section 9.2.2. Recall that we wished to heat cold air ($\dot{m}c_p = C_C = 60(0.25) = 15$) from 80 to 800°F, using hot exhaust gas ($\dot{m}c_p = C_H = 360(0.28) = 101$) at 1000°F. Let us rework the entire problem as expeditiously as possible. First we find the effectiveness required. From Eq. (9.54)

$$\epsilon = \frac{(15)(800 - 80)}{(15)(1000 - 80)} = 0.783$$

We see we need a highly effective unit. We find also

$$R_C = \frac{15}{101} = 0.148$$

For design of new equipment, as opposed to analysis of existing equipment, Eq. (9.57) should be written

$$\bullet \qquad N_{tu} = \frac{1}{1 - R_C} \ln \frac{1 - \epsilon_{cf}R_C}{1 - \epsilon_{cf}} \tag{9.59}$$

$$N_{tu} = 1.173 \ln \frac{0.884}{0.217} = 1.647$$

From the definition of N_{tu}, Eq. (9.55),

$$U\mathscr{P}L = C_{min}N_{tu} = 15(1.647) = 24.7$$

As before, for $U = 10.5$ (see Section 9.5.5), $\mathscr{P}L = 2.35$ ft².

9.9 REGENERATORS

9.9.1 Types of Regenerators

There is a type of heat exchanger which is two-dimensional in nature and therefore beyond the scope of this text, but which is so important technically that not to describe it briefly would be amiss. It is the regenerator in which heat is first transferred from the hot stream to elements in the exchanger. The exchanger stores the heat until later, when it is transferred to a cold stream.

In intermittent regenerators a fixed bed experiences first a flow of hot gas, then a flow of cold gas. The human nose is an example. The principle was first applied to technology in 1816 by Stirling. Checkerboard arrays of bricks were used to preheat air used for combustion in steel blast furnaces.

The rotary regenerator developed by Ljungström contains a bed which is rotated. It is this type which is used for gas turbines and in air conditioning. In the original Ljungström form, stacks of wire mesh are used for the heat transfer elements. Recently R. V. Dunkle has employed parallel plate laminar flow units made by winding spaced sheets into a coil. The ratio of Stanton number to friction factor is three times more favorable than for a packed bed (see Exercise 3 at the end of Chapter 8), and the porosity squared times Stanton number divided by friction factor is still more favorable.

9.9.2 Governing Equations for Balanced Flow

Since regenerators are usually used for balanced flow, $(\dot{m}c_p)_C = C_C = C_H = (\dot{m}c_p)_H$, we consider only this case. We are often justified in neglecting temperature gradients normal to the regenerator walls. Let the half thickness of the wall be δ. The heat capacity of a length dx is then $\rho c \delta \mathscr{P} dx$. Suppose the elements are exposed to a cold flow from $\theta = 0$ to $\theta = \Phi$ and to the hot flow for $\theta = \pi$ to $\theta = \pi + \Phi$. During time dt the wheel (see Figure 9.3) rotates $d\theta = \omega \, dt$, where ω is 2π times the revolutions per second.

As in the case of the counterflow exchanger, we make a heat balance on the cold stream between x and $x + dx$

$$CT_C|_x + h\mathscr{P}dx(T_M - T_C) = CT_C|_{x+dx}$$

Similarly for the hot stream

$$CT_H|_{x+dx} = h\mathscr{P} \, dx(T_H - T_M) + CT_H|_x$$

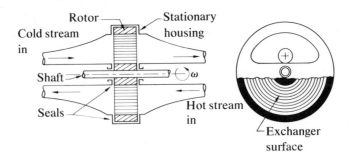

Figure 9.3 A counterflow rotary regenerator.

When hot fluid flows over the matrix we have

$$h \, \mathscr{P} dx (T_H - T_M) = \rho c \delta \, \mathscr{P} dx \, \frac{\partial T_M}{\partial t} = \rho c \delta \, \mathscr{P} dx \, \frac{\partial T_M}{\partial \theta} \omega$$

When cold fluid flows

$$h \, \mathscr{P} dx (T_M - T_C) = - \rho c \delta \, \mathscr{P} dx \, \frac{\partial T_M}{\partial t} = - \rho c \delta \, \mathscr{P} dx \, \frac{\partial T_M}{\partial \theta} \omega$$

Expanding in Taylor's series and rearranging gives for $\theta = 0$ to Φ

$$h \mathscr{P} L (T_M - T_C) = C \, \frac{\partial T_C}{\partial (x/L)} = C \, \frac{\partial T_C}{\partial x^*} \tag{9.60}$$

$$h \mathscr{P} L (T_M - T_C) = - C_R \, \frac{\partial T_M}{\partial \theta} \tag{9.61}$$

where

$$C_R = \rho c \delta \, \mathscr{P} L \omega = c M_T \omega \tag{9.62}$$

The quantity M_T is the total mass in the transfer elements. Similarly for $\theta = \pi$ to $\pi + \Phi$

$$h \mathscr{P} L (T_H - T_M) = C \, \frac{\partial T_H}{\partial (x/L)} = C \, \frac{\partial T_H}{\partial x^*} \tag{9.63}$$

$$h \mathscr{P} L (T_H - T_M) = C_R \, \frac{\partial T_M}{\partial \theta} \tag{9.64}$$

As before, boundary conditions are

$$\begin{array}{lll} x = 0: & \theta = 0 \text{ to } \Phi & T_C = T_{C,i} \\ x = L: & \theta = \pi \text{ to } \pi + \Phi & T_H = T_{H,i} \end{array} \tag{9.65}$$

Dividing Eqs. (9.60 to 9.64) by C shows that the dimensionless equations depend only upon N_{tu} and the ratio

$$R_R = \frac{C_R}{C} \tag{9.66}$$

In general the effectiveness will depend upon only these two quantities. Since the general case is a complex two-dimensional problem, only two special cases will be treated here.

9.9.3   Special Case of $R_R \geqslant N_{tu}$

When the rate of rotation is high, Eqs. (9.61) and (9.64) show that $\partial T_M / \partial \theta$ goes to zero. Consequently, T_M is a function of x only, and no θ variations are forced upon T_C and T_H. By adding Eqs. (9.60) and (9.63) we obtain

$$h \mathscr{P} L (T_H - T_C) = C \left(\frac{dT_H}{dx^*} + \frac{dT_C}{dx^*} \right)$$

But, since for balanced flow $dT_H/dx^* = dT_C/dx^*$, we have

$$h \mathscr{P} L (T_H - T_C) = 2C \frac{dT_C}{dx^*}$$

$$h \mathscr{P} L (T_H - T_C) = 2C \frac{dT_H}{dx^*}$$

These two equations are identical to Eqs. (9.30b) and (9.29b) when

$$U = \frac{h}{2} \quad \text{and} \quad C_H = C_C = C$$

The solution for ϵ as a function of N_{tu} is therefore the same as for a counterflow heat exchanger with balanced flow. When h_H is not the same as h_C, it can be shown that

$$U = \left[\frac{1}{h_H} + \frac{1}{h_C} \right]^{-1} \tag{9.67}$$

$$N_{tu} = U \mathscr{P} L / C \tag{9.68}$$

Note that the transfer area $\mathscr{P} L$ is that of a *single stream* and is therefore one-half of the total area in a rotary unit.

9.9.4   Special Case of $N_{tu} \gg 1$

When h is large there is little temperature difference between T_M and T_H or T_C. In this case the right-hand sides of Eqs. (9.60) and (9.61) may be equated. The result is $T_M = T_C$ is a function of not $x^* = x/L$ and θ independently, but only the combination $x^* - \theta/R_R$. Similarly, Eqs. (9.63) and (9.64) can be equated with $T_M = T_H$, a function of $x^* + \theta/R_R$.

The requirement that T_C is a function of $x^* - \theta/R_R$ and T_H is a function of $x^* + \theta/R_R$ shows that a front moves like a shock wave through the regenerator. In actuality, of course, axial conduction and the finite value of h will smooth the temperature front somewhat, but our simple model overlooks this unnecessary complication. What the simple solution shows is that a cold front advances linearly with increasing θ as the cold gas flows through the regenerator. This cold front retreats as the hot gases flow back the other way.

What is of interest to the designer is whether the cold front has sufficient time to pass all the way through the bed. This passage will occur if the cold cycle angle Φ divided by R_R exceeds unity.

$$\Phi/R_{R,min} = 1, \qquad R_{R,min} = \Phi \qquad (9.69)$$

Equation (9.69) gives the combination of the length and rate of rotation for the cold wave (or hot wave) to just pass through the regenerator. If an R_R much greater than $R_{R,min}$ is used, the bed is unnecessarily long so that there is unnecessary pressure drop, or the rate of rotation is too fast so that there is unnecessary mixing of the streams, or the elements have been made unnecessarily heavy.

9.9.5   *EXAMPLE* **A Cold Weather Mask**

Rough out a design for an Arctic face shield regenerator to relieve some of the thermal stress on the nose. Neglect mass transfer for now. Assume 0.0025 lb of air is inhaled in 2 sec. Suppose the outside air is $-40°F$ and the exhaled breath is $+90°F$. Suppose the inhaled air should be $+50°F$ into the nose.

The effectiveness required is, for balanced flow,

$$\epsilon = \frac{T_{C,o} - T_{C,i}}{T_{H,i} - T_{C,i}} = \frac{50 - (-40)}{90 - (-40)} = \frac{90}{130} = 0.692$$

For a counterflow heat exchanger with $C_{max}/C_{min} = 1$, we would need

$$\epsilon = \frac{N_{tu}}{1 + N_{tu}} = 0.692, \qquad N_{tu} \geqslant 2.24$$

For a fixed bed regenerator we use Φ/t in place of ω, where t is the half period (2 sec)—the time it takes for the wheel to pass through one stream. From Eq. (9.69) we require

$$R_R = \frac{\rho c \delta \mathscr{P} L (\Phi/t)}{\dot{m} c_p} \geqslant \Phi$$

$$\frac{\rho c \delta \mathscr{P} L}{(0.0025)(0.25)} \geqslant 1$$

$$\rho c \delta \mathscr{P} L \geqslant 6.25 \times 10^{-4}$$

If plastic sheet material is used, there is no possibility of heating more than a thickness on the order of $\sqrt{\alpha t}$ (see Chapter 2, transient heat condition). This value is an upper limit of the half thickness

$$\delta \ll (3 \times 10^{-3}[\text{ft}^2/\text{hr}]\, 2/3600[\text{hr}])^{1/2}$$

$$\delta \ll 1.3 \times 10^{-3}\,\text{ft} = 15.5\,\text{mils}$$

So use, say $2\delta = 12$ mils. For $c \doteq 0.3$, $\rho \doteq 55$ lb/ft^3

$$\mathscr{P} L \geqslant \frac{6.25 \times 10^{-4}}{(55)(0.3)(0.006/12)} = 0.075\,\text{ft}^2 = 10.8\,\text{in.}^2$$

If laminar flow is used and a spacing is selected, a value of h may be estimated from the fact that Nu based on twice the plate spacing is 7.5. If $h \simeq 10$ Btu/hr ft² °F and an N_{tu} of 2.24 is used, we find $\mathscr{P}L$ must be

$$N_{tu} = \frac{U \mathscr{P}L}{C} = \frac{(h/2) \mathscr{P}L}{C}, \qquad \mathscr{P}L = \frac{2CN_{tu}}{h}$$

$$\mathscr{P}L = 2\frac{(0.0025)(0.25)}{(2/3600)}\frac{2.24}{10} = 0.5 \text{ ft}^2$$

This value is 6.7 times as much area as is needed to make R_R equal to Φ.

It is left to the ingenuity of the reader to contrive a way of rolling up the requisite area (the size of an ordinary sheet of paper) into a convenient cartridge and connecting it to the nose. The Eskimo merely breathes through the fur of his parka. Provision should be made for quickly replacing the cartridge with a thawed one should it plug with ice. Because of the condensation and evaporation of moisture, more than the calculated amount of heat capacity would be needed, but we have chosen R_R appreciably greater than $R_{R,\min}$ so that our rough design is probably not badly in error.

9.10   SUMMARY

Equations (9.5) and (9.6) give the overall heat balances for two streams. Equation (9.25) shows how the overall convective transfer coefficient for an exchanger is determined. Equation (9.54) then defines effectiveness which is a function of N_{tu} and R_C, as shown in Eqs. (9.55 to 9.58).

Section 9.9 introduces some of the basic ideas about a counterflow regenerator. A quantity R_R is defined, Eqs. (9.62) and (9.66), and when R_R is greater than N_{tu}, it is found that the device acts as a counterflow exchanger.

The designer's problem of balancing size, weight, capital cost, and pumping power requirements has not been mentioned explicitly, but only hinted at. Some of the exercises at the end of Chapter 8, if not done previously, should be done now in order to obtain a feel for the possible trade-offs between size-weight-volume and pressure drop in heat exchanger design.

EXERCISES

1. Steam is being condensed on a tube bundle through which cold water flows. Each tube is $1\frac{1}{4}$ in. outside diameter, and U based on outside area is 40 Btu/ft² hr °F. Water flows in each tube at 0.2 lb/sec and enters at 60°F, and leaves at 160°F. If the condenser is operating at atmospheric pressure, how much steam is condensed on each tube and how long is the tube bundle?

2. A 9 ft long stage of a multistage flash vaporization desalination plant operates at 4.52 psia. The condenser tube bundle is composed of 110 1 in. outside diameter titanium tubes with a 0.035 in. thick wall, through which cooling water

flows at 100,000 gallons per hour. If the cooling water (sea water) enters the stage at 115°F, at what temperature does it exit? Take an average value of the outside heat transfer coefficient for film condensation on horizontal tubes to be 1770 Btu/ft² hr °F (to obtain $U\mathscr{P}$ you must find the inside convective coefficient, etc.). At what rate does the stage produce fresh water?

3. Saturated liquid ammonia at 5°F flows in a $\frac{3}{4}$ in. schedule 40 steel pipe which is coated with polystyrene foam insulation 2 in. thick. The line is suspended in air at 70°F and 30% relative humidity. Some data which pertain to the situation are: $h_i = 200$ Btu/hr ft² °F, $\epsilon_{\text{polystyr}} \doteq 0.85$. (The other data necessary are found in Chapter 5 and Appendix B.) What is $U\mathscr{P}$ for this line, and what is $\dot{Q}/L$ from the air to the ammonia? If water vapor diffuses through the polystyrene foam over a long period of time and condenses, how much of the insulation would ultimately be filled with ice and how much with water? What is $U\mathscr{P}$ under these circumstances?

4. A power station operating on a mercury-steam binary cycle has mercury condensing at 620°F on the outside of schedule 40 2 in. nominal diameter steel pipes. The film coefficient for condensation is 6500 Btu/ft² hr °F. Water boils inside the pipes at 558°F, where the film coefficient may be taken to be $0.475(\Delta T)^3$ Btu/ft² hr °F. How many pipes 2 ft long are required if the turbine steam consumption is 50 lb/sec?

5. A stack 200 ft high, 17 ft in inside diameter, 18 in. thick at the bottom, and 9 in. thick at the top is made of concrete. Exhaust gas at a rate of 7.2×10^5 lb/hr enters the base of the stack at a temperature of 650°F. Air at 65°F blows at 10 mph across the stack. Find the exit temperature of the stack gas.

6. An aircraft oil cooler is to be designed to reduce the oil temperature from 240°F to 195°F. The oil (SAE 50) flow rate is 12,000 lb/hr. If the overall heat transfer coefficient U is 26 Btu/ft² hr °F, find the necessary transfer area for a counter-current exchanger. Take the entering air temperature to be 100°F. Assume balanced flow $C_H = C_C$.

7. For a gas turbine, exhaust gases are used to heat the working fluid after compression and prior to combustion. In such a case it is reasonable to take $C_C = C_H$. Use L'Hospital's rule to find the effectiveness versus N_{tu} from Eq. (9.57).

8. Design a rotary regenerator for preheating air supplied to an automotive catalytic afterburner. Discuss whether an ordinary heat exchanger or a regenerator should be used in such an application.

9. Propose a finite difference scheme for the analysis of a rotary regenerator.

REFERENCES

Heat Exchangers

The literature on heat exchangers is well represented by:

Kays, W. M., and A. L. London, *Compact Heat Exchangers*, 2nd Ed. New York: McGraw-Hill, Inc., 1964.

A text and handbook which has a large number of excellent photographs is:

Fraas, A. P., and M. N. Ozisik, *Heat Exchanger Design*. New York: John Wiley, 1965.

A paper which gives a concise review of the literature on rotary regenerators is:

Dunkle, R. V., and I. L. Maclaine-Cross, "Theory and Design of Rotary Regenerators for Air Conditioning," *Mechanical and Chemical Engineering Transactions of the Institution of Engineers, Australia*, Vol. MC6, No. 1, May 1970.

Much information can be found in journals and the proceedings of symposia published by the American Institute of Chemical Engineers and the American Society of Mechanical Engineers.

CHAPTER 10

MASS EXCHANGERS

10.1 INTRODUCTION

In this final chapter we combine our knowledge of mass transfer gained in Chapters 3, 4, 5, 7, and 8 with our understanding of heat exchangers developed in Chapter 9. We consider mass transfer processes in a single flowing stream or between two such streams. For example, we might be interested in transfer of a given species from a gas mixture to a liquid solution; thus two phases would be involved. As was the case for heat exchangers, a large transfer area is desirable. Large surface areas are obtained by spraying liquid into a gas, by bubbling gas through a liquid, or by forcing both phases through a packed bed. Such processes are illustrated in Figure 10.1.

The equipment used to effect mass exchange between two streams exhibits, accordingly, a variety of geometrical arrangements. Often gravity is used to propel the liquid phase; the equipment is therefore vertical and is called a *tower* or a *column*. The type of flow is then characterized by adjectives that is, *spray* towers, *packed* columns, and *bubble* or *sieve plate* columns.

The principles used to analyze the operation of mass exchangers have already been presented. They are conservation of mass, momentum, energy, and species. Just as in Chapter 9, where we made heat balances on each stream, we make mass balances on each stream. These balances are of two types: an overall balance encompassing the entire exchanger and differential balances over infinitesimal elements within the unit.

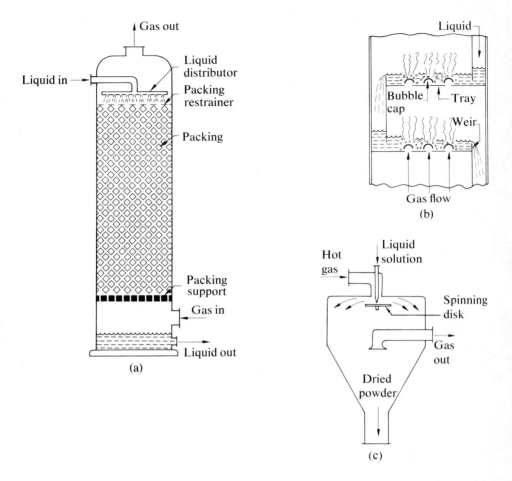

Figure 10.1   Mass exchangers: (a) packed column; (b) bubble plate column; (c) spray drier.

10.2   MASS BALANCES FOR A SINGLE STREAM

10.2.1   An Overall Balance for an Urban Air Volume

For a single flowing stream at steady-state an overall species balance is obtained simply by equating the production rate of a given species within the system to the rate of net outflow of that species from the system. The net outflow is the flow of the species in the outgoing stream minus the inflow, if any, in the entering stream:

{rate of species out} − {rate of species in} = {rate of internal production}

In this section we consider an overall mass balance for an urban air volume. We choose to do so in order that the student may appreciate the nature of the air pollution problem in urban areas. We could, of course, make similar balances for industrial processes. The principles involved and the techniques employed are the same in making a mass balance on a stream of a distillation column for separating gasoline from crude oil as in following the course of a pollutant such as sulfur dioxide (SO_2) in a power plant stack gas.

Consider an urban basin whose area is 1250 sq. mi. Suppose the basin is of such nature that the temperature distribution in the atmosphere is as depicted in Figure 10.2. Such atmospheres are said to possess an *inversion layer*. We assume this layer to be 1500 ft thick, and that the average residence time for air in the layer is one day (that is, we assume that a mass of air equal to all of the air contained in the layer flows into, mixes with, and flows out of the layer each day). The amount of air in the layer is

$$(1250 \text{ sq mi})(5280^2 \text{ ft}^2/\text{sq mi})(1500 \text{ ft}) = 5.22 \times 10^{13} \text{ ft}^3$$

or

$$(5.22 \times 10^{13} \text{ ft}^3)(0.0735 \text{ lb/ft}^3)(1/2000 \text{ tons/lb}) = 1.92 \times 10^9 \text{ tons}$$

of air at 80°F and 14.7 psia. The total mass flow rate ($\dot{m}$) through this volume is thus on the order of 1.92×10^9 tons/day. Now, if a pollutant were to enter this stream at the rate of 1920 tons/day, it would produce in the outlet stream a mass fraction of 10^{-6}. This estimate is to be compared with the values in Table 10.1, which were obtained from observations of the Los Angeles urban area. The pollutant NO_x is actually emitted largely as NO with a little NO_2. Because the NO oxidizes photochemically after entering the atmosphere, the mass flow $\dot{m}_{NO}$ is multiplied by 46/30, the ratio of molecular weights, and added to $\dot{m}_{NO_2}$. This sum is then referred to as the mass flow of pollutant NO_x, $\dot{m}_{NO_x}$.

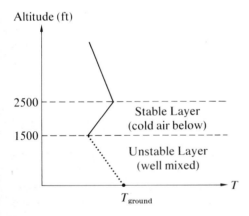

Altitude (ft)

2500 ----

Stable Layer
(cold air below)

1500 ----

Unstable Layer
(well mixed)

T_{ground}

Figure 10.2   Characteristics of an inversion layer.

Since the average of the values in the last column is somewhat larger than our estimated value of 1920 tons/day, we conclude that our assumptions with respect to residence time and/or control volume size are a little crude. This disagreement is not surprising, however, since we know that these quantities are highly dependent on local meteorological conditions. We also observe that the

Table 10.1 Mass Fractions of Pollutants in an Urban Area

Species i	Average Production tons/day	Average Mass Fraction Average of Four Worst 24 hr Periods $m_i \times 10^6$	Peak Mass Fraction Average of Four Highest Maxima $m_i \times 10^6$	Production Equivalent to $m_i = 10^{-6}$ tons/day
NO_x	1050	0.32	0.86	1050/0.32 = 3270
CO	9105	9.7	24.0	9105/9.7 = 950
SO_2	250	0.086	0.42	250/0.086 = 2910
				Average value 2380

peak concentration is roughly three times as great as the average value (compare the third and fourth columns).

These observations are seen to be significant in view of the fact that damage to plant life has been observed to occur with chronic exposure to as little as 0.03 ppm SO_2 (ppm stands for parts per million by volume, that is, a mole fraction multiplied by 10^6). For these dilute solutions one merely multiplies the mass fraction by the molecular weight ratio M_{air}/M_i to obtain the mole fraction of species i. For example, the mass fraction of 0.086×10^{-6} of SO_2 in Table 10.1 amounts to 0.039 ppm. Although the effects of various pollutants on the well-being of animal life are as yet unavailable, there is some preliminary evidence which indicates that prolonged exposure to modest concentrations of such species as CO, NO_x, SO_x, and O_3 may be deleterious to health. It is well established that *episodes* of high concentrations can kill susceptible segments of a population. For this latter reason, one air pollution control district in the United States (Los Angeles County APCD) has established what are termed *first stage alert* pollutant levels for CO (50 ppm), NO_x (3 ppm), SO_x (3 ppm), and O_3 (0.5 ppm). The *third stage alert* levels at which a serious emergency is considered to exist are 150, 10, 10, and 1.5 ppm, respectively.

10.2.2 An Overall Balance for an Oil Fired Power Plant

The major sources of air pollutants are by-products of combustion processes. These by-products arise in three principal ways: (a) from incomplete combustion (yielding CO and unburned hydrocarbons), (2) from secondary chemical reactions involving trace contaminants (such as free sulfur) in the fuel (yielding SO_2), and (3) from side reactions between N_2 and O_2 (yielding oxides of nitrogen). In the following discussion, we illustrate again the principles of overall balances, this time using a combustion process which produces SO_2 as a by-product.

In one urban area of the size considered in Section 10.2.1, the installed generating capacity is presently about 8.7×10^6 kilowatts (kw). However, with anticipated growth of the area, it is likely that the generating capacity will increase to about 20×10^6 kw. Assuming a subregion peak mass fraction of three times the average value, and using the conservative estimate that 2000 tons/day causes an average mass fraction of 10^{-6} we conclude that not more than $2000/3 = 667$ tons/day $= 28$ tons/hr of a given pollutant may be produced if peak regional mass fractions are to be held to 1×10^{-6}. We see that for each 800,000 kw power plant to be added, not more than $(28 \text{ tons/hr})(8 \times 10^5 \text{ kw}/20 \times 10^6 \text{ kw}) = 1.12$ tons/hr $= 2240$ lb/hr of SO_2 should be produced, provided that emissions from the existing plants are also reduced proportionately.

Consider an 8×10^5 kw power plant which is assumed to burn a fuel oil containing 1% free sulfur by weight. (Typically, fuel oils have free sulfur contents ranging from 0.2 to 3%). Assuming the fuel oil is a hydrocarbon with chemical formula $C_{21}H_{44}$, the main combustion reaction is

$$C_{21}H_{44} + 32O_2 \rightarrow 21CO_2 + 22H_2O$$

The equation for the overall chemical reaction contains a great deal of information. First, it shows at a glance the molecular weight of each of the major species. For example, the molecular weight of the fuel is $21 \times 12 + 44 \times 1 = 296$. Second, it defines the mole (and mass) flow rates of the incoming stream (left-hand side) and outgoing stream (right-hand side) when one flow rate is known. Knowing the heating value of the fuel, approximately 18,000 Btu/lb, and the overall efficiency of the plant, about 40 %, we can immediately find the fuel flow rate:

$$\text{power} = \text{efficiency} \times \text{heat rate}, \quad P = \eta \dot{Q}, \quad \dot{Q} = P/\eta$$

$$\text{heat rate} = \text{mass flow} \times \text{heating value}, \quad \dot{Q} = \dot{m}\,\Delta\hat{h}, \quad \dot{m} = \dot{Q}/\Delta\hat{h} = P/(\eta\,\Delta\hat{h})$$

$$\dot{m}_{HC} = \frac{(8 \times 10^5 \text{ kw})(3412 \text{ Btu/kw hr})}{(0.4)(1.8 \times 10^4 \text{ Btu/lb})} = 3.79 \times 10^5 \text{ lb/hr}$$

$$\dot{M}_{HC} = \frac{\dot{m}}{M_{HC}} = \frac{3.79 \times 10^5 \text{ lb/hr}}{296 \text{ lb/lb-mole}} = 1280 \text{ lb-mole/hr}$$

From this flow rate we can at once find the sulfur flow rate, since we have assumed that there is 0.01 lb of sulfur for each pound of hydrocarbon.

$$\dot{m}_S = (0.01 \text{ lb of S/lb of HC})(3.79 \times 10^5 \text{ lb of HC/hr}) = 3790 \text{ lb of S/hr}$$

$$\dot{M}_S = \frac{\dot{m}_S}{M_S} = \frac{(3790 \text{ lb/hr})}{(32 \text{ lb/lb-mole})} = 118.5 \text{ lb-mole/hr}$$

The chemical reaction equation tells us that for each mole of hydrocarbon 32 moles of oxygen are required for a stoichiometric mixture.

$$\dot{M}_{O_2} = (1280 \text{ lb-mole of HC/hr})(32 \text{ moles of } O_2/\text{mole HC}) = 41,000 \text{ lb-mole/hr}$$

Typically 10% excess oxygen is used to reduce CO emissions from the stack. The required flow of oxygen is then

$$\dot{M}_{O_2} = (1.1)(41,000) = 45,000 \text{ lb-mole/hr}$$

For each 20.9 moles of O_2 there are 79.1 moles of nitrogen, approximately, in air. Knowing the oxygen flow rate we can find at once the nitrogen flow rate

$$\dot{M}_{N_2} = (79.1/20.9)\dot{M}_{O_2} = (79.1/20.9)(45,000) = 170,200 \text{ lb-mole/hr}$$

The stack gas flow rates of the major constituents in the effluent may then be summarized as follows:

$$
\begin{array}{lllll}
CO_2\colon & (1280)(21) & = & 26{,}900 \text{ lb-mole/hr} & = 1.182 \times 10^6 \text{ lb/hr} \\
H_2O\colon & (1280)(22) & = & 28{,}200 \text{ lb-mole/hr} & = 0.508 \times 10^6 \text{ lb/hr} \\
N_2\colon & & = & 170{,}200 \text{ lb-mole/hr} & = 4.760 \times 10^6 \text{ lb/hr} \\
O_2\colon & (1/11)(45{,}000) = & & 4{,}090 \text{ lb-mole/hr} & = 0.130 \times 10^6 \text{ lb/hr} \\
SO_2\colon & & = & 118.5 \text{ lb-mole/hr} & = 0.00759 \times 10^6 \text{ lb/hr} \\
& \text{Totals} & & 229{,}508 \text{ lb-mole/hr} & = 6.589 \times 10^6 \text{ lb/hr} \\
& & & & (7.90 \times 10^4 \text{ tons/day})
\end{array}
$$

The total mole flow rate of gas from the stack $\dot{M}_G$ is found to be 229,508 or, to our accuracy, 2.3×10^5 lb-mole/hr. The mass flow rates are found by multiplying the mole flow rates by the molecular weights. Note that Table B.17 contains molecular weights, heats of combustion, and other useful information.

We find that our model plant produces $7.59 \times 10^3/2 \times 10^3 = 3.80$ tons/hr of SO_2. If this rate is to be reduced to 1.12 tons/hr, the mole fraction SO_2 in the stack gas must be reduced from

$$
y_{SO_2,\text{in}} = \frac{\dot{M}_{SO_2}}{\dot{M}_G} = \frac{118.5}{2.3 \times 10^5} = 5.15 \times 10^{-4}
$$

to

$$
y_{SO_2,\text{out}} = \left(\frac{1.12}{3.80}\right)(5.15 \times 10^{-4}) = 1.52 \times 10^{-4}
$$

If this reduction is to occur by absorbing SO_2 in a mass exchanger, we will have to process

$$
\dot{M}_G = 2.3 \times 10^5 \text{ lb-moles/hr}
$$

or

$$
\dot{m}_G = 6.6 \times 10^6 \text{ lb/hr}
$$

of stack gas. Such an exchanger might have to operate at elevated temperatures (800°F for the process described in Section 10.6).

In addition to the major constituents, including the major pollutant SO_2, cited above, minor pollutants such as CO and NO are formed in the combustion process. The amount of CO may range up to a few percent but can be reduced to less than 1% by using excess air and insuring satisfactory mixing of unburned fuel and air. Assuming that not more than 0.5% CO appears in the 79,200 tons/day of effluent produced in our plant, we see that the total number of plants currently operating produce

$$
(0.005)(79{,}200 \text{ tons/day})\left(\frac{8.7 \times 10^6}{8 \times 10^5}\right) = 4300 \text{ tons/day}
$$

of CO, or about two-fifths of the production rate shown in Table 10.1. Actual production rates by power plants are possibly much lower than this figure.

Oxides of nitrogen are formed when air is heated to high temperatures and then quickly cooled. The current economic situation, wherein no direct incentive exists for eliminating NO, dictates that high combustion temperatures be used to minimize gas flow rates and heat transfer surface areas. At present in our model basin, approximately 1.1 g of NO are produced for each kilowatt-hour of electrical energy which is generated. Approximately 15% of the 1050 tons/day shown in Table 10.1 has been attributed to power plants. Starting in 1975 in Los Angeles County, no more than 125 ppm of NO is to leave the stacks of large gas fired units, and no more than 225 ppm for oil or coal fired units. Good practice by utility operators has achieved these values, whereas formerly values of up to 1000 ppm were not uncommon.

10.2.3   An Overall Balance for Automobile Engines

In most urban areas of the United States, the principal source of nitric oxide is automobile engine exhaust. The problem with NO, apart from the fact that it is itself harmful, is that it reacts with O_2 (in the presence of sunlight) to produce nitrogen dioxide (NO_2), which also is harmful. Although ozone (O_3) also is produced (as an intermediate), it reacts very rapidly in the presence of excess NO to produce still more NO_2. However, when most of the available NO has been oxidized, some miles from the source, ozone builds up in concentration and produces the characteristic irritating property of photochemical smog. The precise mechanism for subsequent reactions is unknown. One view is that as ozone builds up, it oxidizes unburned hydrocarbons from automobile and industrial emissions. The ozone is regenerated by action of photons on NO_2,

$$NO_2 + photon \rightarrow NO + O$$
$$O + O_2 \rightarrow O_3$$

Some of the resulting NO then further reacts with the oxidized hydrocarbons to form particularly obnoxious molecules such as peroxyacylnitrate (PAN). These molecules have been found to be injurious to plants and are highly irritating (and presumably injurious) to man as well. The remainder of the NO is reoxidized by O_3 to NO_2, and the cycle repeats itself. When sunlight is plentiful, O_3 levels in a polluted atmosphere tend to be approximately one-half of the NO_2 levels.

Let us suppose that there are 4 million automobiles in our "model" basin. Such a population consumes about 8×10^6 gallons of gasoline (25,000 tons) per day. If carburetors are set approximately at the stoichiometric fuel-air mixture, and gasoline (taken as octane, C_8H_{18}) is burned, giving carbon dioxide (CO_2) and water (H_2O), we can write

$$C_8H_{18} + 12.5O_2 \rightarrow 8CO_2 + 9H_2O$$

We therefore have the following:

1. Fuel consumption (molecular weight $= 8 \times 12 + 18 = 114$),

    $(25,000 \text{ tons/day}) (2000 \text{ lb/ton})/(114 \text{ lb/lb-mole}) = 4.39 \times 10^5$ lb-mole/day

2. Oxygen consumption,

$$(12.5)(4.39 \times 10^5) = 5.48 \times 10^6 \text{ lb-mole/day} = 0.877 \times 10^5 \text{ tons/day}$$

3. Nitrogen throughput,

$$\left(\frac{79.1}{20.9}\right)(5.48 \times 10^6) = 2.08 \times 10^7 \text{ lb-mole/day} = 2.91 \times 10^5 \text{ tons/day}$$

4. Total exhaust flow (flow in = flow out),

$$(0.25 + 0.88 + 2.91) \times 10^5 = 4.04 \times 10^5 \text{ tons/day}$$

Actually, automobile engines run, on the average, slightly rich (1 to 5% excess fuel); although some engines run rich only during acceleration, running lean (1 to 2% excess air) during cruising. The leaner running engines reduce incomplete combustion but probably increase production of NO. There are studies under way to assess the feasibility of operating with up to 30% excess air. Such operation would significantly reduce hydrocarbons. Successful operation will require that effective heat and mass transfer rates be achieved in order that fuel and air be well mixed. Without proper mixing, ignition may not occur and the engine will run rough.

The nitric oxide mass fraction, reported as NO_2, in exhaust emissions typically ranges from 1.2×10^{-3} (large engines) to 3.5×10^{-3} (small engines), depending on operating conditions. (Although larger engines emit lower concentrations of NO and other pollutants, they nonetheless discharge more NO on a total emissions basis.) These values result in approximately 4 grams of NO_x per mile. Based on an average mass fraction of 2.0×10^{-3}, we can estimate

$$(2.0 \times 10^{-3} \text{ tons } NO_x/\text{ton emission})(4.04 \times 10^5 \text{ tons emission/day})$$
$$= 808 \text{ tons } NO_x/\text{day}$$

are discharged into our model basin from automobiles. (Table 10.1 shows that 1050 tons/day are produced in our reference basin, of which 740 tons/day has been attributed to automobiles.) Thus our estimated NO_x emissions rate would result in

$$\frac{808 \text{ tons } NO_x/\text{day}}{1 \cdot 92 \times 10^9 \text{ tons air/day}} = 0.42 \times 10^{-6} \text{ tons } NO_x/\text{ton air}$$

or about 0.26 ppm NO_x on the average. However, during peak periods we could expect concentrations in excess of 0.8 ppm NO_x. The 1976 Federal Standards, if left unaltered, call for a 90% reduction in NO_x automotive emissions.

As mentioned above, present engines run fuel rich; hence HC and CO emissions are high. Over the range of vehicle operating conditions, the CO content of the exhaust varies from 1 to 7%. Using a value of 2% for the average vehicle, we see that

$$(0.02)(4.04 \times 10^5) = 8080 \text{ tons/day}$$

of CO are produced in our model basin, which is somewhat smaller than the total of 9105 tons/day cited in Table 10.1 (of which 8945 tons/day has been attributed to automobile engines).

10.3   SINGLE-STREAM MASS EXCHANGERS

10.3.1   The Catalytic Reactor

In Section 9.3 we introduced the concept of a single-stream heat exchanger. We noted that the essential simplifying feature of the problem was that the temperature varied in only one stream. Similar single-stream situations arise in mass transfer. For example, suppose a gaseous mixture flows through a duct which is packed with irregularly shaped solid particles whose outer surfaces act as a catalyst for chemically converting a reactant species in the mixture to a product species. Such an arrangement is called a *fixed bed catalytic reactor*. As the mixture flows through the bed, the given species undergoes a heterogeneous chemical reaction at the solid surfaces and is removed from the stream. Since the species does not accumulate in an adjacent phase (as would be the case if the species were to dissolve in an adjacent stream of liquid), the analysis of this problem is rather simple. In the following section we analyze this *analog* of the simple heat exchanger problem considered in Section 9.3.

10.3.2   The Governing Differential Equation

As depicted in Figure 10.3, a single-stream mass exchanger involves flow of a single stream with continuous removal of a given chemical species along the flow path. Let $\dot{m}_G$ be the mass flow rate of the stream, m_1 the *bulk* mass fraction of the transferred species 1, $n_{1,s}$ the mass flux of species 1 at the s-surface (the transfer surface), and $\mathscr{P} \Delta x$ the area of the s-surface between positions x and $x + \Delta x$. Here, $\mathscr{P}$ is the total perimeter of the s-surface. From conservation of species 1, on neglecting axial diffusion,

$$(\dot{m}_G m_1)_x - (\dot{m}_G m_1)_{x+\Delta x} - n_{1,s}\, \mathscr{P} \Delta x|_{\bar{x}} = 0$$

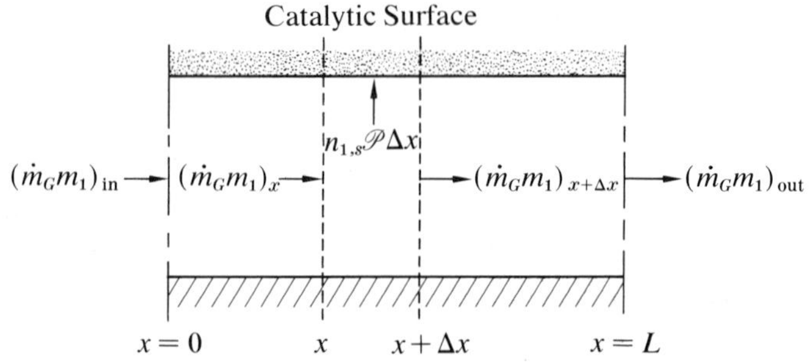

Figure 10.3   Schematic of a single-stream mass exchanger.

which becomes, on letting $\Delta x \to 0$ with $\dot{m}_G$ taken to be constant,

$$\frac{dm_1}{dx} + \frac{n_{1,s}\mathscr{P}}{\dot{m}_G} = 0 \tag{10.1}$$

Further progress requires an expression for the mass flux $n_{1,s}$. Recall Eq. (3.28) in Section 3.6.1. It may be rewritten

$$j_{1,s} = n_{1,s} = K_{oa}m_1 \tag{10.2}$$

where K_{oa}, the overall mass transfer coefficient, is given by $1/K_{oa} = 1/\rho k'' + \delta/\rho\mathscr{D}$. Equation (3.87), when converted to mass units, is also of this form with $1/K_{oa} = 1/(V_p a/S_p)\eta k''\rho + 1/K$. In either case Eq. (10.1) becomes

$$\frac{dm_1}{dx} + \frac{K_{oa}\mathscr{P}}{\dot{m}_G} m_1 = 0 \tag{10.3}$$

10.3.3 Solutions — Number of Transfer Units and Effectiveness Factor

Given the mass fraction of species 1 in the incoming stream, $m_{1,\text{in}}$, Eq. (10.3) may be integrated to obtain

$$\frac{m_1}{m_{1,\text{in}}} = e^{-(K_{oa}\mathscr{P}L/\dot{m}_G)(x/L)} = e^{-N_{tu}x/L} \tag{10.4}$$

where we introduce the symbol N_{tu} (number of transfer units) to denote the dimensionless grouping $K_{oa}\mathscr{P}L/\dot{m}_G$. From a design viewpoint, the mass fraction in the outgoing stream, $m_{1,\text{out}}$, is of interest. For $x = L$, Eq. (10.4) gives the ratio $m_{1,\text{out}}/m_{1,\text{in}}$. If the value of $m_{1,\text{out}}$ were reduced to zero, the unit would be viewed as being 100% effective. For this reason, the *effectiveness* is defined to be

$$\epsilon = 1 - \frac{m_{1,\text{out}}}{m_{1,\text{in}}} = 1 - e^{-N_{tu}} \tag{10.5}$$

Note that this equation and its derivation are entirely parallel to the results of Section 9.3.

10.3.4 *EXAMPLE* Catalytic Removal of CO from Automobile Exhaust

Suppose we wish to design a 99% effective fixed bed catalytic converter for removing CO from automobile engine exhaust. A suitable catalyst for the converter is cupric oxide (CuO). We might examine the feasibility of using a packed bed of copper foil or copper clad metal wound in a spiral or stacked in wafer form with a plate spacing of $d = 0.060$ in. To keep the pressure drop low, we would make the cross-sectional area high enough to maintain laminar flow.

In applying Eq. (10.5), we need to calculate K_{oa}. From Eq. (10.2) and the text immediately following it, we write

$$K_{oa} = \frac{j_1}{m_1} = \frac{1}{1/(\rho k'') + \delta/(\rho \mathcal{D}_{1m})} = \frac{1}{1/(\rho k'') + 1/K_G}$$

where δ is the thickness of an equivalent *stagnant film* for the convective situation involved. In Chapter 5 we saw that an alternative, and more convenient, procedure for obtaining K_G was to employ the appropriate correlation for the mass transfer Nusselt number. We have the relation

$$K_G = Nu_{D_h} \frac{\rho \mathcal{D}_{1m}}{D_h}$$

where, in general, Nu_{D_h} is a function of the Reynolds and Schmidt numbers. However, as shown in Table 5.1, the Nusselt number for fully developed ($L \to \infty$) laminar flow with uniform boundary conditions at the wall ($m_{1,s} =$ constant) is a pure number,

$$Nu_{D_h} = 7.54$$

Now the hydraulic diameter D_h is just

$$D_h = \frac{4A_c}{\mathcal{P}} = \frac{4 \cdot d \cdot W}{2 \cdot W} = 2d$$

where d is the foil spacing and W is the perimeter of one side of the foil. Thus

$$K_G = 7.54 \, \rho \mathcal{D}_{1m}/2d$$

To evaluate K_G we use properties of air at 1500°F and 1 atm. From Appendix B, Table B.5, we have $\rho = 0.0203$ lb/ft³, and from Table B.12b we have $\mathcal{D}_{CO,air} = 6.89$ ft²/hr:

$$K_G = \frac{(7.54)(0.0203 \text{ lb/ft}^3)(6.89 \text{ ft}^2/\text{hr})}{(2)(0.06 \text{ in.})(1/12 \text{ ft/in.})} = 105.5 \text{ lb/ft}^2 \text{ hr}$$

From the Arrhenius relation in Example 3.11.3 we estimate k'' to be 0.93 cm/sec $= 110$ ft/hr. The value of K_{oa} is then

$$K_{oa} = \frac{1}{1/(110)(0.0203) + 1/(105.5)} = 2.19 \text{ lb/ft}^2 \text{ hr}$$

Since K_{oa} is seen to be reaction rate limited, one should explore using porous as well as solid surfaces.

From the value of the effectiveness, $\epsilon = 0.99$, we find

$$N_{tu} = K_{oa} \mathcal{P} L/\dot{m}_G = \ln(1/(1-\epsilon)) = \ln 100 = 4.61$$

Now consider a compact-sized automobile traveling 60 mph and getting 25 mi/gallon. Assuming an air to fuel ratio of 14 lb air/lb fuel,

$$\dot{m}_G = \frac{(60 \text{ mi/hr})}{(25 \text{ mi/gallon})}(0.1337 \text{ ft}^3/\text{gallon})(44 \text{ lb/ft}^3)(15 \text{ lb fuel and air/lb fuel})$$

$$= 212 \text{ lb/hr}$$

But to operate properly, the catalytic converter requires additional oxygen. Assuming 10% excess air is added to the exhaust from the engine gives

$$\dot{m}_G = 1.1\,(212) = 233\ \text{lb/hr}$$

Thus

$$\mathscr{P}L = (N_{tu})\left(\frac{\dot{m}_G}{K_{oa}}\right) = \frac{(4.61)\,(233\ \text{lb/hr})}{(2.19\ \text{lb/ft}^2\ \text{hr})} = 490\ \text{ft}^2$$

It is left to the student to consider how the 490 ft² of copper oxide-coated foil spaced 0.06 in. apart may be packaged and how the desired temperature level could be maintained. In an actual design several alternatives would have to be considered and an optimal design selected.

10.4   INTERPHASE MASS TRANSFER

10.4.1   Transfer between a Gas and a Liquid Stream

In Section 10.3 we introduced our treatment of mass exchangers by considering flow of a single stream with prescribed boundary conditions at the contact surface. Thus the problem was decoupled from the transfer problem in an adjacent phase. Usually, however, engineers are concerned with mass exchangers in which the mass transfer problem must be solved in two adjacent phases, both of which are moving. Although in principle such problems could be solved using the techniques of Chapters 3 and 4, the engineering situation, more often than not, involves complexities in apparatus and/or mechanics of two-phase flow which make rigorous solutions impractical. For these reasons it is customary to express rates of mass transfer across the phase boundaries in terms of overall driving forces, much as was done for heat transfer when dealing with heat exchangers (see Chapter 9). For simplicity in developing this approach for mass exchangers, we consider here only transfer of a single chemical species from a gas phase to a liquid phase.

10.4.2   Interfacial Behavior

Consider the physical situation depicted in Figure 10.4 where a chemical species A, in a gaseous mixture of A and nontransferred species B, is being absorbed in a nonvolatile liquid C (an example of such a system would be the NH_3, N_2, H_2O system at low temperatures). The actual concentration profiles are indicated by solid lines. However, for convenience in what is to follow we assume that the bulk phases are at the uniform concentrations (that is, mole fractions) y_A and x_A as indicated by the dotted lines, the resistances to mass transfer being assigned to narrow regions on either side of the interface — our view of a boundary layer as presented in Chapter 5. Although the phases are not in equilibrium, we may safely assume that the interface presents negligible resistance to mass transfer and locally is in thermodynamic equilibrium. From the principles of thermo-

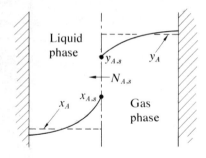

Liquid
phase
$y_{A,s}$ y_A
$-N_{A,s}$
x_A $x_{A,s}$
Gas
phase

Figure 10.4   Concentration distributions of transferred species A in the neighborhood of a gas-liquid interface.

dynamics, the number of thermodynamic variables which may be specified in an *a priori* manner (the *degrees of freedom* or F) is just

$$F = P + C - 2 \qquad \text{(10.6)}$$

where P is the number of phases and C the number of distinguishable chemical species. For the situation in Figure 10.4, therefore,

$$F = 2 + 3 - 2 = 3$$

Since it is usual practice to carry out a mass transfer process at specified temperature and pressure, F effectively equals one. Thus if we specify either $x_{A,s}$ or $y_{A,s}$ (the interfacial concentrations), the other is known. These observations are embodied in the equilibrium relation

$$y_{A,s} = f(x_{A,s}) \qquad (P, T \text{ specified}) \qquad \text{(10.7)}$$

In many practical situations, species A is only sparingly soluble in solvent C and Eq. (10.7) assumes the form, on assuming ideal gas behavior,

$$y_{A,s} = He\, x_{A,s} \qquad \text{(10.8)}$$

where He is called the **Henry number** and is defined as

$$He \equiv \frac{C_{He}(T)}{P} \qquad \text{(10.9)}$$

The quantity $C_{He}(T)$ is called **Henry's constant** and can be seen to be related to the vapor pressure of species A, $P_{A,s}$, and $x_{A,s}$ by

$$C_{He}(T) = \frac{P_{A,s}(T)}{x_{A,s}} \qquad \text{(10.10)}$$

Actually, Henry's constant is found to vary with P and $x_{A,s}$; however, for dilute solutions the variation is small. For our purposes here, we shall assume it to be independent of total pressure and the liquid side composition. Values of Henry's constant for selected liquid solutions are tabulated in Table B.18. (For a more thorough treatment of phase equilibria, the reader is referred to textbooks on chemical thermodynamics.) As a final point, we emphasize that in contrast to heat transfer, where the temperatures in adjacent phases are equal at the interface, the mole fractions $y_{A,s}$ and $x_{A,s}$ are *not* equal.

10.4.3 *EXAMPLE* Vapor-Liquid Equilibria for the N_2-NH_3-H_2O System

A gaseous mixture of ammonia (NH_3) and nitrogen (N_2) is in equilibrium with an aqueous solution of NH_3 at 80°F. Given that Henry's constant is 30.2 atm, compute the Henry number and the composition of the liquid phase when $P = 0.1, 1.0,$ and 10.0 atm and $y_{NH_3} = 0.302$.

From Eq. (10.9)

$$He = \frac{C_{He}(80°)}{P} = \frac{30.2 \text{ atm}}{P(\text{atm})}$$

and from Eq. (10.8)

$$x_{NH3} = \frac{y_{NH_3}}{He} = \frac{0.302}{He}$$

(Since the phases are in equilibrium, $x_{A,s} = x_A$ and $y_{A,s} = y_A$.) The results are tabulated as follows:

P(atm)	0.1	1.0	10.0
He	302	30.2	3.02
x_{NH_3}	0.001	0.01	0.1

Thus, for a given temperature, x_{NH_3} is seen to increase with increasing pressure; that is, the equilibrium is said to shift toward the liquid phase with increasing pressure. (Advantage of this feature, which is characteristic of most sparingly soluble gases, is taken when carbonating soft drinks.)

10.4.4 Overall Mass Transfer Coefficients

As suggested in Section 10.4.1, we choose to represent rates of mass transfer between adjacent phases in terms of the driving forces $(y_A - y_{A,s})$ and $(x_{A,s} - x_A)$. Thus as outlined in Section 3.10.1 of Chapter 3 we take

$$N_{A,s} = K_G(y_A - y_{A,s}) \equiv K_L(x_{A,s} - x_A) \tag{10.11}$$

where we now introduce the subscripts G and L on the mass transfer coefficient K to distinguish between transfer in each phase. Units for the quantities $N_{A,s}$ and the K's are lb-moles per unit area per unit time. A more useful form for $N_{A,s}$ results on eliminating $y_{A,s}$ (or $x_{A,s}$) from Eq. (10.11) by introducing the equilibrium expression, Eq. (10.8):

$$N_{A,s} = K_G(y_A - He\, x_{A,s}) = K_G\left(y_A - \frac{He\, N_{A,s}}{K_L} - He\, x_A\right)$$

Thus

$$N_{A,s} = \frac{1}{\dfrac{1}{K_G} + \dfrac{He}{K_L}}(y_A - He\, x_A) = K_{Goa}(y_A - He\, x_A) \tag{10.12}$$

where

$$K_{Goa} = \frac{1}{\dfrac{1}{K_G} + \dfrac{He}{K_L}} \tag{10.13}$$

is called the overall mass transfer coefficient for a gas side driving force ($y_A -$ $He\, x_A$). (Note that the product $He\, x_A$ may be viewed as a pseudo gas phase mole fraction; it is the gas phase mole fraction which would exist at the interface if the liquid phase were of uniform concentration x_A.)

The task remains of developing procedures for calculating the mass transfer coefficients K_G and K_L. As discussed in Chapters 3 and 5 this usually is done by means of correlations which involve the Nusselt number for mass transfer,

$$Nu_m = \frac{KL}{c\mathscr{D}_A} = \frac{KL \cdot Sc}{cv}$$

and the Reynolds (Re) and Schmidt (Sc) numbers. For example, consider film flow of a liquid down the interior surface of a vertical tube with countercurrent flow of a gas up the center. If the interphase mass transfer rate $N_{A,s}$ is low, we may, as discussed in Section 5.5.3 of Chapter 5, assume a strict analogy between heat and mass transfer and write, for turbulent flow of the gas phase,

$$\frac{K_G D}{cv} Sc = 0.023\, Re_D^{0.8} Sc^{1/3}$$

that is,

$$K_G Sc^{2/3} = \left(\frac{0.023\, cv}{D}\right)(Re_D)^{0.8} \tag{10.14}$$

where D is the diameter of the tube. Equation (10.14) applies equally well when the cross section open to gas flow is noncircular, in which event we merely replace D by the hydraulic diameter $D_h = 4A_c/\mathscr{P}$, where A_c is the cross-sectional area and $\mathscr{P}$ is the total interfacial perimeter.

For the liquid film, the problem is somewhat less straightforward. We might, as discussed in Section 4.7 of Chapter 4, use penetration theory which results in an expression for the average value of K_L over a film of height H,

$$K_L = 2c\left(\frac{\mathscr{D}_A v_0}{\pi H}\right)^{1/2} \tag{10.15}$$

where v_0 is the surface velocity of the falling film. Equation (10.15) may be rearranged to obtain

$$K_L Sc^{1/2} = 1.128c(vv_0/H)^{1/2} \tag{10.16}$$

For laminar flow, v_0 is just

$$v_0 = g\delta^2/2v; \qquad \delta = (3v\Gamma/\rho g)^{1/3} \tag{10.17}$$

as developed in Section 4.6 of Chapter 4. For turbulent flow the student should work Exercise 9 at the end of Chapter 8.

Alternative correlations have been developed for a great variety of other flow patterns such as (1) film flow of liquid over the surfaces of irregularly shaped solid particles in a packed column with gas flow through the interstitial spaces or (2) flow of a gas stream through droplets of liquid in a spray tower. Textbooks by Spalding, Treybal, Levich, and Sherwood and Pigford review and tabulate some of the more useful expressions. *Compact Heat Exchangers* by Kays and London also contains useful data.

10.4.5 *EXAMPLE* Controlling Resistances for Mass Transfer– Gas Absorption in a Falling Film Column

A gaseous mixture of nitrogen and a trace amount of an inert chemical species A flows counter-current to a falling film of water in a vertical tube. The velocity of the gas is 1.0 ft/sec, the diameter of the tube is 1.0 ft, the average thickness of the liquid film is 0.012 in., and the column is 10 ft high. The column operates at a temperature of 80°F and a pressure of 1 atm. The water enters saturated with N_2 but contains no species A. Determine the relative resistances to mass transfer in each phase ($1/K_G$ and He/K_L), and the overall mass transfer coefficient K_{Goa} when the trace gas is (1) ammonia (NH_3), (2) chlorine (Cl_2), and (3) carbon dioxide (CO_2). The Henry numbers for these gases may be taken as (1) 30.2, (2) 620, and (3) 1710, respectively.

Consider first the gas phase, assuming its physical properties to be those of pure N_2 at 80°F and 1 atm:

$$\rho = 0.0713 \text{ lb/ft}^3, \qquad \nu = 16.8 \times 10^{-5} \text{ ft}^2/\text{sec}$$

$$c = \frac{\rho}{M} = \frac{(0.0713 \text{ lb/ft}^3)}{(28 \text{ lb/lb-mole})} = 2.54 \times 10^{-3} \text{ lb-mole/ft}^3$$

Thus the Reynolds number based on column diameter is

$$Re_D = \frac{(1.0 \text{ ft})(1.0 \text{ ft/sec})}{(16.8 \times 10^{-5} \text{ ft}^2/\text{sec})} = 5950$$

and we see that the flow is just turbulent. For low Re turbulent flows Eq. (10.14) is none too accurate, but we use it for an estimate.

$$K_G Sc^{2/3} = 0.023 \, c\nu Re_D^{0.8}/D$$

$$= \frac{(0.023)(2.54 \times 10^{-3} \text{ lb-mole/ft}^3)(16.8 \times 10^{-5} \text{ ft}^2/\text{sec})(5950)^{0.8}}{(1.0 \text{ ft})}$$

$$= 1.027 \times 10^{-5} \text{ lb-moles/ft}^2 \text{ sec}$$

$$= 0.0370 \text{ lb-moles/ft}^2 \text{ hr}$$

For the liquid phase, assuming pure water at 80°F,

$$\rho = 62.2 \text{ lb/ft}^3, \qquad \nu = 0.930 \times 10^{-5} \text{ ft}^2/\text{sec}$$

$$c = \frac{\rho}{M} = \frac{(62.2 \text{ lb/ft}^3)}{(18 \text{ lb/lb-mole})} = 3.46 \text{ lb-mole/ft}^3$$

The film surface velocity is

$$v_0 = \frac{g\delta^2}{2\nu} = \frac{(32.2 \text{ ft/sec}^2)(10^{-6} \text{ ft}^2)}{(2)(0.930 \times 10^{-5} \text{ ft}^2/\text{sec})} = 1.73 \text{ ft/sec}$$

and, hence, from Eq. (10.16)

$$K_L Sc^{1/2} \simeq 1.1 c (v_0 \nu/H)^{1/2}$$

$$= (1.1)(3.46 \text{ lb-mole/ft}^3) \left[\frac{(1.73 \text{ ft/sec})(0.930 \times 10^{-5} \text{ ft}^2/\text{sec})}{(10 \text{ ft})} \right]^{1/2}$$

$$= 4.83 \times 10^{-3} \text{ lb-moles/ft}^2 \text{ sec} = 17.35 \text{ lb-moles/ft}^2 \text{ hr}$$

Summarizing the results, we have the values in Table 10.2.

Table 10.2   Summary of Results for Example 10.4.5

Derived Properties	Chemical Species		
	NH_3	Cl_2	CO_2
Sc (gas, N_2 taken as air)	0.61	1.42	1.0
$K_G = 0.0370/Sc^{2/3}$, lb-moles/ft² hr	0.0514	0.0293	0.037
Sc (liquid)	409.0	592.0	402.0
$K_L = 17.35/Sc^{1/2}$, lb-moles/ft² hr	0.858	0.713	0.865
He	30.2	620.0	1710.0
Gas side resistance, $1/K_G$	19.45	34.1	27.0
Liquid side resistance, He/K_L	35.2	870.0	1977.0
Total resistance, $1/K_G + He/K_L$	54.7	904.0	2004.0
$K_{Goa} = 1/(1/K_G + He/K_L)$, lb-moles/ft² hr	1.83×10^{-2}	1.11×10^{-3}	4.99×10^{-4}

We conclude from the above results that the liquid side resistance controls the rates of absorption of Cl_2 and CO_2, but that neither resistance controls for NH_3. The liquid side resistance could be reduced, for example, by increasing the velocity v_0. However, quite the opposite is the case, as will be shown shortly, when the absorbed species reacts with the solvent. When reactions occur in the liquid phase, the gas side resistance usually controls.

10.4.6   *EXAMPLE*   Overall Mass Transfer Coefficients for Packed Columns

Since the overall rate of mass transfer in a mass exchanger increases with interfacial area, it is common practice to *break up* the flow of the incoming streams by packing the column with, for example, irregularly shaped solid particles. Typically, these particles range in size from about ½ to 3 in., are made of chemical stoneware, and are loaded into the column in a random fashion. Some common types are shown in Figure 10.5.

Alternatively, regular packings are used. These packings result in lowered pressure drops for a given throughput of fluid, usually at the expense of increased installation cost. An example of such a packing is illustrated in Figure

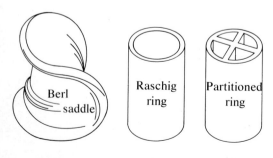

Berl saddle   Raschig ring   Partitioned ring

Figure 10.5   Column packings.

10.6. The column itself is usually circular in cross section, contains an open space at the bottom beneath the packing support to ensure good distribution of the incoming gas stream, and provides for uniform distribution of incoming liquid at the top of the packing. Typically, column sizes range from 1 to 50 ft in diameter and from five to hundreds of feet in height. From a fluid flow standpoint, problems arise with respect to flooding (gas stream flow rate so high that liquid down flow is reversed because of interfacial drag), channeling (nonuniform down flow of liquid), entrainment (droplets of liquid carried off at top in a mist), and prediction of pressure drops. Here, however, we are interested in predicting mass transfer coefficients.

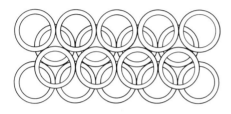

Figure 10.6   Raschig rings, staggered.

Consider counter-current flow of aqueous ammonia and a gaseous mixture of nitrogen and trace amounts of ammonia in a packed column. The packing is 1.5 in. Berl saddles whose shape is such that the diameter d_s of a sphere having the same surface area is 0.155 ft. The flow rates of gas and liquid are 2800 lb/hr ($\dot{m}_G$) and 15,000 lb/hr ($\dot{m}_L$), respectively. The effective operating void space in the packing, $\epsilon_v = A_c/(\pi D^2/4)$, may be taken as 0.7, while the geometric area of the column is $\pi D^2/4 = 10$ ft². The individual mass transfer coefficients are correlated as

$$K_L = 25.1 \frac{cv}{d_s} \left(\frac{d_s \dot{m}_L}{\rho v A} \right)^{0.45} (Sc)^{-1/2}$$

$$K_G = 1.195 \frac{\dot{m}_G}{AM} \left(\frac{d_s \dot{m}_G}{\rho v (1 - \epsilon_v) A} \right)^{-0.36} (Sc)^{-2/3}$$

Taking data from Section 10.4.5, with $T = 80°F$, there results for the liquid phase:

$$\rho = 62.2 \text{ lb/ft}^3, \nu = 0.930 \times 10^{-5} \text{ ft}^2/\text{sec} = 0.0335 \text{ ft}^2/\text{hr}, Sc = 409$$
$$c = 3.46 \text{ lb-mole/ft}^3$$

$$\frac{cv}{d_s} = \frac{(3.46 \text{ lb-mole/ft}^3)(0.0335 \text{ ft}^2/\text{hr})}{(0.155 \text{ ft})} = 0.748 \text{ lb-mole/ft}^2 \text{ hr}$$

$$K_L = (25.1)(0.748 \text{ lb-mole/ft}^2 \text{ hr}) \left[\frac{(0.155 \text{ ft})(15,000 \text{ lb/hr})}{(62.2 \text{ lb/ft}^3)(0.0335 \text{ ft}^2/\text{hr})(10 \text{ ft}^2)} \right]^{0.45} (409)^{-1/2}$$
$$= 7.75 \text{ lb-mole/ft}^2 \text{ hr}$$

For the gas phase

$$\rho = 0.0713 \text{ lb/ft}^3, \quad \nu = 16.8 \times 10^{-5} \text{ ft}^2/\text{sec} = 0.605 \text{ ft}^2/\text{hr}, \quad Sc = 0.61$$

$$\frac{\dot{m}_G}{AM} = \frac{2800 \text{ lb/hr}}{(10 \text{ ft}^2)(28 \text{ lb/lb-mole})} = 10 \text{ lb-mole/ft}^2 \text{ hr}$$

$$K_G = (1.195)\,(10\,\text{lb-mole/ft}^2\,\text{hr})\left[\frac{(0.155\,\text{ft})\,(2800/10\,\text{lb/ft}^2\,\text{hr})}{(0.0713\,\text{lb/ft}^3)\,(0.605\,\text{ft}^2/\text{hr})\,(1-0.7)}\right]^{-0.36}$$

$$\times\,(0.61)^{-2/3} = 0.894\,\text{lb-mole/ft}^2\,\text{hr}$$

For a Henry number of 30.2

$$K_{Goa} = \frac{1}{\dfrac{1}{K_G}+\dfrac{He}{K_L}} = \frac{1}{\dfrac{1}{0.894}+\dfrac{30.2}{7.75}} = 0.199\,\text{lb-moles/ft}^2\,\text{hr}$$

10.5　COUNTER-CURRENT MASS EXCHANGERS

10.5.1　The Governing Differential Equation

As depicted in Figure 10.7, a counter-current mass exchanger has streams flowing in opposite directions. In many applications the liquid stream flows down under the influence of gravity with counter-current flow of the gas stream. For visual convenience, we separate the gas and liquid flows into two distinct streams. However, it should be kept in mind that the usual engineering situation involves multiple paths for the flow of each phase in order to increase the contact area $\mathscr{P}\Delta z$ and hence the mass transfer rate $N_{A,s}\,\mathscr{P}\Delta z$. The overall liquid and gas flow rates are denoted, respectively, by $\dot{M}_L$ and $\dot{M}_G$ (both taken positive) with units moles per unit time.

For an element of length Δz, the species balance for the gas stream between planes at z and $z+\Delta z$ gives

$$(\dot{M}_G y_A)_z = (\dot{M}_G y_A)_{z+\Delta z} + (N_{A,s}\,\mathscr{P}\Delta z)_{\bar{z}}$$

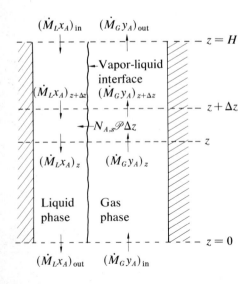

Figure 10.7　Schematic of a counter-current mass exchanger.

Assuming $\dot{M}_G$ constant and introducing for $N_{A,s}$ the expression defined in Eq. (10.12), there results

$$\frac{dy_A}{dz} = -\frac{K_{Goa}\mathscr{P}}{\dot{M}_G}(y_A - He\,x_A) \tag{10.18}$$

For gas absorption, $y_A > He\,x_A$ and y_A is seen to decrease with increasing z. A similar balance for the liquid stream gives

$$\frac{dx_A}{dz} = -\frac{K_{Goa}\mathscr{P}}{\dot{M}_L}(y_A - He\,x_A)$$

Likewise, x_A decreases with increasing z.

Since Eq. (10.18) contains the grouping $He\,x_A$, both sides of the above equation are multiplied by He, and the quantity $\dot{M}_L/He$ is written $\dot{M}_L{}^\dagger$. There results

$$\frac{d(He\,x_A)}{dz} = -\frac{K_{Goa}\mathscr{P}}{\dot{M}_L{}^\dagger}(y_A - He\,x_A) \tag{10.19}$$

Subtracting Eq. (10.19) from Eq. (10.18) then gives

$$\frac{d}{dz}(y_A - He\,x_A) = -K_{Goa}\mathscr{P}\left(\frac{1}{\dot{M}_G} - \frac{1}{\dot{M}_L{}^\dagger}\right)(y_A - He\,x_A) \tag{10.20}$$

Equations (10.18) to (10.20) are equivalent forms of the governing differential equation for local mass exchange in the column. Solutions to these equations and interpretations of the results follow. Note the complete parallelism between these equations and those of Section 9.5.1.

10.5.2   Boundary Conditions and Overall Balance Constraints

In addition to the governing differential equation, analysis of a counter-flow mass exchanger involves boundary conditions (that is, compositions of entering and leaving streams) and an overall balance. The boundary conditions are

$$\text{at } z = 0: \quad y_A(0) = y_{A,\text{in}}, \quad x_A(0) = x_{A,\text{out}}$$

$$\text{at } z = H: \quad y_A(H) = y_{A,\text{out}}, \quad x_A(H) = x_{A,\text{in}} \tag{10.21}$$

The overall balance is

$$(\dot{M}_G y_A)_{\text{in}} - (\dot{M}_L x_A)_{\text{out}} = (\dot{M}_G y_A)_{\text{out}} - (\dot{M}_L x_A)_{\text{in}}$$
$$= (\dot{M}_G y_A) - (\dot{M}_L x_A) \tag{10.22}$$

as is evident by inspecting Figure 10.7.

Equation (10.22) and any one of the Eqs. (10.18) to (10.20) constitute two equations in essentially eight unknowns: $\dot{M}_L$, $\dot{M}_G$, H, $\mathscr{P}$, $x_{A,\text{in}}$, $x_{A,\text{out}}$, $y_{A,\text{in}}$, and $y_{A,\text{out}}$. (The quantities K_{Goa} and He are not regarded as unknowns since they implicitly are functions of the other variables as well as temperature and pressure, which are assumed to be known.) Thus a total of six unknowns must be specified in advance (eight including temperature and pressure) in order that a given problem may be solved. As will be seen shortly, particular choices will have a bearing on the method of solution.

10.5.3   Effectiveness Factors and NTU Concepts

Frequently engineers are confronted with a mass transfer problem for which equipment is available whose geometry (essentially $\mathscr{P}$ and H) is already fixed. Thus we would be interested in the effectiveness of the equipment in removing a given chemical species from, say, a gaseous stream by absorbing it in a liquid stream. In such an event we ordinarily would know (1) the rate at which gas must be processed ($\dot{M}_G$), (2) the inlet composition of the gas ($y_{A,\text{in}}$), (3) a reasonable estimate of the rate at which liquid may be charged to the equipment ($\dot{M}_L$), and (4) the inlet composition of the liquid ($x_{A,\text{in}}$). So, how do we determine how well the equipment will perform? In answering this question, it will prove instructive to integrate Eq. (10.20):

$$\int_{(y_{A,\text{in}} - He\,x_{A,\text{out}})}^{(y_{A,\text{out}} - He\,x_{A,\text{in}})} d\ln(y_A - He\,x_A) = -\left(\frac{1}{\dot{M}_G} - \frac{1}{\dot{M}_L^{\dagger}}\right)(K_{Goa}\mathscr{P}H) \tag{10.23}$$

That is,

$$y_{A,\text{out}} = He\,x_{A,\text{in}} + (y_{A,\text{in}} - He\,x_{A,\text{out}})\exp\left[-\left(\frac{1}{\dot{M}_G} - \frac{1}{\dot{M}_L^{\dagger}}\right)(K_{Goa}\mathscr{P}H)\right] \tag{10.24}$$

Now, from the viewpoint of effectiveness, we would like to establish an expression for $\dot{M}_G(y_{A,\text{in}} - y_{A,\text{out}})$ or equivalently for $\dot{M}_L(x_{A,\text{out}} - x_{A,\text{in}})$. Let $\gamma = (1/\dot{M}_G - 1/\dot{M}_L^{\dagger}) \times (K_{Goa}\mathscr{P}H)$. Then,

$$(y_{A,\text{in}} - y_{A,\text{out}}) = y_{A,\text{in}} - [He\,x_{A,\text{in}} + (y_{A,\text{in}} - He\,x_{A,\text{out}})e^{-\gamma}]$$

$$= (y_{A,\text{in}} - He\,x_{A,\text{in}}) - (y_{A,\text{in}} - He\,x_{A,\text{out}})e^{-\gamma}$$

$$= (y_{A,\text{in}} - He\,x_{A,\text{in}}) - \left[y_{A,\text{in}} - He\,x_{A,\text{in}} - \frac{\dot{M}_G}{\dot{M}_L^{\dagger}}(y_{A,\text{in}} - y_{A,\text{out}})\right]e^{-\gamma}$$

$$= (y_{A,\text{in}} - He\,x_{A,\text{in}})(1 - e^{-\gamma}) + (y_{A,\text{in}} - y_{A,\text{out}})\frac{\dot{M}_G}{\dot{M}_L^{\dagger}}e^{-\gamma}$$

where we have eliminated $x_{A,\text{out}}$ by means of the overall balance, Eq. (10.22). Solving for $(y_{A,\text{in}} - y_{A,\text{out}})$ and dividing the result by $(y_{A,\text{in}} - He\,x_{A,\text{in}})$ gives

$$\frac{y_{A,\text{in}} - y_{A,\text{out}}}{y_{A,\text{in}} - He\,x_{A,\text{in}}} = \frac{1 - e^{-\gamma}}{1 - \frac{\dot{M}_G}{\dot{M}_L^{\dagger}}e^{-\gamma}} \tag{10.25}$$

Now suppose that $\gamma > 0$ (that is, $1/\dot{M}_G > 1/\dot{M}_L^{\dagger}$). In such an event, Eq. (10.25) is seen to exhibit the following limiting behavior: as $\gamma \to \infty$ (that is, as $K_{Goa}\mathscr{P}H \to \infty$), the right-hand side is seen to approach 1. From this result we see that $y_{A,\text{out}} \to He\,x_{A,\text{in}}$; that is, the bulk streams at the top of the column are in thermodynamic equilibrium. Thus the left side of Eq. (10.25) may be interpreted as the ratio of the rate at which species A is actually removed from the gas stream by the exchanger, $\dot{M}_G(y_{A,\text{in}} - y_{A,\text{out}})$, to the maximum possible rate, $\dot{M}_G(y_{A,\text{in}} - He\, x_{A,\text{in}})$. This ratio is called the effectiveness factor, ϵ_{cf}, for a counter-current mass exchanger,

$$\epsilon_{cf} \equiv \frac{\dot{M}_G(y_{A,\text{in}} - y_{A,\text{out}})}{\dot{M}_G(y_{A,\text{in}} - y_{A,\text{out}})_{\text{max}}} = \frac{\dot{M}_G(y_{A,\text{in}} - y_{A,\text{out}})}{\dot{M}_G(y_{A,\text{in}} - He\, x_{A,\text{in}})} \tag{10.26}$$

or from Eq. (10.25)

$$\epsilon_{cf} = \frac{1 - e^{-\gamma}}{1 - \dfrac{\dot{M}_G}{\dot{M}_L{}^\dagger} e^{-\gamma}} \tag{10.27}$$

When $\gamma < 0$ $(1/\dot{M}_G < 1/\dot{M}_L{}^\dagger)$, we can deduce from Eq. (10.24) that $y_{A,\text{in}} \to He\, x_{A,\text{out}}$ as $|\gamma| \to \infty$. In such an event, it is more useful to express ϵ_{cf} in the form

$$\epsilon_{cf} = \frac{\dot{M}_G(y_{A,\text{in}} - y_{A,\text{out}})}{\dot{M}_L(x_{A,\text{out}} - x_{A,\text{in}})_{\text{max}}}$$

$$= \frac{\dot{M}_G(y_{A,\text{in}} - y_{A,\text{out}})}{\dfrac{\dot{M}_L}{He}(y_{A,\text{in}} - He\, x_{A,\text{in}})}$$

$$= \frac{\dot{M}_G}{\dot{M}_L{}^\dagger} \frac{(y_{A,\text{in}} - y_{A,\text{out}})}{(y_{A,\text{in}} - He\, x_{A,\text{in}})} \tag{10.28}$$

from which it may easily be shown that

$$\epsilon_{cf} = \frac{1 - e^{\gamma}}{1 - \dfrac{\dot{M}_L{}^\dagger}{\dot{M}_G} e^{\gamma}} \tag{10.29}$$

The details are left as an exercise. Actually, Eqs. (10.26) to (10.29) may be written in the more compact forms

● $$\epsilon_{cf} = \frac{\dot{M}_G}{C_{\text{min}}}\left(\frac{y_{A,\text{in}} - y_{A,\text{out}}}{y_{A,\text{in}} - He\, x_{A,\text{in}}}\right) = \frac{\dot{M}_L}{C_{\text{min}}}\left(\frac{x_{A,\text{out}} - x_{A,\text{in}}}{y_{A,\text{in}} - He\, x_{A,\text{in}}}\right) \tag{10.30}$$

and

● $$\epsilon_{cf} = \frac{1 - e^{-N_{tu}(1 - R_C)}}{1 - R_C e^{-N_{tu}(1 - R_C)}} \tag{10.31}$$

where N_{tu} is the number of transfer units and is given by

$$N_{tu} = K_{Goa}\, \mathscr{P}H/C_{\text{min}} \tag{10.32}$$

and

$$R_C = \frac{C_{\text{min}}}{C_{\text{max}}} = \frac{\min(\dot{M}_G, \dot{M}_L{}^\dagger)}{\max(\dot{M}_G, \dot{M}_L{}^\dagger)} \tag{10.33}$$

We note from Eq. (10.31) that $0 < \epsilon_{cf} < 1$ for $0 < N_{tu} < \infty$.

Returning to the discussion in the first paragraph in this section, we observe that, given an existing piece of equipment with items (1) through (4) specified, C_{min}, C_{max}, and R_C are known from the definitions in Eq. (10.33); N_{tu} may be calculated from Eq. (10.32); and ϵ_{cf} may be extracted from Eq. (10.31).

Once the effectiveness is known, the terminal compositions ($y_{A,\text{out}}$ and $x_{A,\text{out}}$) may be obtained from Eq. (10.30).

Now suppose instead that we wish to design a mass exchanger which will realize a specified effectiveness. This is the corollary of the case discussed above. With ϵ_{cf} given, we extract N_{tu} from Eq. (10.31):

$$N_{tu} = \frac{1}{1 - R_C} \ln \frac{1 - R_C \epsilon_{cf}}{1 - \epsilon_{cf}} \tag{10.34}$$

With N_{tu} known we can then assign the required column parameters $\mathscr{P}$ and H as constrained by such side issues as pressure drop, space limitations, and overall mass transfer coefficient (K_{Goa}). We illustrate the utility of the preceding expressions with the following examples:

10.5.4   *EXAMPLE* **Effectiveness of a Falling Film Absorption Column**

Suppose that ammonia is to be removed from a nitrogen stream in the falling film column described in Section 10.4.5. Given that $x_{\text{NH}_3,\text{in}} = 0$ and $y_{\text{NH}_3,\text{in}} = 0.028$, calculate the effectiveness, ϵ_{cf}, of the column and the exit compositions, $y_{A,\text{out}}$ and $x_{A,\text{out}}$.

From the data of Section 10.4.5, we have

$$\dot{m}_G = \rho A_c V = \rho(\pi D^2/4)V$$
$$= (0.0713 \text{ lb/ft}^3)(\pi/4 \text{ ft}^2)(1.0 \text{ ft/sec})(3600 \text{ sec/hr})$$
$$= 202 \text{ lb/hr}$$

$$\dot{M}_G = \frac{(202 \text{ lb/hr})}{(28 \text{ lb/lb-mole})} = 7.2 \text{ lb-moles/hr}$$

From Eq. (4.42) of Chapter 4

$$\dot{m}_L = \mathscr{P}\Gamma = \frac{\pi D \rho g \delta^3}{3\nu}$$

$$\dot{m}_L = \frac{(3.14 \text{ ft})(62.2 \text{ lb/ft}^3)(32.2 \text{ ft/sec}^2)(10^{-9} \text{ ft}^3)(3600 \text{ sec/hr})}{(3)(0.930 \times 10^{-5} \text{ ft}^2/\text{sec})},$$
$$= 812 \text{ lb/hr}$$

$$\dot{M}_L = \frac{(812 \text{ lb/hr})}{(18 \text{ lb/lb-mole})} = 45.1 \text{ lb-moles/hr}$$

Thus for $He = 30.2$

$$\dot{M}_L^\dagger = \dot{M}_L/He = 45.1/30.2 = 1.49 \text{ lb-moles/hr}$$

and, therefore, from Eq. (10.33),

$$C_{\min} = \min(\dot{M}_G, \dot{M}_L^\dagger) = \min(7.2, 1.49) = 1.49 \text{ lb-moles/hr}$$
$$C_{\max} = \max(\dot{M}_G, \dot{M}_L^\dagger) = 7.2 \text{ lb-moles/hr}$$
$$R_C = C_{\min}/C_{\max} = 1.49/7.2 = 0.207$$

The number of transfer units is, from Eq. (10.32),

$$N_{tu} = K_{Goa} \mathscr{P} H / C_{min}$$
$$= K_{Goa} \pi D H / C_{min}$$
$$= \frac{(0.0183 \text{ lb-moles/ft}^2 \text{ hr})(10\pi \text{ ft}^2)}{1.49 \text{ lb-moles/hr}}$$
$$= 0.386$$

giving an effectiveness factor, from Eq. (10.31),

$$\epsilon_{cf} = \frac{1 - e^{-N_{tu}(1-R_C)}}{1 - R_C e^{-N_{tu}(1-R_C)}} = \frac{1 - e^{-0.386(1-0.207)}}{1 - 0.207 e^{-0.386(1-0.207)}} = 0.311$$

Finally, from Eq. (10.30), noting that $x_{A,in} = 0$, we find

$$x_{A,out} = \frac{C_{min} y_{A,in} \epsilon_{cf}}{M_L} = \frac{y_{A,in} \epsilon_{cf}}{He} = \frac{(0.028)(0.311)}{30.2} = 2.88 \times 10^{-4}$$

$$y_{A,out} = y_{A,in}\left(1 - \frac{\epsilon_{cf} M_L}{M_G He}\right) = (0.028)\left(1 - \frac{(0.311)(45.1)}{(7.2)(30.2)}\right) = 0.0263$$

Thus we see that the column is rather ineffective in removing NH$_3$ from the gas stream. This is due to the fact that the overall resistance is very high. We conclude that mass exchangers for absorption of inert species require large interfacial areas. In practice this area is provided by contacting the phases in a packed column.

10.5.5   *EXAMPLE*   Number of Transfer Units Required for a Falling Film Column

Let us redo the previous example, this time specifying that the column be 95% effective ($\epsilon_{cf} = 0.95$). From Eq. (10.34), the number of transfer units required is

$$N_{tu} = \frac{1}{1 - R_C} \ln \frac{1 - R_C \epsilon_{cf}}{1 - \epsilon_{cf}} = \frac{1}{1 - 0.207} \ln \frac{1 - (0.207)(0.95)}{1 - 0.95} = 3.5$$

Assuming for the moment that K_{Goa} is unchanged, the contact area required for a 95% effective column is, from Eq. (10.32)

$$\mathscr{P} H = N_{tu} \frac{C_{min}}{K_{Goa}} = \frac{(3.5)(1.49 \text{ lb-moles/hr})}{(0.0183 \text{ lb-moles/ft}^2 \text{ hr})} = 285 \text{ ft}^2$$

As stated above, a packed column is the practical way to obtain the requisite area. The value of K_{Goa} would be calculated for the type of packing used.

10.6   PRELIMINARY DESIGN OF AN SO$_2$ SCRUBBER

10.6.1   The Process: Absorption by Molten Alkali-Carbonates

To reduce SO$_2$ emissions (recall Section 10.2.2) one might gasify fossil fuel and remove H$_2$S before burning or remove SO$_2$ from a conventional flue gas stream. When a water solution is used in the latter process, the *wet scrubbing* must be done below 212°F. Reheating the flue gases to make them rise may then be necessary. Oldenkamp has described *dry scrubbing* with a molten mixture of carbonates of lithium (32%), sodium (33%), and potassium (35%) at 800°F. The absorption step was followed by a complex sequence of operations to recover the carbonates in essentially pure form; we consider here only the design of the SO_2 absorber.

Consider flow of molten carbonate (hereafter M_2CO_3) down the walls of closely packed concentric cylinders with counter-current flow of gas up through the annular spaces of the absorber, as illustrated in Figure 10.8. Assume, as discussed in Section 10.2.2, that we are to reduce the mole fraction of SO_2 in the stack gas from 5.15×10^{-4} to 1.52×10^{-4}. Equation (10.34) is seen to apply in this case, and hence the number of transfer units required is

$$N_{tu} \equiv \frac{K_{Goa}\,\mathscr{P}H}{C_{min}} = \frac{1}{1-R_C}\ln\left(\frac{1-R_C\epsilon_{cf}}{1-\epsilon_{cf}}\right)$$

In applying Eq. (10.34), we shall need:

1. Such geometric quantities as the wetted perimeter ($\mathscr{P}$), the cross-sectional area open to gas flow (A_c), and the hydraulic diameter(s) of the annular spaces (D_h).
2. Equilibrium data for estimating the Henry number (He), compositions of entering and leaving streams (y_{in}, x_{in} and y_{out}, x_{out}), and stream flow rates ($\dot{M}_G$ and $\dot{M}_L$).

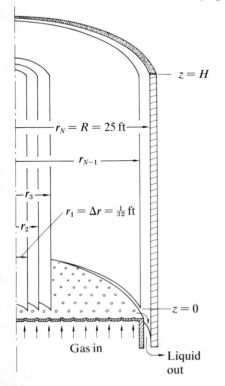

$z = H$

$r_N = R = 25$ ft

r_{N-1}

r_3

$r_1 = \Delta r = \frac{1}{32}$ ft

r_2

$z = 0$

Gas in

Liquid out

Figure 10.8   Schematic of a concentric cylinder counter-current falling film absorber.

3. Physical properties of the gas and liquid phases (ρ, μ, $\mathscr{D}$).
4. The overall mass transfer coefficient (K_{Goa}).

Since our calculations are only illustrative, we shall make a number of assumptions including (1) perfect gas behavior, (2) gas properties evaluated assuming the stack gas is air, and (3) constant molar flow rates $\dot{M}_G$ and $\dot{M}_L$.

10.6.2   Geometrical Considerations

Since we may wish to install our exchanger within the stack of our model plant, we let the radius R of the exchanger be 25 ft (a stack diameter of 50 ft for so large a plant is not unreasonable). Assuming that 0.375 in. spacings (center to center) between the cylinders will not result in an unacceptable pressure drop in the gas flow, we have $\Delta r = \frac{1}{32}$ ft and require $N = R/\Delta r = 800$ cylinders. Assuming further that the combined thickness of cylinder wall and falling liquid films is 2δ, we have in turn the following geometrical values:

Wetted Perimeter ($\mathscr{P}$)
Noting from Figure 10.8 that $r_n = n\,\Delta r = nR/N$, we write

$$\mathscr{P} = \sum_{n=1}^{N} \mathscr{P}_n$$

$$= 2\pi \sum_{n=1}^{N} (r_n - \delta + r_n + \delta) - 2\pi(r_N + \delta)$$

$$= \frac{2\pi R}{N} \sum_{n=1}^{N} 2n - 2\pi(R + \delta)$$

$$\doteq 2\pi R(N - 1)$$

$$\doteq 2\pi RN$$

where we have replaced the sum by an integral,

$$\sum_{n=1}^{N} 2n \doteq \int_{0}^{N} 2n\,dn = N^2$$

Thus, for $R = 25$ ft and $N = 800$,

$$\mathscr{P} = (6.28)(25)(800) = 1.26 \times 10^5 \text{ ft}$$

Cross-Sectional Area (A_c)

$$A_c = \pi(R - \delta)^2 - \sum_{n=1}^{N-1} (2\pi r_n)(2\delta)$$

$$= \pi(R - \delta)^2 - \frac{4\pi\delta R}{N} \sum_{n=1}^{N-1} n$$

$$\doteq \pi R^2 - 2\pi\delta RN$$

$$\doteq \pi R^2 (1 - 2\delta/\Delta r)$$

Assuming that $2\delta = 0.16$ in. is adequate to accommodate the liquid flow (we should check this supposition later), we have $A_c = (3.14)(625)(1-0.16/0.375) = 1125$ ft².

Hydraulic Diameter (D_h)

By definition, the hydraulic diameter D_h for a duct of arbitrary shape is

$$D_h = 4A_c/\mathscr{P}$$

For the annular space between the $(n-1)$st and nth cylinders,

$$D_h = \frac{4\pi[(r_n-\delta)^2 - (r_{n-1}+\delta)^2]}{2\pi[(r_n-\delta)+(r_{n-1}+\delta)]}$$

$$= 2[(r_n-\delta) - (r_{n-1}+\delta)]$$

$$= 2[\Delta r - 2\delta]$$

Thus D_h is the same for all annuli (excepting the core). This finding is important because it implies that the gas stream velocity and transfer coefficients are the same in each annular section. For our problem

$$D_h = 2[0.375 - 0.16]/12 = 0.0358 \text{ ft}$$

10.6.3   Equilibrium Data, Compositions, and Flow Rates

For the air-SO_2-M_2CO_3 system, Oldenkamp reports equilibrium data as shown in Figure 10.9. Dropping the subscript A, we have

$$y_s = \left[\frac{3.12 \times 10^{-4}}{(1-x_s)}\right][x_s]$$

Thus the Henry number is seen to be

$$He = \frac{3.12 \times 10^{-4}}{(1-x_s)}$$

and is seen to be a *constant* only near $x_s = 0$. Table 10.3 summarizes the data.

Table 10.3 Equilibrium Data for the Air-SO_2-M_2CO_3 system

x_s	$He \times 10^4$	$y_s \times 10^4$
0	3.12	0
0.1	3.47	0.347
0.2	3.90	0.780
0.3	4.46	1.34
0.4	5.20	2.08
0.5	6.24	3.12
0.6	7.80	4.68

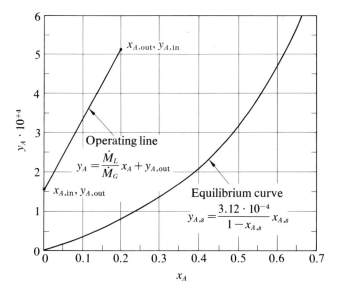

Figure 10.9 Operating and equilibrium lines for the SO_2-molten-carbonate system.

Figure 10.9 also shows the operating line derived from the species balance

$$\dot{M}_G y - \dot{M}_L x = \dot{M}_G y_{\text{out}} - \dot{M}_L x_{\text{in}}$$

We assume pure M_2CO_3 in the feed to the exchanger; that is, we take $x_{\text{in}} = 0$. Then we obtain

$$y = (\dot{M}_L/\dot{M}_G)x + y_{\text{out}} \tag{10.35}$$

Equation (10.35) is called an operating line, because it shows the relationship between x and y during the mass transfer operation. For every point (x, y) on the operating line there is a point (x_s, y_s) on the equilibrium curve such that the second equality is satisfied in Eq. (10.11) at the z-location corresponding to the bulk fractions (x, y). The vertical distance between the two points on Figure 10.9 is then the gas phase driving force $(y - y_s)$, and the horizontal distance is the liquid phase driving force $(x_s - x)$. When an exchanger operates with large values of x_s, so that the curvature present in the equilibrium curve must be accounted for, a constant Henry number cannot be introduced, as was done to derive Eq. (10.18), and the differential conservation of species equation corresponding to Eq. (10.18) must be integrated numerically or graphically. Then an accurate plot such as Figure 10.9 is a great aid. But if the operating line and equilibrium curve are separated widely and x_s is small, we can neglect the curvature in the equilibrium line and employ a constant value of Henry number.

Oldenkamp has shown, for the M_2CO_3 system, that x_s does not differ significantly from x; that is, the liquid phase resistance to mass transfer is negligible compared to ordinary gas side resistances. Hence, if we have a small value of x_{out}, we will never have a large value of x_s. The extreme points on the operating curve can be found from Section 10.2.2 as follows:

$$\dot{M}_G = 2.30 \times 10^5 \text{ lb-moles/hr}$$

$$\dot{m}_G = 6.59 \times 10^6 \text{ lb/hr}$$

$$y_{\text{in}} = 5.15 \times 10^{-4}$$

$$y_{\text{out}} = 1.52 \times 10^{-4}$$

We elect to set

$$x_{\text{out}} = 0.2$$

first, because we wish to assume He constant, second, because we wish to sustain an appreciable driving force throughout the exchanger (note that for $y_{\text{in}} = 5.15 \times 10^{-4}$, $y_{\text{out}} - y_s \to 0$ as $x_{\text{out}} \to 0.623$). Thus we take

$$He = \frac{(3.12 + 3.90) \times 10^{-4}}{2} = 3.51 \times 10^{-4}$$

and from Eq. (10.22)

$$\frac{\dot{M}_G}{\dot{M}_L} = \frac{x_{\text{out}}}{(y_{\text{in}} - y_{\text{out}})} = \frac{0.2}{3.63 \times 10^{-4}} = 551$$

Hence

$$\dot{M}_L = \frac{2.30 \times 10^5}{551} = 417 \text{ lb-moles/hr}$$

$$\dot{M}_L{}^\dagger = \frac{\dot{M}_L}{He} = \frac{417}{3.51 \times 10^{-4}} = 1.19 \times 10^6 \text{ lb-mole/hr}$$

$$C_{\text{min}} = \min (\dot{M}_G, \dot{M}_L{}^\dagger) = 2.30 \times 10^5 \text{ lb-moles/hr}$$

$$C_{\text{max}} = \max (\dot{M}_G, \dot{M}_L{}^\dagger) = 1.19 \times 10^6 \text{ lb-moles/hr}$$

$$R_C = \frac{C_{\text{min}}}{C_{\text{max}}} = \frac{2.30 \times 10^5}{1.19 \times 10^6} = 0.193$$

10.6.4   Physical Properties

Assuming the stack gas to be similar to air, we have at 800°F and 1 atm:

$$\rho = 0.0315 \text{ lb/ft}^3$$

$$\nu = 70.1 \times 10^{-5} \text{ ft}^2/\text{sec} = 2.52 \text{ ft}^2/\text{hr}$$

$$c = \frac{P}{\mathcal{R}T} = \frac{(14.7 \text{ lb}_f/\text{in.}^2)(144 \text{ in.}^2/\text{ft}^2)}{(1545 \text{ ft lb}_f/\text{lb-mole °R})(1260°\text{R})}$$

$$= 0.001087 \text{ lb-mole/ft}^3$$

From Example 7.4.5

$$\mathcal{D}_{\text{SO}_2} \doteq 2.3 \text{ ft}^2/\text{hr}$$

10.6.5   Overall Mass Transfer Coefficient

For gas side controlling, $K_{Goa} = K_G$. From Eq. (5.22) written in molar terms and Table 5.1, we have, since the rate of mass transfer is low,

$$Nu_m = \frac{K_G D_h}{c \mathcal{D}_{SO_2,air}} = 7.54$$

Checking the Reynolds number for the parallel plate geometry,

$$Re = \frac{D_h \rho V}{\mu} = \frac{D_h (\dot{m}_G/A_c)}{\rho \nu} = \frac{(0.0358 \text{ ft})(6.59 \times 10^6 \text{ lb/hr})}{(0.031 \text{ lb/ft}^3)(2.52 \text{ ft}^2/\text{hr})(1125 \text{ ft}^2)}$$

$$= 2680$$

So the flow is laminar as was desired ($Re < 2800$). Hence

$$K_G = 7.54(1.09 \times 10^{-3} \text{ lb-moles/ft}^3)(2.3 \text{ ft}^2/\text{hr})/(0.0358 \text{ ft})$$

$$= 0.528 \text{ lb-moles/ft}^2 \text{ hr}$$

10.6.6 N_{tu} and Exchanger Height

In order to apply Eq. (10.31) we need to calculate the effectiveness ϵ_{cf}. Applying Eq. (10.30) we find, since $C_{min} = \dot{M}_G$,

$$\epsilon_{cf} = \frac{y_{in} - y_{out}}{y_{in}} = \frac{5.15 - 1.52}{5.15} = 0.705$$

Thus

$$N_{tu} = \frac{1}{1 - 0.193} \ln \frac{1 - (0.193)(0.705)}{1 - 0.705}$$

$$= 1.33$$

and the height H required is

$$H = \frac{C_{min} N_{tu}}{K_{Gou} \mathcal{P}} = \frac{(2.30 \times 10^5)(1.33)}{(0.528)(1.26 \times 10^5)} = 4.60 \text{ ft}$$

The results of Section 4.6 for the falling laminar film may be used to show that the liquid film thickness is negligible. It is also left to the student to use the relations of Table 5.1 and Chapter 8 to estimate pressure drop by finding the friction factor from the calculated value of Reynolds number. The result, a ΔP of 0.9 in. of water, is somewhat high for a power plant. Increasing the diameter of the stack or using a very tall one or providing forced draft might prove to be necessary. As the student can well imagine, many iterations and much careful thought must go on before a final design can be fixed and the large amount of capital required for the project be committed. Such is the nature of engineering.

10.7   SUMMARY

The design problem in the preceding section summarized not only the salient points of this chapter but those of earlier sections of the text as well. For example, we obtained such quantities as wetted perimeter for interphase mass transfer, cross-sectional area open to gas flow, hydraulic diameter for a flow passage, and Henry number for relating interfacial equilibria. These results together with the physical properties of the contacting fluids were then used to calculate such dimensionless groups as Reynolds and Schmidt numbers and hence an overall mass transfer coefficient. Finally, for a specified reduction in gas stream concentration of the transferred species (effectiveness), we obtained the number of transfer units (N_{tu}) required for a counter-current mass exchanger of specified diameter, that is, the required column height. For completeness, reference was made to such considerations as pressure drop and liquid flow cross section, using concepts developed in Chapters 4 and 5.

Although the chapter is lengthy, the only new idea presented is that of interfacial equilibrium in which the mole fraction of a species in the gas phase is related to that in the liquid phase through the Henry number, Eq. (10.8). Once Eq. (10.8) was introduced to eliminate the liquid side equilibrium mole fraction $x_{A,s}$, the concept of an overall mass transfer coefficient, Eq. (10.13), arose naturally in the same way as the overall heat transfer coefficient was defined in Eq. (9.25). Thus the mass balance equations, (10.18) and (10.19), have the same mathematical form as Eqs. (9.29b) and (9.30b); as a result their solutions, Eqs. (10.30), (10.31) or (10.34), and Eq. (10.32), are identical to those obtained earlier in Chapter 9, Eqs. (9.54), (9.57) or (9.59), and Eq. (9.55). The student now is urged to reread the theoretical developments in Chapters 9 and 10 and note the analogous treatment of heat and mass exchangers.

EXERCISES

1. In Section 10.2.1 we obtained an overall balance for an urban air volume under conditions for which pure air flowed into the system at a rate $\dot{m}$, absorbed and mixed with a pollutant species 1 which was generated within the system at a rate $\dot{r}_1$, and then flowed out of the considered volume with a pollutant concentration given by

$$(\dot{m}m_1)_{out} = (\dot{m}m_1)_{in} + \dot{r}_1$$

or since $m_{1,in}$ was presumed to be zero and $\dot{m}$ was essentially constant,

$$m_{1,out} = \frac{\dot{r}_1}{\dot{m}}$$

Consider now a balance on an urban air volume over a community which is downwind of a major polluter (this frequently is the case when a sea breeze carries Los Angeles smog inland over Pasadena, California). Suppose that a layer of air 2000 ft thick advances on the community at 10 mph. If the concentration of CO in the incoming air is 7.5 ppm, calculate the maximum rate at which CO may be generated within the community if the CO concentration within its air volume is not to exceed 10.0 ppm. (Take as the effective width of the community 3.5 mi.)

2. Due to its cleaner burning properties, propane (C_3H_8) is receiving consideration as an alternate fuel for internal combustion engines. Assuming a medium-

sized passenger car gets 15 mi per gallon of liquid propane while cruising at 60 mph:

(a) Write the main combustion reaction

(b) Calculate the rates at which fuel and oxygen are consumed, water vapor and carbon dioxide are produced, and nitrogen enters and leaves

(c) Calculate the rate at which nitric oxide is produced assuming an emissions concentration of 600 ppm for an engine operating at the stoichiometric fuel-air mixture.

3. Calculate the surface areas required for the catalytic converter described in Section 10.3.4 if (a) the foil spacing is 0.03 in., (b) the effectiveness is 0.999, and (c) the vehicle gets only 10 mi per gallon. Compare your results with the result obtained in Section 10.3.4 and sketch qualitatively the effects of the three parmeters cited in (a), (b), and (c).

4. Determine the length of a 5 in. diameter catalytic reactor necessary to effect a 20-fold reduction in CO emission from an automobile engine as it develops 45 bhp at a specific fuel consumption of 0.8 lb/bhp hr and an air-fuel ratio of 12.5:1. The reactor is to operate at 1100°F, the concentration of CO in the exhaust is 3% by volume, and the exhaust back pressure is 16 psia. The catalyst is in the form of $\frac{1}{8}$ in. diameter spherical pellets of copper oxide on alumina. The pellet density is 0.712 gm/cm³ and has 100 m²/cm³ catalytic surface area. The pellets are packed to give a bed volume void fraction of 25%. Assume a pellet effectiveness of 2.0%.

5. Determine the pressure of pure CO_2 required to produce an equilibrium mole fraction of 0.001 in a carbonated soft drink at 60°F. What would be the concentration of CO_2 in the same beverage in equilibrium with the atmosphere? Why does the carbonated effect persist for some time after opening a soft drink?

6. Show that in deriving Eq. (10.12) the same result would be obtained if $x_{A,s}$ rather than $y_{A,s}$ were eliminated from Eq. (10.11). Show that this result is also true if the equilibrium relation is nonlinear in form, say $y_{A,s} = f(x_{A,s})x_{A,s}$.

7. Calculate the average value of the overall mass transfer coefficient K_{Goa} for a gas mixture of SO_2 in trace amounts in N_2 flowing counter-current to a falling film of aqueous SO_2 on the walls of a vertical rectangular channel. The gas velocity is 10 ft/sec. The channel is 6 ft high, 12 in. by 1 in. in cross section, and operates at 60°F and 1 atm. The average thickness of the liquid film is 0.015 in.

8. Paralleling the derivations in Sections 10.5.1, 10.5.2, and 10.5.3, develop the equations for a co-current mass exchanger.

9. Redesign the mass exchanger of Section 10.6 for co-current flow.

10. An existing counterflow unit to recover SO_2 from industrial exhaust gas is a packed tower 40 ft tall. The gas contains 6% SO_2 by volume and is scrubbed at atmospheric pressure by water at 60°F running down over the packing at a rate 50% greater than that which would be required by an infinite counterflow unit to recover 85% of the SO_2. Experimental data indicate that the

existing unit recovers 85% of the SO_2 in the entering gas stream. A study is being made of ways to reduce the air pollution still further. One proposal is to build a second tower alongside the first and have gas from the top of the first tower enter the top of the second tower and flow co-current with the falling water in the new unit. It is proposed to continue the same gas and water flow rates, but the water will enter the top of the new tower, be collected at its bottom, and pumped to the top of the existing one. The entering gas will continue to flow into the base of the existing unit and will exit from the base of the new one.

(a) Find y_{out} and the percent recovery which would be obtained if the proposal were adopted.

(b) Discuss an alternative proposal to use the existing unit as it stands, but to add ammonia to the incoming water.

REFERENCES

Mass Exchangers

A comprehensive text on chemical engineering mass transfer operations is that of:

Treybal, R. E., *Mass Transfer Operations*, 2nd Ed. New York: McGraw-Hill, 1968.

It covers at some length humidification, gas absorption, distillation, liquid extraction, adsorption and ion exchange, drying, and leaching, and devotes a few pages to other operations such as fractional crystallization, flotation, and reverse osmosis. In chemical engineering operations the interfacial equilibria relations are often nonlinear, and the techniques presented show how chemical engineers handle such nonlinearity.

A text with a somewhat narrower scope, but perhaps more unity is:

Spalding, D. B., *Convective Mass Transfer*. London: Edward Arnold, 1963.

In the application of mass transfer theory to mass exchangers, Section 2.7, this text assumes linear interfacial equilibria and, within this limitation, develops a unified treatment. Section 2.7 and the earlier ones take up such processes as evaporative cooling, combustion, deaeration, and transpiration cooling. The text shows how to treat combined heat and mass transfer and gives simplified methods for cooling tower design.

CONSERVATION EQUATIONS

A.1 INTRODUCTION

Each time an engineer or scientist studies a new problem of heat, mass, or momentum transfer in a fluid, he does not need to start from scratch. He knows that the principles which apply to any problem (which does not involve nuclear reactions) are as follows:

1. Conservation of mass
2. Conservation of momentum
3. Conservation of energy
4. Conservation of atomic species

For this reason it is profitable to formulate general governing equations based on these principles; for a specific problem the engineer then need only delete terms in the equations which are zero or negligibly small.

In the first half of this appendix the equations embodying these principles are so formulated. They will be found to be quite complicated. The second half of the appendix will show that, despite their complexity, they are very valuable for the following reasons: (1) With no attempt whatever to solve them, we can find how to model. (2) With a little arithmetic we can discover orders of magnitude, for example, whether viscous dissipation is important or not, whether gravity forces are important or not, or whether streamwise gradients are important compared to transverse gradients. (3) In simple cases most of the terms may drop out leaving a simple equation which may be solved analytically. (4) In other less simple cases numerical solutions may still be feasible.

The procedure used to derive the conservation equations is quite straightforward. It is one of accounting. Just as an accountant sets up a procedure to follow the flow of money, goods, and services in an enterprise, the engineer sets up a system to account for the flow of mass, momentum, energy, and chemical species. Applying the accounting principle is done as follows:

1. Set up a control volume.
2. Write down a statement of the conservation principle for the quantity for which you have to account. In general this statement has the form: rate of storage within the volume equals net rate of inflow across the volume boundaries plus rate of production within the volume.
3. Decide how the quantity is stored within the control volume. Then formulate the rate of storage in terms of a time derivative.
4. Identify whether or not the quantity can be produced within the control volume; if it can be, write down the rate of generation.
5. Recognize the mechanisms by which the quantity can cross the control volume boundaries.
6. Formulate the net rate of inflow in terms of rates of convections and diffusive transport laws.
7. Write down the resulting conservation equation.

A.2   CONSERVATION OF MASS

A.2.1   Cartesian Coordinates

For simplicity we first consider a Cartesian control volume element dx by dy by dz, fixed in space, as shown in Figure A.1. Mass can be stored within the volume by a change in density. Mass can cross the boundaries of the control volume by convection at the mass average velocity. The conservation principle is simply that the rate of storage of mass within the control volume equals the net rate of inflow of mass across the control volume boundaries.

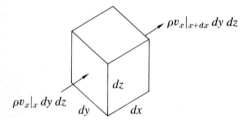

Figure A.1   Control volume for conservation of mass.

The rate of storage is

$$\frac{\partial \rho}{\partial t}\, dx\, dy\, dz$$

The gross rate of inflow is

$$\rho v_x|_x\, dy\, dz + \rho v_y|_y\, dx\, dz + \rho v_z|_z\, dx\, dy$$

Similarly the gross rate of outflow is

$$\rho v_x|_{x+dx}\, dy\, dz + \rho v_y|_{y+dy}\, dx\, dz + \rho v_z|_{z+dz}\, dx\, dy$$

The net rate of inflow is found by subtracting the outflow from the inflow, expanding a quantity evaluated at $x + dx$, $y + dy$, or $z + dz$ in a Taylor series, and canceling terms which subtract out. For example, the x-direction convection terms become

$$\rho v_x|_x\, dy\, dz - \rho v_x|_{x+dx}\, dy\, dz$$

$$= \rho v_x|_x\, dy\, dz - \left[\rho v_x|_x + \frac{\partial}{\partial x}(\rho v_x)\, dx + \frac{1}{2}\frac{\partial^2}{\partial x^2}(\rho v_x)\, dx^2 + \cdots\right] dy\, dz$$

$$= -\frac{\partial}{\partial x}(\rho v_x)\, dx\, dy\, dz - \frac{1}{2}\frac{\partial^2}{\partial x^2}(\rho v_x)\, dx^2\, dy\, dz - \cdots$$

The convection in the other directions results in similar expressions. Now substitute into our statement of conservation, divide throughout by the volume $dx\, dy\, dz$, and then let dx, dy, and dz approach zero. Only first order terms remain as the convective contribution, and the final result is

$$\frac{\partial \rho}{\partial t} = -\frac{\partial}{\partial x}(\rho v_x) - \frac{\partial}{\partial y}(\rho v_y) - \frac{\partial}{\partial z}(\rho v_z) \qquad (\text{A.1})$$

which may be rewritten

$$\frac{\partial \rho}{\partial t} + \frac{\partial}{\partial x}(\rho v_x) + \frac{\partial}{\partial y}(\rho v_y) + \frac{\partial}{\partial z}(\rho v_z) = 0 \qquad (\text{A.2})$$

A.2.2   Arbitrary Coordinates, Use of Vector Calculus

Now let our control volume be any nonreentrant region V. Denote an outward directed unit vector normal to the boundary surface by $\mathbf{n}$. The rate of storage is simply the volume integral of the time rate of change of the density

$$\frac{\partial}{\partial t}\int_V \rho\, dV = \int_V \frac{\partial \rho}{\partial t}\, dV$$

The flow *out* across a surface area element dS is $\rho \mathbf{v} \cdot \mathbf{n}\, dS$. The net inflow is then the surface integral

$$-\int_S \rho \mathbf{v} \cdot \mathbf{n}\, dS$$

Substituting into our statement of conservation of mass we find

$$\int_V \frac{\partial \rho}{\partial t} \, dV = - \int_S \rho \mathbf{v} \cdot \mathbf{n} \, dS \tag{A.3}$$

The surface integral in Eq. (A.3) can be converted into a volume integral by use of the Gauss divergence theorem,

$$\int_V \frac{\partial \rho}{\partial t} \, dV = - \int_V \nabla \cdot (\rho \mathbf{v}) \, dV$$

or rearranging,

$$\int_V \frac{\partial \rho}{\partial t} \, dV + \int_V \nabla \cdot (\rho \mathbf{v}) \, dV = 0$$

$$\int_V \left(\frac{\partial \rho}{\partial t} + \nabla \cdot (\rho \mathbf{v}) \right) dV = 0 \tag{A.4}$$

But the volume V is arbitrary; therefore the integrand in Eq. (A.4) must be identically zero,

$$\frac{\partial \rho}{\partial t} + \nabla \cdot (\rho \mathbf{v}) = 0 \tag{A.5}$$

This equation is the general form of our previous result, Eq. (A.2).

A.3   CONSERVATION OF MOMENTUM

A.3.1   The Control Volume

In setting up a control volume we must decide whether to use a volume fixed in space, in which case fluid flows through the boundaries, or a volume containing a fixed mass of fluid and which moves with the fluid. The former is known as the Eulerian viewpoint and was used in Section A.2; the latter is the Lagrangian viewpoint. Both approaches will, of course, yield equivalent results. Here we choose to take the Lagrangian view because we can regard the principle of conservation of momentum in its simplest form, that is, time rate of change of momentum equals the sum of forces. A Cartesian coordinate system will again be used.

A.3.2   The Substantial Derivative

We need to determine the *time rate of change of momentum.* Consider the flow of water through a fireman's hose; in a time Δt two things can happen. First, a fireman can be cranking open a valve so that the velocity everywhere is changing with time. As a result the velocity component v_x at a particular spot changes an amount

$$\frac{\partial v_x}{\partial t} \Delta t$$

Second, the fluid can leave the hose and enter the nozzle; it then speeds up even if the valve is already fully open and the hydrant pressure is steady. The fluid speeds up because in time Δt it moves a distance $\Delta x = v_x \Delta t$, $\Delta y = v_y \Delta t$, $\Delta z = v_z \Delta t$, and the velocity at the new position is higher. The change in the velocity component v_x due to this second cause is

$$\frac{\partial v_x}{\partial x} \Delta x + \frac{\partial v_x}{\partial y} \Delta y + \frac{\partial v_x}{\partial z} \Delta z$$

$$= \frac{\partial v_x}{\partial x} v_x \Delta t + \frac{\partial v_x}{\partial y} v_y \Delta t + \frac{\partial v_x}{\partial z} v_z \Delta t$$

Since our control surface is wrapped around a fixed mass, the time rate of change of momentum is

$$\rho \, dx \, dy \, dz \frac{D\mathbf{v}}{Dt}$$

where, for v_x,

$$\frac{Dv_x}{Dt} = \frac{\partial v_x}{\partial t} + v_x \frac{\partial v_x}{\partial x} + v_y \frac{\partial v_x}{\partial y} + v_z \frac{\partial v_x}{\partial z} \tag{A.6}$$

Corresponding expressions hold for the other two components of velocity. The operator D/Dt is called the substantial (or "total") derivative. It accounts for the fact that a fluid element, moving through a flow field, can experience a velocity change both due to the flow field changing with time and the element changing its position in the flow field.

A.3.3   Surface Forces in a Pure Fluid

Surface forces per unit area, surface stresses, act within a fluid. The forces are of two kinds, the viscous stresses and the static pressure. We need to adopt a convenient notation and sign convention for these stresses. In Chapter 7 it was shown, from kinetic theory, that for simple Couette-type shear flow

$$\tau_{zx} = -\mu \frac{\partial v_x}{\partial z}$$

This shear stress is felt on a plane parallel to the xy plane, that is, a plane of constant z. The negative sign indicates that the stress exerts a drag on the faster moving fluid above. We adopt the following conventions:

1. The first subscript of τ_{zx} denotes the plane on which the stress acts, a plane of constant z.
2. The convention is adopted that the stress is applied to the fluid of greater z from below.
3. The second subscript denotes the direction in which the force acts.
4. The term τ_{zz} does not include the pressure but only the viscous contribution to the stress exerted in the z-direction on the fluid of greater z on a surface of constant z, a surface perpendicular to the z-axis.

A.3.4   The Stokes Hypothesis

A Newtonian fluid is one in which the shear stress is linearly related to the velocity gradient in a simple Couette-type shear flow. Stokes generalized the concept to a fluid in which there is more than one velocity gradient by assuming that the shear stress is linearly related to the rate of angular deformation. When a real fluid obeys the Stokes hypothesis it is said to be Newtonian. Most common fluids such as water and air are Newtonian, but many important fluids are not (tomato ketchup, strawberry jam, and blood, to name just three).

Figure A.2 shows how angular deformation occurs in simple shear and in the more general case. In time dt the upper left corner of what was originally a square of fluid will move distance $(\partial v_x/\partial y)\, dy\, dt$ farther than the lower left corner.

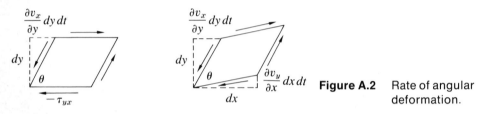

Figure A.2   Rate of angular deformation.

This distance divided by the lever arm, length dy, gives the change in angle θ. Similarly the movement of the lower right corner causes an angular change of $(\partial v_y/\partial x)\, dt$. The combined time rate of angular change in the angle θ is then $-\Delta\theta/\Delta t$, which is equal to

$$\frac{\partial v_x}{\partial y} + \frac{\partial v_y}{\partial x}$$

This angular rate of change was brought about by (or gives rise to) the negative shear stresses shown in Figure A.2. By the Stokes hypothesis

$$\bullet \qquad\qquad \tau_{yx} = -\mu\left(\frac{\partial v_x}{\partial y} + \frac{\partial v_y}{\partial x}\right) \qquad\qquad \text{(A.7a)}$$

$$\bullet \qquad\qquad \tau_{zx} = -\mu\left(\frac{\partial v_x}{\partial z} + \frac{\partial v_z}{\partial x}\right) \qquad\qquad \text{(A.7b)}$$

$$\bullet \qquad\qquad \tau_{zy} = -\mu\left(\frac{\partial v_y}{\partial z} + \frac{\partial v_z}{\partial y}\right) \qquad\qquad \text{(A.7c)}$$

It can be shown, by some rather involved proofs involving coordinate transformations, that the compatible relations for τ_{xx}, τ_{yy}, and τ_{zz} are

$$\bullet \qquad\qquad \tau_{xx} = -\mu\left(2\frac{\partial v_x}{\partial x} - \frac{2}{3}\nabla\cdot\mathbf{v}\right) \qquad\qquad \text{(A.8a)}$$

$$\bullet \qquad\qquad \tau_{yy} = -\mu\left(2\frac{\partial v_y}{\partial y} - \frac{2}{3}\nabla\cdot\mathbf{v}\right) \qquad\qquad \text{(A.8b)}$$

$$\tau_{zz} = -\mu\left(2\frac{\partial v_z}{\partial z} - \frac{2}{3}\nabla\cdot\mathbf{v}\right) \tag{A.8c}$$

where

$$\nabla\cdot\mathbf{v} = \frac{\partial v_x}{\partial x} + \frac{\partial v_y}{\partial y} + \frac{\partial v_z}{\partial z}$$

In some fluids, such as liquids with small gas bubbles, a second coefficient of viscosity is needed. It adds a term proportional to $\nabla\cdot\mathbf{v}$, to each normal stress. But such situations are rare and will be ignored here. In fact, there are very few physical situations for which we cannot simply approximate Eqs. (A.8a, b, c) by

$$\tau_{xx} = -2\mu\frac{\partial v_x}{\partial x} \tag{A.9a}$$

$$\tau_{yy} = -2\mu\frac{\partial v_y}{\partial y} \tag{A.9b}$$

$$\tau_{zz} = -2\mu\frac{\partial v_z}{\partial z} \tag{A.9c}$$

A.3.5 Net Surface Forces

Figure A.3 shows all the x-direction forces acting on a cube dx by dy by dz. The net x-direction surface forces, neglecting Taylor series expansion terms of order higher than the first, are

$$\left(-\frac{\partial P}{\partial x} - \frac{\partial \tau_{xx}}{\partial x} - \frac{\partial \tau_{yx}}{\partial y} - \frac{\partial \tau_{zx}}{\partial z}\right) dx\,dy\,dz$$

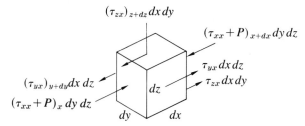

Figure A.3 Surface forces in the x-direction.

A.3.6 Volume Forces

These forces are easy to write down. In general, gravitational and electromagnetic forces may act on a fluid. For example, if the gravitational force per unit mass is denoted by vector $\mathbf{g}$, then the resulting force acting on the fluid within the control volume is $\mathbf{g}\rho\,dx\,dy\,dz$. Electromagnetic forces may be regarded as acting on the fluid as a whole, for example, on a liquid metal or, in the case of a plasma or electrolyte, the forces may differ for the various chemical species. To include the effect of these forces we would write $\Sigma_i\,\rho_i\mathbf{f}_i\,dx\,dy\,dz$, where $\mathbf{f}_i$ is the force per unit mass acting on the ith species. In what follows only gravitational volume sources will be considered.

A.3.7   The Momentum Conservation Equation

We now equate the time rate of change of momentum to the sum of the forces. For the x-direction

$$\rho \frac{Dv_x}{Dt}\, dx\, dy\, dz = \left(-\frac{\partial P}{\partial x} - \frac{\partial \tau_{xx}}{\partial x} - \frac{\partial \tau_{yx}}{\partial y} - \frac{\partial \tau_{zx}}{\partial z} + \rho g_x\right) dx\, dy\, dz$$

$$\rho \frac{Dv_x}{Dt} = -\frac{\partial P}{\partial x} - \frac{\partial \tau_{xx}}{\partial x} - \frac{\partial \tau_{yx}}{\partial y} - \frac{\partial \tau_{zx}}{\partial z} + \rho g_x$$

The equations for the components of the momentum equation in the y- and z-directions are similar. We can summarize all three equations in matrix form

$$\bullet \quad \rho \frac{D}{Dt}\begin{bmatrix} v_x \\ v_y \\ v_z \end{bmatrix} = -\begin{bmatrix} P+\tau_{xx} & \tau_{yx} & \tau_{zx} \\ \tau_{xy} & P+\tau_{yy} & \tau_{zy} \\ \tau_{xz} & \tau_{yz} & P+\tau_{zz} \end{bmatrix}\begin{bmatrix} \dfrac{\partial}{\partial x} \\ \dfrac{\partial}{\partial y} \\ \dfrac{\partial}{\partial z} \end{bmatrix} + \rho\begin{bmatrix} g_x \\ g_y \\ g_z \end{bmatrix} \quad \textbf{(A.10)}$$

The convention for matrix multiplication of the operator is that the ith row of the desired one column vector results from operating on the term in the ith row and jth column of the stress tensor by the differential operator in the jth row of the operator vector and summing the terms for $j = 1, 2, 3$. Notice that the 3×3 stress tensor is symmetric, because $\tau_{xy} = \tau_{yx}$, and so on, as may be seen from Eq. (A.7).

For a constant-density constant-viscosity flow, the momentum equation can be written in the compact form

$$\bullet \qquad \rho \frac{D\mathbf{v}}{Dt} = -\nabla P + \mu \nabla^2 \mathbf{v} + \rho \mathbf{g} \qquad \textbf{(A.11)}$$

Equations (A.10) and (A.11) are often referred to as the Navier-Stokes equations.

A.4   CONSERVATION OF ENERGY

A.4.1   The Control Volume and Conservation Statement

We now return to the Eulerian viewpoint and consider a Cartesian coordinate volume element, dx by dy by dz, fixed in space. For simplicity we will initially assume a pure fluid. Conservation of energy requires that the energy stored within the volume equal the net inflow of heat across the boundary plus the work done on the fluid within the volume plus the energy produced within the volume.

A.4.2   Energy Storage

Energy can be stored within the volume as kinetic energy of the mass as a whole and as internal energy. Since we are formulating our equations on a *macroscopic* basis, the internal energy is to be regarded simply as a thermodynamic property. The energy storage per unit time is thus

$$\frac{\partial}{\partial t}(\rho\hat{u}+\tfrac{1}{2}\rho v^2)\,dx\,dy\,dz$$

where

$$v^2 = \mathbf{v}\cdot\mathbf{v} = v_x{}^2+v_y{}^2+v_z{}^2$$

A.4.3 Heat Inflow

Heat can flow into the control volume by convection and conduction. Consider the convective contribution first. As mass flows across the volume boundaries it brings with it energy $(\hat{u}+\tfrac{1}{2}v^2)$ per unit mass. For example, the convection in across the x-face per unit time is

$$\rho v_x(\hat{u}+\tfrac{1}{2}v^2)\,dy\,dz$$

while that out across the $x+dx$ face is

$$[\rho v_x(\hat{u}+\tfrac{1}{2}v^2)]_{x+dx}\,dy\,dz$$

The net inflow for the two faces is

$$-\frac{\partial}{\partial x}[\rho v_x(\hat{u}+\tfrac{1}{2}v^2)]\,dx\,dy\,dz$$

Similar terms arise for each of the other two pairs of faces. The sum of all these terms can be written

$$-\nabla\cdot[\rho\mathbf{v}(\hat{u}+\tfrac{1}{2}v^2)]\,dx\,dy\,dz$$

The conduction contribution is given by Fourier's law. Heat crosses the x-face by conduction at the rate

$$-k\frac{\partial T}{\partial x}\,dy\,dz$$

while the outflow across the $x+dx$ face is

$$-\left(k\frac{\partial T}{\partial x}\right)_{x+dx}\,dy\,dz$$

The net inflow for the two faces is

$$+\frac{\partial}{\partial x}\left(k\frac{\partial T}{\partial x}\right)\,dx\,dy\,dz$$

Again the other two pairs of faces give similar terms. The sum for all faces can be written

$$\nabla\cdot k\nabla T\,dx\,dy\,dz$$

A.4.4 Work Done

The rate at which work is done by the body force $\mathbf{g}$ is simply the force times the velocity,

$$\rho\mathbf{g}\cdot\mathbf{v}\,dx\,dy\,dz = (\rho g_x v_x+\rho g_y v_y+\rho g_z v_z)\,dx\,dy\,dz$$

Work can also be done by the surface forces acting on the volume element. The vector force acting on the x-face is the first column of the stress tensor times the area of the face. The rate at which work is done *on* the element at this face is the vector dot product of velocity and force. Denote the first, second, and third columns of the stress tensor $\mathbf{F}_x$, $\mathbf{F}_y$, and $\mathbf{F}_z$, respectively. The subscripts denote the direction of the normal to the face on which the force acts. Then, for example,

$$\mathbf{F}_x = \mathbf{i}(P+\tau_{xx}) + \mathbf{j}(\tau_{xy}) + \mathbf{k}(\tau_{xz}) \tag{A.12}$$

and the rate at which work is done on the element by this force is

$$\mathbf{v} \cdot \mathbf{F}_x \, dy \, dz = [(P+\tau_{xx})v_x + \tau_{xy}v_y + \tau_{xz}v_z] \, dy \, dz \tag{A.13}$$

At the $x+dx$ face work is done *by* the element, on the fluid of greater x, at the rate

$$(\mathbf{v} \cdot \mathbf{F}_x)_{x+dx} \, dy \, dz$$

The *net* work done *on* the element for these two faces is

$$-\frac{\partial}{\partial x}(\mathbf{v} \cdot \mathbf{F}_x) \, dx \, dy \, dz$$

The net work done on the element for all six faces is then

$$-\left[\frac{\partial}{\partial x}(\mathbf{v} \cdot \mathbf{F}_x) + \frac{\partial}{\partial y}(\mathbf{v} \cdot \mathbf{F}_y) + \frac{\partial}{\partial z}(\mathbf{v} \cdot \mathbf{F}_z) \right] dx \, dy \, dz$$

A.4.5   Volume Heat Sources

It is often easy to extend the applicability of our simple pure fluid model by including an equivalent volumetric heat source term, $\dot{Q}_V$. Sources of energy which can be so represented are resistive (I^2R) heating, neutron and fission fragment slowing or gamma ray absorption, photon absorption and emission, and so on. Since $\dot{Q}_V$ is defined as an energy production per unit volume, the production rate for the volume element is

$$\dot{Q}_V \, dx \, dy \, dz$$

A.4.6   The Total Energy Equation

Substituting into our statement of conservation of energy and dividing by the volume $dx \, dy \, dz$ gives

$$\frac{\partial}{\partial t}\rho\,(\hat{u}+\tfrac{1}{2}v^2) = -\nabla \cdot [\rho\mathbf{v}(\hat{u}+\tfrac{1}{2}v^2)] + \nabla \cdot k\nabla T + \rho\mathbf{g} \cdot \mathbf{v}$$

$$-\frac{\partial}{\partial x}(\mathbf{v} \cdot \mathbf{F}_x) - \frac{\partial}{\partial y}(\mathbf{v} \cdot \mathbf{F}_y) - \frac{\partial}{\partial z}(\mathbf{v} \cdot \mathbf{F}_z) + \dot{Q}_V \tag{A.14}$$

Since Eq. (A.14) represents conservation of both thermal and mechanical energy, it is not usually the most convenient starting point for the solution of heat transfer problems. We now proceed to derive, from Eq. (A.14), a simpler conservation equation which isolates the thermal effects.

A.4.7 The Thermal Energy Equation

The equation accounting for *mechanical* energy can be obtained by taking the dot product of the momentum equation with velocity $\mathbf{v}$. When subtracted from the total energy equation there is left an equation governing conservation of thermal energy. In Cartesian coordinates we take v_x times the x-momentum equation, v_y times the y-momentum equation, and v_z times the z-momentum equation, and subtract them all from the total energy equation, Eq. (A.14). The result can be rearranged through use of the mass conservation equation, Eq. (A.5), to yield

$$\rho\left(\frac{\partial \hat{u}}{\partial t}+\mathbf{v}\cdot\nabla\hat{u}\right)+P\nabla\cdot\mathbf{v}=\nabla\cdot k\nabla T+\mu\Phi+\dot{Q}_V \tag{A.15}$$

where the quantity $\mu\Phi$ represents the work irreversibly converted into heat by the viscous stresses.

$$\Phi = 2\left[\left(\frac{\partial v_x}{\partial x}\right)^2+\left(\frac{\partial v_y}{\partial y}\right)^2+\left(\frac{\partial v_z}{\partial z}\right)^2\right]-\frac{2}{3}(\nabla\cdot v)^2$$

$$+\left(\frac{\partial v_x}{\partial y}+\frac{\partial v_y}{\partial x}\right)^2+\left(\frac{\partial v_x}{\partial z}+\frac{\partial v_z}{\partial x}\right)^2+\left(\frac{\partial v_y}{\partial z}+\frac{\partial v_z}{\partial y}\right)^2 \tag{A.16}$$

The quantity $\mu\Phi$ is called the **viscous dissipation** or **frictional heating**; Φ is often simply called the **dissipation function**.

We often prefer to work with enthalpy $\hat{h}$, rather than internal energy $\hat{u}$. Substituting the thermodynamic relation

$$\hat{h} = \hat{u}+\frac{P}{\rho}$$

into Eq. (A.15) and again rearranging using the mass conservation equation gives

$$\rho\left(\frac{\partial \hat{h}}{\partial t}+\mathbf{v}\cdot\nabla\hat{h}\right)=\frac{\partial P}{\partial t}+\mathbf{v}\cdot\nabla P+\nabla\cdot k\nabla T+\mu\Phi+\dot{Q}_V$$

or

$$\rho\frac{D\hat{h}}{Dt}=\frac{DP}{Dt}+\nabla\cdot k\nabla T+\mu\Phi+\dot{Q}_V \tag{A.17}$$

For an ideal gas Eq. (A.17) becomes

$$\rho c_p\frac{DT}{Dt}=\frac{DP}{Dt}+\nabla\cdot k\nabla T+\mu\Phi+\dot{Q}_V \tag{A.18}$$

For an incompressible liquid with specific heat $c=c_p=c_v$, we go back to Eq. (A.15) to obtain

$$\rho c\frac{DT}{Dt}=\nabla\cdot k\nabla T+\mu\Phi+\dot{Q}_V \tag{A.19}$$

A.4.8 The Thermal Energy Equation for a Mixture

The preceding derivation of the energy equations assumed a pure fluid. But if a situation involving simultaneous heat and mass transfer is to be considered, we require a thermal energy equation valid for a mixture of chemical species.

Fortunately we do not have to redo the complete derivation, since we can argue as follows: For a pure fluid, conduction is the only diffusive mechanism of heat flow; hence Fourier's law was used which resulted in the term $\nabla \cdot k\nabla T$; see Eq. (A.17), for example. More generally this term may be written $-\nabla \cdot \mathbf{q}$, where $\mathbf{q}$ is the diffusive heat flux, that is, the heat flux relative to the mass average velocity. To extend our results to a mixture of chemical species we simply must recognize that there are now three contributions to $\mathbf{q}$. First there is ordinary conduction described by Fourier's law, $-k\nabla T$, where k is the *mixture* thermal conductivity. Second, there is a contribution due to interdiffusion of species, given by $\Sigma_i \mathbf{j}_i \hat{h}_i$. This contribution is easy to see if we imagine a binary mixture where the species diffuse in opposite directions and $\mathbf{j}_1 = -\mathbf{j}_2$. Then the interdiffusion contribution becomes $\mathbf{j}_1(\hat{h}_1 - \hat{h}_2)$ and a net transport of energy occurs whenever the species enthalpies differ. Third, there is diffusional conduction (also called the diffusion-thermo effect or Dufour effect). This effect is a second order contribution which is negligible for most physical situations. We will ignore diffusional conduction. Thus for a mixture we can write $\mathbf{q} = -k\nabla T + \Sigma_i \mathbf{j}_i \hat{h}_i$ and substitute in, for example, Eq. (A.17), to obtain

$$\rho \frac{D\hat{h}}{Dt} = \frac{DP}{Dt} + \nabla \cdot k\nabla T - \nabla \cdot \sum_i \mathbf{j}_i \hat{h}_i + \mu\Phi + \dot{Q}_V \qquad \textbf{(A.20)}$$

Finally we note that, for a *nonreacting* mixture, the term we have added, $\nabla \cdot (\sum_i \mathbf{j}_i \hat{h}_i)$ is often of minor significance. But when endothermic or exo-thermic reactions occur, the term can play a dominant role. For a reacting mixture the species enthalpies,

$$\hat{h}_i = \hat{h}_i^\circ + \int_{T^\circ}^{T} c_{p,i}\, dT$$

must be written with a consistent set of heats of formation $\hat{h}_i^\circ$.

A.5 CONSERVATION OF CHEMICAL SPECIES

A.5.1 The Control Volume and Conservation Statement

Again we take the Eulerian viewpoint and consider a Cartesian coordinate volume element dx by dy by dz, fixed in space. Our conservation statement is that the time rate of storage of species i within the volume equals the net rate of inflow of species i across the boundary plus the production rate of species i within the volume due to chemical reactions.

A.5.2 Species Storage

The time rate of storage of species i within the control volume is simply

$$\frac{\partial \rho_i}{\partial t}\, dx\, dy\, dz$$

A.5.3 Species Inflow

Species i flows into the control volume by convection and diffusion. First consider the convective contribution. The mass average velocity sweeps species i into and out of the volume. The net rate may be found as was done in the derivation of the mass conservation equation and is

$$-\nabla \cdot (\rho_i \mathbf{v}) \, dx \, dy \, dz$$

Similarly, the net rate of diffusion into the volume is

$$-\nabla \cdot \mathbf{j}_i \, dx \, dy \, dz$$

In fact, this term and the previous one could have been written together as

$$-\nabla \cdot (\rho_i \mathbf{v} + \mathbf{j}_i) = -\nabla \cdot \mathbf{n}_i$$

A.5.4 Species Production

Species i may be produced at the rate $\dot{r}_i$ per unit volume due to homogeneous chemical reactions. The rate of production of i within the volume is then

$$\dot{r}_i \, dx \, dy \, dz$$

A.5.5 The Conservation Equation for Species i

Substitution in a statement of conservation and dividing by the volume $dx \, dy \, dz$ gives

$$\frac{\partial \rho_i}{\partial t} = -\nabla \cdot (\rho_i \mathbf{v}) - \nabla \cdot \mathbf{j}_i + \dot{r}_i = -\nabla \cdot \mathbf{n}_i + \dot{r}_i$$

This relation may be written in terms of mass fraction m_i by substituting $\rho_i = m_i \rho$. The terms containing ρ are simplified via the mass conservation equation as follows:

$$\frac{\partial \rho_i}{\partial t} + \nabla \cdot (\rho_i \mathbf{v}) = \frac{\partial}{\partial t}(m_i \rho) + \nabla \cdot (m_i \rho \mathbf{v})$$

$$= m_i \left(\frac{\partial \rho}{\partial t} + \nabla \cdot \rho \mathbf{v} \right) + \rho \left(\frac{\partial m_i}{\partial t} + \mathbf{v} \cdot \nabla m_i \right)$$

$$= \rho \left(\frac{\partial m_i}{\partial t} + \mathbf{v} \cdot \nabla m_i \right)$$

$$= \rho \frac{D m_i}{D t}$$

and the species conservation equation becomes

$$\bullet \qquad \rho \frac{D m_i}{D t} = -\nabla \cdot \mathbf{j}_i + \dot{r}_i \qquad \qquad (A.21)$$

The conservation of species equation can also be derived in terms of mole concentrations and mole fractions. When dealing with gas mixtures at constant pressure and temperature, it is usually preferred to use molar quantities.

A.6   USE OF THE CONSERVATION EQUATIONS TO SET UP PROBLEMS

A.6.1   Introduction

The conservation equations, summarized in Table A.1 at the bottom of this page, are seen to be a formidable set of simultaneous partial differential equations. There are four independent variables, three space coordinates and a time coordinate, say x, y, z, t. Consider first a pure fluid: then there are five equations; mass conservation, three momentum equations, and conservation of energy. The accompanying five dependent variables are pressure, three components of velocity, and temperature. Also, a thermodynamic equation of state serves to relate density to the pressure, temperature, and composition. For a mixture of n chemical species, there are n species conservation equations, but one is redundant,

Table A.1   Summary of Conservation Equations

Conservation of Mass

$$\frac{\partial \rho}{\partial t} + \nabla \cdot (\rho \mathbf{v}) = 0$$

Conservation of Momentum

$$\rho \frac{D}{Dt} \begin{bmatrix} v_x \\ v_y \\ v_z \end{bmatrix} = - \begin{bmatrix} P + \tau_{xx} & \tau_{yx} & \tau_{zx} \\ \tau_{xy} & P + \tau_{yy} & \tau_{zy} \\ \tau_{xz} & \tau_{yz} & P + \tau_{zz} \end{bmatrix} \begin{bmatrix} \frac{\partial}{\partial x} \\ \frac{\partial}{\partial y} \\ \frac{\partial}{\partial z} \end{bmatrix} + \rho \begin{bmatrix} g_x \\ g_y \\ g_z \end{bmatrix}$$

Conservation of Energy

$$\rho \frac{D\hat{u}}{Dt} + P\nabla \cdot \mathbf{v} = \nabla \cdot k\nabla T + \mu \Phi + \dot{Q}_V$$

$$\rho \frac{D\hat{h}}{Dt} = \frac{DP}{Dt} + \nabla \cdot k\nabla T + \mu \Phi + \dot{Q}_V$$

Conservation of Species

$$\rho \frac{Dm_i}{Dt} = -\nabla \cdot \mathbf{j}_i + \dot{r}_i$$

since they sum to yield the mass conservation equation already listed. Therefore we use only $(n-1)$ species conservation equations together with the relation that the sum of the mass fractions equals unity.

General solution, even by numerical means, of the full equations in the four independent variables is much beyond present capabilities. Fortunately, however, many problems of engineering interest are adequately described by simplified forms of the full conservation equations, and these forms can often be solved. Chapters 2 through 5 contained examples of such problems, and there are a large number more. Formerly we set up the governing equations for these problems directly from first principles; now we are in a position to obtain the governing equations by simply deleting superfluous terms in the full conservation equations. The following remarks apply directly to laminar flows. In the case of turbulent flows some caution must be exercised. For example, on an average basis a flow may be two-dimensional and steady, but if it is unstable and as a result turbulent, fluctuations in the three components of velocity may be occurring with respect to time and the three spatial coordinates. Then the remarks about dropping terms apply only to the time-averaged equations.

A.6.2   Deleting Superfluous Terms

Often the manner in which a particular problem is posed immediately implies that certain terms are identically zero. For example, a timewise steady-state solution might be sought; thus time derivatives would be set equal to zero. Additional terms may be judged to be negligibly small based on physical intuition or on experimental evidence. Some resulting classes of simplified problems are:

1. Constant transport properties
2. Constant density
3. Timewise steady flow (or quasi-steady flow)
4. Two-dimensional flow
5. One-dimensional flow
6. Established flows (no dependence on the streamwise coordinate)
7. Rigid bodies or stagnant fluids

Terms may also be shown to be negligibly small by order of magnitude estimates. Some classes of flow which result are:

1. Creeping flows: inertia terms are negligible
2. Forced flows: gravity forces are negligible
3. Natural convection: gravity forces predominate
4. Low speed gas flows: viscous dissipation and compressibility terms are negligible
5. Boundary layer flows: streamwise diffusion terms are negligible

The student is advised to return to Chapters 2 through 5 and set up governing equations to the various problems in the manner described above.

A.6.3   Boundary and Initial Conditions

A complete mathematical statement of a problem requires specification of boundary conditions around the domain under consideration. Each boundary condition is based on a physical statement or principle; each physical fact gives rise to one, and only one, boundary condition at a particular location. Satisfactory boundary conditions are of three general types.

1. Specification of the dependent variable along the boundary, for example, the *no slip* condition for viscous flow which requires that the component of velocity parallel to a stationary surface be zero.
2. Specification of the normal component of the gradient of a dependent variable along the boundary, for example, for an insulated wall we require that the derivative of temperature normal to the wall be zero.
3. Provision of an algebraic relation which relates the value of the dependent variable and its normal gradient to the velocity component normal to the boundary. Such a specification is often required when there is mass transfer across the boundary.

More complicated formulations arise when the conservation equations must be simultaneously solved in two adjacent domains. Then the boundary conditions take the form of continuity statements, for example, continuity of mass flow normal to the interface; or compatibility constraints, for example, the requirement of thermodynamic equilibrium in the form of Henry's law.

But the student already knows about boundary conditions; they were extensively used in Chapters 2 through 5, 9, and 10. Thus he can peruse these chapters and study the variety of types that were used. One type which does not appear there is that applying to the pressure difference across a curved interface between two phases. For completeness we give

$$\Delta P = \sigma_t \left(\frac{1}{r_1} + \frac{1}{r_2} \right) \tag{A.22}$$

where the higher pressure is on the concave side of the interface, σ_t is the surface tension, and r_1 and r_2 are the principal radii of curvature of the interface.

Initial conditions are required if the problem cannot be treated as quasi-steady and the time derivatives are retained. We must specify values of each dependent variable at some point in time before a solution is commenced. Prediction of where the system is going is then accomplished by a forward integration in time from these initial values.

A.7   SIMILARITY

A.7.1   Dimensionless Equations and Boundary Conditions

In order to introduce the concept of *similarity* we see how the equations governing a problem can be made dimensionless. For simplicity we consider a binary mixture with constant transport properties and density flowing at low speed. Body forces, volume heat sources, and chemical reactions are neglected. Since compressibility effects will prove to be negligible under these conditions, the reasoning below applies to gases at low speeds as well as to liquids. The conservation equations are, from Eqs. (A.5), (A.11), (A.18), and (A.21),

$$\text{Mass:} \qquad \nabla \cdot \mathbf{v} = 0 \qquad \text{(A.23)}$$

$$\text{Momentum:} \qquad \rho \frac{D\mathbf{v}}{Dt} = -\nabla P + \mu \nabla^2 \mathbf{v} \qquad \text{(A.24)}$$

$$\text{Thermal energy:} \quad \rho c \frac{DT}{Dt} = +k\nabla^2 T + \mu \Phi \qquad \text{(A.25)}$$

$$\text{Species:} \qquad \frac{Dm_1}{Dt} = \mathscr{D}_{12} \nabla^2 m_1 \qquad \text{(A.26)}$$

Let a characteristic length of the system be L, for example, tube diameter or plate length. Suppose that the flow is forced, say by a blower, so that we can identify a characteristic velocity V. For flow transverse to a cylinder we would choose the free stream velocity; for flow inside a tube the bulk velocity would be suitable. In a stirred tank the speed of the stirrer tip would be used. In order to render Eqs. (A.23) through (A.26) dimensionless we simply scale all lengths with L, all velocities with V, and as a consequence, time with L/V. We define the dimensionless variables

$$x^* = \frac{x}{L}, \qquad y^* = \frac{y}{L}, \qquad z^* = \frac{z}{L} \qquad \text{(A.27)}$$

$$v_x{}^* = \frac{v_x}{V}, \quad v_y{}^* = \frac{v_y}{V}, \quad v_z{}^* = \frac{v_z}{V} \qquad \left(\mathbf{v}^* = \frac{\mathbf{v}}{V} \right) \qquad \text{(A.28)}$$

$$t^* = \frac{t}{L/V} \qquad \text{(A.29)}$$

And in anticipation of how we are going to rearrange the equations, we choose to scale pressure with ρV^2. The dimensionless pressure is

$$P^* = \frac{P}{\rho V^2} \qquad \text{(A.30)}$$

When the boundary layer concept is discussed in Section A.8, we will see how this scaling can be arrived at in a less arbitrary manner. The required transformation is therefore

$$\frac{\partial}{\partial x} = \frac{\partial}{\partial x^*} \frac{dx^*}{dx} = \frac{1}{L} \frac{\partial}{\partial x^*}, \quad \frac{\partial}{\partial y} = \frac{1}{L} \frac{\partial}{\partial y^*}, \quad \frac{\partial}{\partial z} = \frac{1}{L} \frac{\partial}{\partial z^*}$$

$$\frac{\partial}{\partial t} = \frac{\partial}{\partial t^*} \frac{dt^*}{dt} = \frac{V}{L} \frac{\partial}{\partial t^*}$$

$$v_x = V v_x{}^*, \quad v_y = V v_y{}^*, \quad v_z = V v_z{}^*, \quad P = \rho V^2 P^*$$

Consider, for example, the x-momentum equation, which becomes

$$\rho\left(\frac{V^2}{L}\frac{\partial v_x{}^*}{\partial t^*}+\frac{V^2}{L}v_x{}^*\frac{\partial v_x{}^*}{\partial x^*}+\frac{V^2}{L}v_y{}^*\frac{\partial v_x{}^*}{\partial y^*}+\frac{V^2}{L}v_z{}^*\frac{\partial v_x{}^*}{\partial z^*}\right)=-\rho\frac{V^2}{L}\frac{\partial P^*}{\partial x^*}$$

$$+\mu\left(\frac{V}{L^2}\frac{\partial^2 v_x{}^*}{\partial x^{*2}}+\frac{V}{L^2}\frac{\partial^2 v_x{}^*}{\partial y^{*2}}+\frac{V}{L^2}\frac{\partial^2 v_x{}^*}{\partial z^{*2}}\right)$$

Dividing through by $\rho V^2/L$ there results

$$\frac{\partial v_x{}^*}{\partial t^*}+v_x{}^*\frac{\partial v_x{}^*}{\partial x^*}+v_y{}^*\frac{\partial v_x{}^*}{\partial y^*}+v_z{}^*\frac{\partial v_x{}^*}{\partial z^*}=-\frac{\partial P^*}{\partial x^*}+\frac{\mu}{VL\rho}\left(\frac{\partial^2 v_x{}^*}{\partial x^{*2}}+\frac{\partial^2 v_x{}^*}{\partial y^{*2}}+\frac{\partial^2 v_x{}^*}{\partial z^{*2}}\right)$$

The parameter $VL\rho/\mu = Re$ is the Reynolds number, a measure of the magnitude of the inertia forces $\rho V^2(L^2)$ compared with the viscous forces $\mu(V/L)(L^2)$. The parameter Re can be seen to be the only one appearing in the dimensionless x-momentum equation.

Returning to the vector form of the momentum equation, we can write the mass conservation equation and momentum conservation equation as simply

- $$\nabla^* \cdot \mathbf{v}^* = 0 \tag{A.31}$$

- $$\frac{D\mathbf{v}^*}{Dt^*}=-\nabla^*P^*+\frac{1}{Re}\nabla^{*2}\mathbf{v}^* \tag{A.32}$$

We now define a dimensionless temperature $T^* = (T-T_e)/(T_s-T_e)$ and dimensionless concentration $m^* = (m_1-m_{1,e})/(m_{1,s}-m_{1,e})$. The subscript e refers to external free stream, or average conditions, and the subscript s refers to conditions adjacent to a bounding surface, across which transfer of heat and mass might occur. Transformation of Eqs. (A.25) and (A.26) yields, respectively,

- $$\frac{DT^*}{Dt^*}=\frac{1}{Re\ Pr}\nabla^{*2}T^*+\frac{2Ec}{Re}\Phi^* \tag{A.33}$$

- $$\frac{Dm^*}{Dt^*}=\frac{1}{Re\ Sc}\nabla^{*2}m^* \tag{A.34}$$

The additional dimensionless parameters which appear (Pr, Sc, and Ec) have all been encountered earlier in the text.

Prandtl number

$$Pr=\frac{\nu}{\alpha}=\frac{\mu c_p}{k},\qquad \frac{\text{diffusivity for momentum (vorticity)}}{\text{diffusivity for heat}}$$

Schmidt number

$$Sc=\frac{\nu}{\mathscr{D}}=\frac{\mu}{\rho\mathscr{D}},\qquad \frac{\text{diffusivity for momentum (vorticity)}}{\text{diffusivity for mass species}}$$

Eckert number

$$Ec=\frac{V^2}{2c_p(T_s-T_e)},\qquad \frac{\text{kinetic energy of flow}}{\text{enthalpy difference}}$$

Dimensionless boundary conditions are best discussed in terms of a specific situation. For example, consider transverse flow over a cylinder. Appropriate boundary conditions would be as follows

In the free stream far from the cylinder,

$$\mathbf{v} = \mathbf{V}, \quad T = T_e, \quad m_1 = m_{1,e} \tag{A.35}$$

Adjacent to the surface of the cylinder,

$$\mathbf{v} \cdot \mathbf{t} = 0, \quad \rho \mathbf{v} \cdot \mathbf{n} = n_s, \quad T = T_s, \quad m_1 = m_{1,s} \tag{A.36}$$

In Eq. (A.36) the first condition states that the tangential component of the velocity is zero (no-slip), while the second condition relates the normal component of the velocity to the net rate of mass transfer. The boundary conditions transform into the following set:

Free stream

$$\mathbf{v}^* = 1, \quad T^* = 0, \quad m^* = 0 \tag{A.37}$$

Surface

$$\mathbf{v}^* \cdot \mathbf{t} = 0, \quad \mathbf{v}^* \cdot \mathbf{n} = n_s^*, \quad T^* = 1, \quad m^* = 1 \tag{A.38}$$

Notice if the cylinder were of radius R and were spinning with angular velocity Ω, then the no-slip condition would be replaced by $\mathbf{v} \cdot \mathbf{t} = \Omega R$, which transforms into $\mathbf{v}^* \cdot \mathbf{t} = \Omega^* R^*$, where $\Omega^* = \Omega/(V/L)$.

A.7.2   Dimensionless Wall Transfer Rates

Our primary purpose is the prediction of drag, heat transfer, and mass transfer at surfaces bounding the flow. Thus these quantities are usually the end result of solving the governing conservation equations, and, too, must be made dimensionless. For simplicity assume that the flow is two dimensional (x, y) and the surface is lined up in the x-direction.

Drag

$$\tau_s = -\mu \frac{\partial v_x}{\partial y}\bigg|_s$$

$$= -\mu \frac{V}{L} \frac{\partial v_x^*}{\partial y^*}\bigg|_{y^* = 0}$$

neglecting any $\partial v_y/\partial x$ caused by mass transfer. We define the dimensionless skin friction coefficient c_f such that

$$\frac{c_f}{2} = \tau_s^* = \frac{\tau_s}{\rho V^2} \qquad \text{(disregarding, again, the sign of } \tau_s)$$

$$\frac{c_f}{2} = \frac{1}{Re} \frac{\partial v_x^*}{\partial y^*}\bigg|_{y^* = 0} \tag{A.39}$$

Heat transfer

$$q_s = -k\frac{\partial T}{\partial y}\bigg|_{y=0} = \frac{k(T_s - T_e)}{L}\left(-\frac{\partial T^*}{\partial y^*}\bigg|_{y^*=0}\right)$$

and we define the dimensionless Nusselt number for heat transfer as

$$Nu = q_s{}^* = \frac{q_s}{(k/L)(T_s - T_e)} = -\frac{\partial T^*}{\partial y^*}\bigg|_{y^*=0} \tag{A.40}$$

Mass transfer

$$j_{1,s} = -\rho\mathscr{D}_{12}\frac{\partial m_1}{\partial y}\bigg|_{y=0} = \frac{\rho\mathscr{D}_{12}(m_{1,s} - m_{1,e})}{L}\left(-\frac{\partial m^*}{\partial y^*}\bigg|_{y^*=0}\right)$$

and we define the dimensionless Nusselt number for mass transfer as

$$Nu_m = j_s{}^* = \frac{j_{1,s}}{(\rho\mathscr{D}_{12}/L)(m_{1,s} - m_{1,e})} = -\frac{\partial m^*}{\partial y^*}\bigg|_{y^*=0} \tag{A.41}$$

A.7.3   Similarity in Low Speed Forced Flows

First, what do we mean by a low speed flow? Equation (A.33) may be written in the form

$$\frac{DT^*}{Dt^*} = \frac{1}{Re\,Pr}(\nabla^{*2}T^* + 2\,Ec\,Pr\,\Phi^*) \tag{A.42}$$

By restricting our attention to low speed flows we hope to be able to disregard viscous dissipation effects. From Eq. (A.42) it is clear that if the product $Ec\,Pr$ is small then viscous dissipation will be negligible compared with conductive heat transport. For gases a pressure gradient term $2\,Ec\,DP^*/Dt^*$ appears, and for gases Pr is of order unity. So if $Ec\,Pr$ is small, so will be Ec. For forced flows, where P scales with ρV^2, the consequence is that the compressibility effects are negligible whenever viscous dissipation effects are negligible. (In Section A.7.4 we will see that for *natural* convection compressibility effects in tall systems may be significant.) The dimensionless thermal energy equation is then

$$\frac{DT^*}{Dt^*} = \frac{1}{Re\,Pr}\nabla^{*2}T^* \tag{A.43}$$

We are now in a position to discuss the concept of "similarity." We have shown that the dimensionless mass and momentum equations contain only one parameter, the Reynolds number. If the walls are impermeable and stationary, the dimensionless boundary conditions introduce no further parameters. It follows that the dimensionless velocity and hence dimensionless wall drag are functions only of the dimensionless independent variables and the Reynolds number. The wall averaged drag $\bar{\tau}_s$ is therefore given by

•

$$\bar{\tau}_s{}^* = \frac{\bar{C}_f}{2} = F(Re) \tag{A.44}$$

If we have the complex situation of net mass transfer from a spinning cylinder (mentioned in Section A.7.1), then the dimensionless mass transfer rate and the dimensionless angular velocity are additional parameters,

$$\frac{\overline{C_f}}{2} = F(Re, n_s{}^*, \Omega^*) \qquad \text{(A.45)}$$

The dimensionless energy equation contains only two parameters, the Reynolds and Prandtl numbers. Although these appear as the product $Re\, Pr = Pe$, the Peclet number, the Reynolds number itself is an independent parameter through the coupling to the momentum equation via the velocity field $\mathbf{v}^*$. The wall averaged heat transfer is therefore given by

- $$\overline{Nu} = F(Re, Pr) \qquad \text{(A.46)}$$

Likewise, for mass transfer at low rates of net mass transfer,

- $$\overline{Nu}_m = F(Re, Sc) \qquad \text{(A.47)}$$

We can now see how to run a proper model test. If the same dimensionless differential equations subject to the same dimensionless boundary conditions govern the modeling experiment, as govern the full-scale prototype, the dimensionless solutions are identical. We can thus solve the conservation equations by making careful measurements of the pressure, velocity, temperature, and species concentration fields around, or inside, a model and reporting these data in dimensionless form. The wall friction, heat transfer, and mass transfer are likewise measured, or in some cases calculated. For example, the heat transfer may be calculated from the temperature field near the wall using Fourier's law. These too are reported in dimensionless form.

The solution of problems by model testing is essential to obtain information about flows in three dimensions over bodies of complex shapes or about turbulent flows. Analytical and numerical skills are not yet sufficiently advanced to treat these complex technological problems without introducing at least some experimental observation. Perhaps the most well-known application of modeling is the use of wind tunnels for aerodynamic testing. The advance of the art of aeronautics would have been severely handicapped had wind tunnels not been widely used for model testing.

Notice particularly that our results hold true for both laminar and turbulent flow. The same value of Re for a model as exists for the prototype will duplicate the turbulent flow (unless very extreme efforts are made to eliminate *disturbances* which trigger the instability). We have, however, made the restrictive assumption of constant properties. When property variations are important, we must introduce $\mu^* = \mu/\mu_0$ and so on. For similarity we require μ^*, ρ^*, k^*, and so on, to vary with dimensionless temperature, pressure, and composition in the same manner for the prototype as they do for the model. This requirement often necessitates using the same fluid and the same temperatures in the thermal boundary conditions.

A.7.4 Similarity in Natural Convection

In Chapter 5 it was shown that the dimensionless parameter governing free convection arising in a gravitational field due to density differences is the Grashof number

$$Gr = \frac{(\Delta\rho/\rho)gL^3}{\nu^2} \tag{A.48}$$

The Grashof number may be interpreted as a Reynolds number squared for buoyancy-driven convection. If there is mixed free and forced convection (a small blower is added to the electronic gear of our example in Chapter 5), then both the Reynolds number characterizing the forced flow and the Grashof number are important. It was shown in Chapter 5 that for an order of magnitude estimate as to whether forced or free convection is more important, we compare the ratio

$$\frac{\rho V^2}{\rho V_0^2} = \frac{\rho V^2}{\Delta\rho gL} = \frac{Re^2}{Gr} = Fr \tag{A.49}$$

This ratio is sometimes called the *Froude number*. When Fr is large, forced convection is dominant, and we do not care what value Gr has. When Fr is small, free convection is dominant, and we do not care what value Re is. When Fr is of order unity, we must duplicate both Re and Gr in a model test. This modeling is somewhat difficult as it requires a wind tunnel operating at elevated pressures, if the model size is to be less than the prototype size.

The Grashof number scales the effect of gravity on the momentum equation. For compressible flow there is also an effect of gravity on the energy equation. In Eq. (A.18), for steady flow, there is the term $\mathbf{v} \cdot \nabla P$. For free convection, taking z in the vertical direction, $\nabla P = -\mathbf{k}\rho g$. Hence the dimensionless parameter scaling the pressure gradient term is obtained by replacing V^2 in the Eckert number by gL to yield the dimensionless parameter $gL/c_p \Delta T$. In most engineering problems this term is negligible. However, in situations which verge upon being meteorological in scale, such as tall smoke stacks or plumes from these stacks, the pressure gradient term in quite important. It is this effect which causes the Santa Ana winds, which blow occasionally in Los Angeles during the winter, to be hot. The high plateaus where these winds originate are relatively cool, but the gas is compressionally heated in changing elevation. Such winds also blow in New Zealand and South Africa.

A.8 THE LAMINAR BOUNDARY LAYER EQUATIONS

A.8.1 Order of Magnitude Estimates

Our discussion of similarity has shown that it is valuable to make the governing equations and their boundary conditions dimensionless. Before carrying out an analytical or numerical solution, one should also make an order of magnitude assessment of all terms in the equations. Often a problem can be greatly simplified by discovering that a term, which would be very difficult to handle if large, is in fact negligibly small. Even if the primary thrust of the investigation is experimental, making the equations dimensionless and estimating the orders of magnitude of the terms is good practice. It is usually not possible for an experimental test to satisfy (simulate) all conditions exactly; a good engineer will focus on the most important conditions.

Making an order of magnitude estimate is best done at the same time the governing equations are rendered dimensionless. What we do is to forsake the simplicity of scaling all lengths by L, all velocities by V, and so on. Instead, for boundary layer flows, allowance is made for the fact that lengths transverse to the main flow scale with a much shorter length than those measured in the direction of the main flow.

A.8.2 The Momentum Equation for Flat Plate Boundary Layer Flow

Consider a flat plate of length L with a fluid flowing parallel to it, as shown in Figure A.4. We expect a layer of fluid, moving slower than the oncoming stream, to build up adjacent to the plate due to the viscous shear stresses. At the distance L down the plate this region might be characterized by a thickness δ. We must choose a suitable δ with which to scale lengths in the y-direction. Assume for simplicity that the flow is steady ($\partial/\partial t = 0$), laminar, two-dimensional ($v_z = 0$, $\partial/\partial z = 0$), incompressible ($\rho = $ constant), and has constant viscosity. Furthermore we wish to investigate situations where the condition $\delta \ll L$ is met; this situation corresponds to Prandtl's concept of a thin **boundary layer** in which the

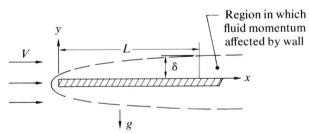

Region in which fluid momentum affected by wall

Figure A.4 A momentum boundary layer.

fluid velocity changes from the free stream value to zero. The dimensional governing equations are the following:

Conservation of mass

$$\frac{\partial v_x}{\partial x} + \frac{\partial v_y}{\partial y} = 0 \tag{A.50}$$

Conservation of momentum

$$\rho v_x \frac{\partial v_x}{\partial x} + \rho v_y \frac{\partial v_x}{\partial y} = -\frac{\partial P}{\partial x} + \mu \left[\frac{\partial^2 v_x}{\partial x^2} + \frac{\partial^2 v_x}{\partial y^2} \right] \tag{A.51}$$

$$\rho v_x \frac{\partial v_y}{\partial x} + \rho v_y \frac{\partial v_y}{\partial y} = -\frac{\partial P}{\partial y} + \mu \left[\frac{\partial^2 v_y}{\partial x^2} + \frac{\partial^2 v_y}{\partial y^2} \right] - \rho g \tag{A.52}$$

We choose to make the equations dimensionless in such a manner that the dimensionless dependent variables are order of magnitude unity or less. The lengths x and y, and the velocity v_x pose no problem.

$$x^* = \frac{x}{L}, \quad y^* = \frac{y}{\delta}, \quad v_x^* = \frac{v_x}{V} \tag{A.53}$$

The velocity v_y does pose a problem, but we resolve it in the same way as we did the problem of scaling y; we assign a scaling value v_S and let the equations themselves tell us what the appropriate order of magnitude is.

$$v_y^* = \frac{v_y}{v_S} \tag{A.54}$$

The pressure terms present a good puzzle. Pressure variations arise due to (1) changes in the hydrostatic pressure that would exist if there were no wall present, and (2) changes in velocity caused by the wall disturbing the main flow. The hydrostatic pressure which would exist if there were no wall is readily found. We simply set v_y identically equal to zero, and the y-momentum equation reduces to

$$0 = \frac{\partial P_0}{\partial y} + \rho g$$

Integrating gives

$$P_0 = P_{0,0} - \rho g y \tag{A.55}$$

where $P_{0,0}$, the constant of integration, is the hydrostatic pressure at $y = 0$. We now choose the dimensionless pressure to be the difference between P and P_0 divided by a scaling value P_S. What is P_S? We will have to wait and see. When we started to solve our problem we thought we had just one unknown, δ. But now we find we have three: δ, v_S, and P_S.

The rules of transformation are quite simple:

$$\frac{\partial}{\partial x} = \frac{\partial}{\partial x^*} \frac{dx^*}{dx} = \frac{1}{L} \frac{\partial}{\partial x^*} \tag{A.56}$$

$$\frac{\partial}{\partial y} = \frac{\partial}{\partial y^*} \frac{dy^*}{dy} = \frac{1}{\delta} \frac{\partial}{\partial y^*} \tag{A.57}$$

$$v_x = V v_x^*, \quad v_y = v_S v_y^*, \quad P = P_0 + P_S P^* \tag{A.58}$$

The mass conservation equation transforms into

$$\frac{V}{L} \frac{\partial v_x^*}{\partial x^*} + \frac{v_S}{\delta} \frac{\partial v_y^*}{\partial y^*} = 0 \tag{A.59}$$

In Figure A.4 visualize measuring the velocity v_x along a horizontal traverse near the plate, starting from the leading edge and progressing along to $x = L$. The velocity v_x will change from the value V to about zero in a length L; thus we conclude that $\partial v_x{}^*/\partial x^*$ is order of magnitude -1. Looking at the second term in Eq. (A.59), we intend $\partial v_y{}^*/\partial y^*$ to be order of magnitude $+1$. Equation (A.59) thus requires that v_S be order of magnitude $V\delta/L$, for the two terms must sum to zero. We are free to specify a precise value of v_S; we take

$$v_S \equiv \frac{\delta}{L}V \qquad\qquad \textbf{(A.60)}$$

so that our equations will be of simple form. Next the x-momentum equation is to be transformed. We make use of our new-found knowledge of v_S,

$$\frac{\rho V^2}{L}v_x{}^*\frac{\partial v_x{}^*}{\partial x^*}+\frac{\rho V^2}{L}v_y{}^*\frac{\partial v_x{}^*}{\partial y^*}=-\frac{P_S}{L}\frac{\partial P^*}{\partial x^*}+\mu\left[\frac{V}{L^2}\frac{\partial^2 v_x{}^*}{\partial x^{*2}}+\frac{V}{\delta^2}\frac{\partial^2 v_x{}^*}{\partial y^{*2}}\right]$$

$$v_x{}^*\frac{\partial v_x{}^*}{\partial x^*}+v_y{}^*\frac{\partial v_x{}^*}{\partial y^*}=-\frac{P_S}{\rho V^2}\frac{\partial P^*}{\partial x^*}+\frac{\mu}{VL\rho}\left[\frac{\partial^2 v_x{}^*}{\partial x^{*2}}+\frac{L^2}{\delta^2}\frac{\partial^2 v_x{}^*}{\partial y^{*2}}\right] \qquad \textbf{(A.61)}$$

Also we have y-momentum

$$\frac{\rho V^2\delta}{L^2}v_x{}^*\frac{\partial v_y{}^*}{\partial x^*}+\frac{\rho V^2\delta}{L^2}v_y{}^*\frac{\partial v_y{}^*}{\partial y^*}=-\frac{P_S}{\delta}\frac{\partial P^*}{\partial y^*}+\mu\left[\frac{V\delta}{L^3}\frac{\partial^2 v_y{}^*}{\partial x^{*2}}+\frac{V}{L\delta}\frac{\partial^2 v_y{}^*}{\partial y^{*2}}\right]$$

$$v_x{}^*\frac{\partial v_y{}^*}{\partial x^*}+v_y{}^*\frac{\partial v_y{}^*}{\partial y^*}=-\frac{P_S}{\rho V^2}\frac{L^2}{\delta^2}\frac{\partial P^*}{\partial y^*}+\frac{\mu}{VL\rho}\left[\frac{\partial^2 v_y{}^*}{\partial x^{*2}}+\frac{L^2}{\delta^2}\frac{\partial^2 v_y{}^*}{\partial y^{*2}}\right] \qquad \textbf{(A.62)}$$

Now we must construct some logical trains. At the outset we said we were going to investigate situations where the condition $\delta/L \ll 1$ is met. We will *assume* $L/\delta \gg 1$ and examine the consequences. Inspecting Eq. (A.61) we observe that $\partial^2 v_x{}^*/\partial y^{*2}$ is order of magnitude -1, since $\partial v_x{}^*/\partial y^*$ is order of magnitude unity near the wall and is near zero at $y = \delta(y^* = 1)$. Therefore $(L^2/\delta^2) \times (\partial^2 v_x{}^*/\partial y^{*2})$ is order of (L^2/δ^2) and we conclude that either $\mu/VL\rho$ is very small *or* $(P_S/\rho V^2)(\partial P^*/\partial x^*)$ is very large. To find out more about the pressure gradient term we can see what form Eq. (A.61) takes in the free stream outside the boundary layer. We can drop the viscous stress terms to obtain

$$v_x{}^*\frac{\partial v_x{}^*}{\partial x^*}+v_y{}^*\frac{\partial v_x{}^*}{\partial y^*}=-\frac{P_S}{\rho V^2}\frac{\partial P^*}{\partial x^*} \qquad\qquad \textbf{(A.63)}$$

If we have chosen the scaling velocity V sensibly, that is, order of magnitude comparable to the highest velocity in the flow field, then the terms on the left-hand side of Eq. (A.63) are of order no greater than 1. Thus $(P_S/\rho V^2)(\partial P^*/\partial x^*)$ is also of order no greater than unity (accordingly we can choose $P_s \equiv \rho V^2$). It follows that $\mu/VL\rho$ is very small. In fact, $VL\rho/\mu = Re$, the Reynolds number, must be order of magnitude L^2/δ^2. We have therefore succeeded in estimating δ.

$$\frac{\delta}{L} = \frac{1}{Re^{1/2}} \qquad\qquad \textbf{(A.64)}$$

Is δ/L often small as we have assumed? Unless the density is very low, as for rocket propelled flight in the upper atmosphere, or unless the length L is quite small, as for a small droplet, the Reynolds number is indeed large, and δ/L is thus small. For example, consider air at normal density flowing at 10 ft/sec over a plate only 1 ft long,

$$Re \doteq \frac{(10 \text{ ft/sec}) (1 \text{ ft})}{(2 \times 10^{-4} \text{ ft}^2/\text{sec})} = 50,000$$

The value of δ/L is then of the order of 1/200.

Turning now to the y-momentum equation, Eq. (A.62), we neglect terms of order $(\delta^2/L^2) = (1/Re)$ and rearrange the others to obtain

$$\frac{P_S}{\rho V^2} \frac{\partial P^*}{\partial y^*} = \frac{1}{Re} \left\{ \frac{\partial^2 v_y^*}{\partial y^{*2}} - \left[v_x^* \frac{\partial v_y^*}{\partial x^*} + v_y^* \frac{\partial v_y^*}{\partial y^*} \right] \right\} \tag{A.65}$$

We conclude that pressure variations across the boundary layer are of order $1/Re$, that is, *the pressure which exists in the free stream outside the boundary layer* ($y > \delta$) *impresses itself through the boundary layer*. The partial derivative $\partial P^*/\partial x^*$ in Eq. (A.61) can be replaced by dP_e^*/dx^*. Outside the boundary layer Eq. (A.63) holds, but if the wall lies approximately along the x-direction, the term $v_y^* (\partial v_x^*/\partial y^*)$ can be dropped. Denoting the velocity external to the boundary layer v_e and returning to dimensional form allows Eq. (A.63) to be written

$$v_e \frac{dv_e}{dx} = -\frac{1}{\rho} \frac{dP}{dx} \tag{A.66}$$

or

$$v_e \, dv_e + \frac{dP}{\rho} = 0, \qquad \text{the Euler Equation} \tag{A.67}$$

The x-momentum equation for a boundary layer can therefore be written as

$$v_x^* \frac{\partial v_x^*}{\partial x^*} + v_y^* \frac{\partial v_x^*}{\partial y^*} = v_e^* \frac{dv_e^*}{dx^*} + \frac{\partial^2 v_x^*}{\partial y^{*2}} \tag{A.68}$$

For the simple situation of a flat plate parallel to the flow, v_e is a constant, and dv_e/dx is zero. Equation (A.68) reduces to

$$v_x^* \frac{\partial v_x^*}{\partial x^*} + v_y^* \frac{\partial v_x^*}{\partial y^*} = \frac{\partial^2 v_x^*}{\partial y^{*2}} \tag{A.69}$$

A.8.3   The Thermal Energy Equation for a Boundary Layer

Now consider our flat plate to be heated such that a surface temperature of T_s is maintained. The free stream is at temperature T_e. Figure A.5 shows the physical situation envisioned.

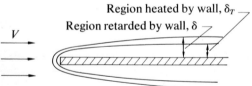

Region heated by wall, δ_T

Region retarded by wall, δ

V

Figure A.5   A thermal boundary layer.

The governing equations are those considered in Section A.8.2 with the addition of the thermal energy equation. For low speed flow, that is, low Eckert number $V^2/(2c_p\,\Delta T)$, we neglect dissipation and compressibility effects. The thermal energy equation for two-dimensional, steady, laminar flow with constant thermal conductivity is, from Eq. (A.25),

$$\rho v_x c_p \frac{\partial T}{\partial x} + \rho v_y c_p \frac{\partial T}{\partial y} = k\left[\frac{\partial^2 T}{\partial x^2} + \frac{\partial^2 T}{\partial y^2}\right] \tag{A.70}$$

We introduce a dimensionless temperature,

$$T^* = \frac{T - T_e}{T_s - T_e} \tag{A.71}$$

and now scale y with δ_T, the thermal thickness,

$$y_T^* = \frac{y}{\delta_T} \tag{A.72}$$

Upon making the energy equation dimensionless there results

$$\frac{\rho c_p (T_s - T_e) V}{L}\left(v_x^* \frac{\partial T^*}{\partial x^*} + \frac{\delta}{\delta_T}\, v_y^* \frac{\partial T^*}{\partial y_T^*}\right)$$
$$= \frac{k(T_s - T_e)}{\delta_T^2}\left[\frac{\partial^2 T^*}{\partial y_T^{*2}} + \frac{\delta_T^2}{L^2}\frac{\partial^2 T^*}{\partial x^{*2}}\right]$$

Dividing through by $\rho c_p (T_s - T_e) V/L$ and dropping the last term in the square brackets,

$$v_x^* \frac{\partial T^*}{\partial x^*} + \frac{\delta}{\delta_T} v_y^* \frac{\partial T^*}{\partial y_T^*} = \frac{1}{Re\,Pr}\frac{L^2}{\delta_T^2}\frac{\partial^2 T^*}{\partial y_T^{*2}} \tag{A.73}$$

We can use order of magnitude arguments to deduce behavior of δ_T/δ as a function of Pr. Using Eq. (A.64) we rewrite Eq. (A.73) as

$$v_x^* \frac{\partial T^*}{\partial x^*} + \frac{\delta}{\delta_T} v_y^* \frac{\partial T^*}{\partial y_T^*} = \frac{1}{Pr}\frac{\delta^2}{\delta_T^2}\frac{\partial^2 T^*}{\partial y_T^{*2}} \tag{A.74}$$

When Pr is very small it follows that δ/δ_T must be small also; from Eq. (A.74).

$$O(1) = \frac{1}{Pr}\frac{\delta^2}{\delta_T^2}$$

or

$$\frac{\delta_T}{\delta} = O\!\left(\frac{1}{Pr^{1/2}}\right) \tag{A.75}$$

When Pr is very large, δ/δ_T must be large. A somewhat more lengthy argument is is now required to establish the Pr dependence. Since δ/δ_T is large, the entire thermal boundary layer is deep within the velocity boundary layer, very close to the wall. The x-momentum equation tells us that in this region $\partial v_x^*/\partial y^*$ is nearly constant so that v_x^* is order of magnitude $y^* \leqslant \delta_T/\delta$. From the mass conservation equation v_y^* is, in this region very close to the wall, of order of magnitude $y^{*2} \leqslant$

δ_T^2/δ^2. Thus the entire left-hand side of Eq. (A.74) is order of magnitude δ_T/δ, and it follows that

$$\frac{\delta_T}{\delta} = O\left(\frac{1}{Pr^{1/3}}\right) \tag{A.76}$$

The significance of the Prandtl number is clear from a comparison of Eqs. (A.68) and (A.73). When the kinematic viscosity $\nu = \mu/\rho$ is larger than the thermal diffusivity $\alpha = k/(\rho c_p)$, the influence of the wall on the fluid momentum penetrates more deeply into the flow than does the thermal effect. Or, in other words, the thermal layer is thin compared to δ when the Prandtl number is large.

A.8.4 The Species Equation for a Boundary Layer

The conservation of species equation for a two-dimensional steady laminar flow of a binary mixture, with constant density, is from Eq. (A.26)

$$\rho v_x \frac{\partial m_1}{\partial x} + \rho v_y \frac{\partial m_1}{\partial y} = \rho \mathcal{D}_{12} \left[\frac{\partial^2 m_1}{\partial x^2} + \frac{\partial^2 m_1}{\partial y^2}\right] \tag{A.77}$$

If the net rate of mass transfer is low, $n_s \ll (\delta/L)\rho_e v_e$, then the momentum and mass conservation equations will be essentially unaffected by the mass transfer. Notice that Eq. (A.77) is the same as the energy equation if $k/(\rho c_p)$ is replaced by $\mathcal{D}_{12}$. We can write the dimensionless equation at once by defining

$$y_M^* = \frac{y}{\delta_M} \quad \text{and} \quad m^* = \frac{m_1 - m_{1,e}}{m_{1,s} - m_{1,e}} \tag{A.78}$$

and replacing T^* with m^* and $k/(\rho c_p)$ with $\mathcal{D}_{12}$ in Eq. (A.73)

$$v_x^* \frac{\partial m^*}{\partial x^*} + \frac{\delta}{\delta_M} v_y^* \frac{\partial m^*}{\partial y_M^*} = \frac{1}{Re\ Sc} \frac{L^2}{\delta_M^2} \frac{\partial^2 m^*}{\partial y_M^{*2}} \tag{A.79}$$

As before, when the Schmidt number is very small,

$$\frac{\delta_M}{\delta} = O\left(\frac{1}{Sc^{1/2}}\right) \tag{A.80}$$

and when Sc is much larger than unity,

$$\frac{\delta_M}{\delta} = O\left(\frac{1}{Sc^{1/3}}\right) \tag{A.81}$$

Table A.2 summarizes the boundary layer equations which result when small terms are dropped but the equations are restated in dimensional form.

A.8.5 Estimating the Order of Magnitude of the Wall Transfer Rates

From our order of magnitude estimates of δ, δ_T, and δ_M we can estimate τ_s, q_s, and j_s within an order of magnitude or two, as follows:

Table A.2   Conservation Equations in the Boundary Layer Approximation

Equations	Steady Two-Dimensional Flow	Steady Axisymmetric Flow
Mass	$\dfrac{\partial}{\partial x}(\rho v_x) + \dfrac{\partial}{\partial y}(\rho v_y) = 0$	$\dfrac{1}{r}\dfrac{\partial}{\partial x}(\rho r v_x) + \dfrac{\partial}{\partial y}(\rho v_y) = 0$
Momentum	$\rho v_x \dfrac{\partial v_x}{\partial x} + \rho v_y \dfrac{\partial v_x}{\partial y} = -\dfrac{dP}{dx}$ $+ \dfrac{\partial}{\partial y}\left(\mu \dfrac{\partial v_x}{\partial y}\right) + \rho g_x$	$\rho v_x \dfrac{\partial v_x}{\partial x} + \rho v_y \dfrac{\partial v_x}{\partial y} = -\dfrac{dP}{dx}$ $+ \dfrac{\partial}{\partial y}\left(\mu \dfrac{\partial v_x}{\partial y}\right) + \rho g_x$
Energy	$\rho v_x \dfrac{\partial \hat{u}}{\partial x} + \rho v_y \dfrac{\partial \hat{u}}{\partial y} + P\nabla \cdot \mathbf{v} =$ $\dfrac{\partial}{\partial y}\left(k\dfrac{\partial T}{\partial y}\right) + \mu\left(\dfrac{\partial v_x}{\partial y}\right)^2 + \dot{Q}_V$	$\rho v_x \dfrac{\partial \hat{u}}{\partial x} + \rho v_y \dfrac{\partial \hat{u}}{\partial y} + P\nabla \cdot \mathbf{v} =$ $\dfrac{\partial}{\partial y}\left(k\dfrac{\partial T}{\partial y}\right) + \mu\left(\dfrac{\partial v_x}{\partial y}\right)^2 + \dot{Q}_V$
Species Dilute or binary mixtures	$\rho v_x \dfrac{\partial m_i}{\partial x} + \rho v_y \dfrac{\partial m_i}{\partial y} = \dfrac{\partial}{\partial y}\left(\rho \mathscr{D}\dfrac{\partial m_i}{\partial y}\right) + \dot{r}_i$	$\rho v_x \dfrac{\partial m_i}{\partial x} + \rho v_y \dfrac{\partial m_i}{\partial y} = \dfrac{\partial}{\partial y}\left(\rho \mathscr{D}\dfrac{\partial m_i}{\partial y}\right) + \dot{r}_i$

$$\tau_s = -\mu \left.\frac{\partial v_x}{\partial y}\right|_{y=0} = -\frac{\mu V}{\delta}\left.\frac{\partial v_x^*}{\partial y^*}\right|_{y^*=0}$$

Thus

$$\tau_s \approx \mu V/\delta \tag{A.82}$$

Similarly

$$q_s \approx \frac{k(T_s - T_e)}{\delta_T} \tag{A.83}$$

$$j_s \approx \frac{\rho \mathscr{D}_{12}(m_{1,s} - m_{1,e})}{\delta_M} \tag{A.84}$$

Introducing the definitions of $c_f/2$, Nu, and Nu_m from Eqs. (A.39), (A.40), and (A.41), we obtain

$$\frac{c_f}{2} \approx \frac{1}{Re^{1/2}} \tag{A.85}$$

$$Nu \approx Re^{1/2}Pr^{1/3}, \qquad Pr > 1 \tag{A.86a}$$

$$Nu \approx Re^{1/2}Pr^{1/2}, \qquad Pr < 1 \tag{A.86b}$$

$$Nu_m \approx Re^{1/2}Sc^{1/3}, \qquad Sc > 1 \tag{A.87a}$$

$$Nu_m \approx Re^{1/2}Sc^{1/2}, \qquad Sc < 1 \qquad\qquad \text{(A.87b)}$$

These results may be compared to exact numerical solutions for laminar boundary layer flow along a flat plate

$$\frac{c_f}{2} = \frac{0.332}{Re_x^{1/2}} \qquad\qquad\qquad \text{(A.88)}$$

$$Nu(x) \doteq 0.332\, Re_x^{1/2}Pr^{1/3}, \qquad Pr \geqslant 1 \qquad\qquad \text{(A.89a)}$$

$$Nu(x) = 0.564\, Re_x^{1/2}Pr^{1/2}, \qquad Pr \ll 1 \qquad\qquad \text{(A.89b)}$$

$$Nu_m(x) \doteq 0.332\, Re_x^{1/2}Sc^{1/3}, \qquad Sc \geqslant 1 \qquad\qquad \text{(A.90a)}$$

$$Nu_m(x) = 0.564\, Re_x^{1/2}Sc^{1/2}, \qquad Sc \ll 1 \qquad\qquad \text{(A.90b)}$$

A.8.6   Self-Similar Solutions

It is beyond the scope of this introductory text to develop exact solutions of the type presented above. However, it may be noted that our order of magnitude estimates point the way for an analytical solution by a similarity transformation. We found for a plate of length L that variations in the y-direction scaled according to

$$y^* = \frac{y}{\delta} = \frac{y}{L\, Re^{-1/2}}$$

For an arbitrary distance x one is thus led to suspect that a scaled length

$$\eta^* = \frac{y}{x\, Re_x^{-1/2}}, \qquad Re_x = \frac{Vx\rho}{\mu} \qquad\qquad \text{(A.91)}$$

may be the only independent variable for some boundary layer transfer problems. If a solution does exist such that v_x^*, T^*, and m^* are functions of only $\eta^*(x, y)$, such a solution is called **self-similar**. For flat plate flow with constant values or power function variations of T_s and $m_{1,s}$ along the surface, self-similar solutions can be obtained. Equations (A.88) through (A.90) are the results of such analyses.

REFERENCES

Conservation Equations

Any but the most elementary fluid mechanics text gives a derivation of the Navier-Stokes equations. A classic presentation for Cartesian coordinates is given by:

Lamb, H., *Hydrodynamics*, 6th Ed. New York: Dover Publications, 1932. (See Articles 1, 2, 30, and 323–328.)

Alternative presentations suitable for the beginning student are given by:

Schlichting, H., *Boundary Layer Theory*, 6th Ed. New York: McGraw-Hill, 1968. (See Chapter 3.)

Whitaker, S., *Introduction to Fluid Mechanics*. Englewood Cliffs, New Jersey: Prentice-Hall, 1968. (Chapters 4 and 5. See particularly pp. 139–145.)

The energy and species equations are presented well by:

Bird, R. B., W. E. Stewart, and E. N. Lightfoot, *Transport Phenomena*. New York: John Wiley, 1960. (See Tables 10.2, 18.2, and 18.3.)

The idea of the boundary layer is developed well by Schlichting, cited above. Boundary layer solutions for heat and mass transfer are given, for example, by Kays, and Chapter F.2 by Ostrach reviews natural convection boundary layers:

Kays, W. M., *Convective Heat and Mass Transfer*. New York: McGraw-Hill, 1966. (See Chapter 10 for laminar boundary layer solutions of the type alluded to in Section A.8.6.)

Ostrach, S., "Laminar Flows with Body Forces," Section F of F. K. Moore's *Theory of Laminar Flows*. Princeton University Press, 1964.

APPENDIX **B**

SELECTED VALUES OF

CONSTANTS, CONVERSION

FACTORS, AND PROPERTIES

LIST OF TABLES

Table B.1   Mathematical and Physical Constants

Mathematical Constants

$$e = 2.7182818$$
$$\pi = 3.1415927$$
$$\ln 10 = 2.3025851$$

Physical constants

Gas constant

$$\mathscr{R} = 1.545 \times 10^3 \text{ [ft lb}_f/\text{lb-mole }°R]$$
$$1.987 \text{ [cal/gm-mole }°K]$$
$$8.3143 \times 10^3 [\text{joules/kg-mole }°K]$$
$$82.05 \text{ [cm}^3 \text{ atm/gm-mole }°K]$$

Standard gravitational acceleration

$$g = 32.174 \text{ [ft/sec}^2]$$
$$980.665 \text{ [cm/sec}^2]$$
$$9.80665 \text{ [m/sec}^2]$$

Avogadro's number

$$N_{Av} = 6.02252 \times 10^{23} \text{ [molecules/gm-mole]}$$

Boltzmann constant

$$k = 1.38054 \times 10^{-16}[\text{erg/molecule }°K]$$
$$1.38054 \times 10^{-23}[\text{joule/molecule }°K]$$

Atomic mass unit

$$\text{amu} = 1.66043 \times 10^{-24} \text{ [gm]}$$
$$1.66043 \times 10^{-27} \text{ [kg]}$$

Planck's constant

$$h = 6.6256 \times 10^{-27}[\text{erg sec}]$$
$$6.6256 \times 10^{-34}[\text{joule sec}]$$

Speed of light

$$c = 2.997925 \times 10^{10}[\text{cm/sec}]$$
$$2.997925 \times 10^8[\text{m/sec}]$$

Table B.2 Conversion Factors

Absolute Temperature

$T[°R] = T[°F] + 459.67$

$T[°K] = T[°C] + 273.15$

$1.8[(°R)/(°K)]$

Temperature Difference

$1.8[(°R)/(°K)]$

$1.8[(°F)/(°K)]$

Length

$(1/12)[(ft)/(in.)]$

$(1/30.48)[(ft)/(cm)]$

$3.281[(ft)/(m)]$

Mass

$(1/32.17)[(slug)/(lb)]$

$(1/453.6)[(lb)/(gm)]$

$2.2046[(lb)/(kg)]$

Force

$2.248 \times 10^{-6}[(lb_f)/(dyne)]$

$0.22481[(lb_f)/(newton)]$

Energy, work

$778.16[(ft\ lb_f)/(Btu)]$

$7.376 \times 10^{-8}[(ft\ lb_f)/(erg)]$

$0.7376[(ft\ lb_f)/(joule)]$

$3.9657 \times 10^{-3}[(Btu)/(cal)]$

$3412.2[(Btu)/(kw\ hr)]$

Power

$3412.2[(Btu/hr)/(kw)]$

$0.7457[(kw)/(hp)]$

Heat Flux

$3170.0[(Btu/ft^2hr)/(w/cm^2)]$

$0.3170[(Btu/ft^2hr)/(w/m^2)]$

$13,263[(Btu/ft^2hr)/(cal/cm^2sec)]$

Mass Flux

$7373[(lb/ft^2hr)/(gm/cm^2\ sec)]$

$737.3[(lb/ft^2hr)/(kg/m^2\ sec)$

Pressure, stress

$144[(lb_f/ft^2)/(lb_f/in.^2)]$

$0.0020886[(lb_f/ft^2)/(dyne/cm^2)]$

$2.0886 \times 10^{-2}[(lb_f/ft^2)/(newton/m^2)]$

$14.696[(lb_f/in.^2)/(atm)]$

$760[(torr)/(atm)]$

Viscosity

$2.0886 \times 10^{-5}[(lb_f sec/ft^2)/(cp)]$

$0.020886[(lb_f sec/ft^2)/(newton\ sec/m^2)]$

$8.6336 \times 10^{-6}[(lb_f sec/ft^2)/(lb/hr\ ft)]$

Thermal Conductivity

$241.75[(Btu/hr\ ft\ °F)/(cal/sec\ cm°K)]$

$0.5778[(Btu/hr\ ft\ °F)/(w/m°K)$

Diffusivity

$(1/3600)[(ft^2/sec)/(ft^2/hr)]$

$3.875[(ft^2/hr)/(cm^2/sec)]$

$3.875 \times 10^4[(ft^2/hr)/(m^2/sec)]$

Table B.3   Thermal Properties of Solid Metals

	Thermal Properties at Normal Temperature			
Selected Metals	k, Btu/hr ft °F	ρ, lb/ft³	c_p, Btu/lb °F	α, ft²/hr
Aluminum, pure	118	169	0.214	3.26
Aluminum alloy, 0.03 to 0.05 Cu, trace Mg	95	174	0.211	2.58
Antimony	10.5	413	0.050	0.51
Beryllium	110	115	0.40	2.39
Bismuth	4.7	612	0.029	0.27
Brass, 0.70 Cu, 0.30 Zn	64	532	0.092	1.31
Cadmium	53	540	0.055	1.78
Constantan, 0.60 Cu, 0.40 Ni	13.1	557	0.10	0.24
Copper, pure	223	559	0.092	4.35
Gold	170	1205	0.031	4.55
Iron	42	493	0.108	0.79
Lead	20	705	0.030	0.95
Magnesium	99	109	0.242	3.76
Molybdenum	71	638	0.060	1.86
Nickel, pure	52	556	0.11	0.85
Platinum	40	1340	0.032	0.93
Platinum-rhodium, 0.10 Rh	17.5	1280	0.032	0.43
Silver, pure	244	657	0.056	6.6
Steel, mild, 0.01 C	28	489	0.11	0.52
Steel, stainless, 0.18 Cr, 0.08 Ni	9.4	488	0.11	0.17
Tin	37	456	0.054	1.51
Titanium	12.7	281	0.11	0.41
Tungsten	94	1208	0.032	2.43
Uranium	14.5	1167	0.028	0.44
Zinc	65	446	0.092	1.59

From E. R. G. Eckert, and R. M. Drake, *Heat and Mass Transfer*. New York: McGraw-Hill, 1959 and *Metallic Elements and Their Alloys*, T.P.R.C. Data Book, Vol. 1, Purdue Research Foundation, 1966.

Table B.4   Thermal Properties of Solid Dielectrics

	Thermal Properties at Normal Temperature (or as noted)		
Selected Materials	k, Btu/hr ft°F	ρ, lb/ft³	c_p, Btu/lb °F
Aluminum oxide (sapphire)	21	231	0.19
Beryllium oxide, pressed	125	185	0.25
Brick, common	0.40	100	0.20
Carbon	3–5	97	0.18
Chrome brick (392°F)	0.82	246	0.20
(1832°F)	0.96		

Table B.4   (Continued)

	Thermal Properties at Normal Temperature (or as noted)		
Selected Materials	k, Btu/hr ft °F	ρ, lb/ft^3	c_p, Btu/lb °F
Concrete	0.5–0.8	140	0.21
Earth, dry	0.30	128	0.44
wet	1.5	150	0.52
Glass			
Borosilicate, pyrex type	0.63	165	0.19
Flint, 0.45 PbO, 0.10 misc.	0.44	169	0.2
Silica, pure	0.78	165	0.19
Soda-lime, 0.25 Na$_2$O, 0.10 CaO	0.51	150	0.2
Ice (32°F)	1.28	57	0.46
Insulations			
Asbestos	0.09	36	0.25
Cork	0.025	10	0.04
Diatomaceous earth (100°F)	0.030	14	0.21
(600°F)	0.046		
Fiberglass (100°F)	0.031	1.5	0.19
(200°F)	0.043		
Polystyrene foam	0.02	1–4	0.27
Magnesium oxide, crystal	35	223	0.22
Paper	0.075	44–72	0.5–0.7
Plastics, solid			
Acrylics	0.12	74	0.35
Cellulose acetate	0.10–0.19	81	0.3–0.42
Neoprene rubber	0.11	78	0.46
Phenolic, filler	0.19–0.39	95–125	0.28–0.32
Polyamide (nylon)	0.14	71	0.4
Polyethylene (high density)	0.19	60	0.46–0.55
Polypropylene	0.10	57	0.46
Polystyrene	0.058–0.090	65–82	0.30–0.35
Polytetrafluoroethylene	0.14	137	0.25
Polyvinylchloride	0.053	107	0.25
Silicon dioxide (quartz)	3.6	165	0.18
Uranium oxide	5.9	680	0.056
Wood			
Oak, perpendicular to grain	0.12	51	0.57
parallel to grain	0.20		
White pine, perpendicular to grain	0.06	31	0.67
parallel to grain	0.14		
Zirconia brick (392°F)	0.84	304	0.13
(1832°F)	1.13		0.15

From various sources. See, for example, W. H. McAdams, *Heat Transmission*, 3d Ed. New York: McGraw-Hill, 1954.

Table B.5   Thermal Properties of Gases (1 atm pressure)

Gas	T, °F	k, Btu/hr ft°F	ρ, lb/ft³	c_p, Btu/lb°F	$\mu \times 10^{7\,a}$, lb$_f$sec/ft²	$\nu \times 10^5$, ft²/sec	Pr
Air	−200	0.0088	0.1527	0.243	2.16	4.55	0.69
	−100	0.0114	0.1103	0.241	2.83	8.27	0.69
	0	0.0138	0.0863	0.241	3.42	12.75	0.69
	80	0.0154	0.0735	0.240	3.89	16.9	0.69
	100	0.0159	0.0709	0.240	3.95	17.9	0.69
	200	0.0179	0.0601	0.241	4.44	23.7	0.69
	400	0.0217	0.0462	0.243	5.32	37.1	0.69
	600	0.0254	0.0374	0.248	6.13	52.7	0.69
	800	0.0290	0.0315	0.253	6.87	70.1	0.69
	1000	0.0327	0.0272	0.260	7.56	89.4	0.695
	1500	0.0415	0.0203	0.274	9.15	145.3	0.70
	2000	0.0497	0.0161	0.284	10.61	211	0.70
	2500	0.0574	0.0134	0.291	11.97	287	0.70
	3000	0.0645	0.0115	0.296	13.24	371	0.70
	3500	0.0712	0.0100	0.299	14.45	464	0.70
Ammonia	0	0.0117	0.0515	0.52	1.81	10.4	0.91
(NH₃)	80	0.0142	0.0439	0.52	2.15	15.9	0.89
	100	0.0148	0.0423	0.52	2.22	17.2	0.88
	200	0.0185	0.0355	0.53	2.63	24.0	0.87
	400	0.0275	0.0270	0.57	3.48	41.2	0.84
Argon	−200	0.0053	0.213	0.126	2.49	3.78	0.68
(A)	−100	0.0072	0.153	0.125	3.38	7.14	0.68
	0	0.0089	0.119	0.125	4.19	11.31	0.68
	100	0.0105	0.0978	0.125	4.92	16.19	0.675
	200	0.0120	0.0829	0.124	5.59	21.7	0.67
	400	0.0148	0.0636	0.124	6.81	34.4	0.66
	600	0.0172	0.0516	0.124	7.90	49.2	0.66
	800	0.0194	0.0434	0.124	8.89	65.9	0.66
	1000	0.0215	0.0375	0.124	9.83	84.3	0.66
	1500	0.0260	0.0279	0.124	11.95	137.8	0.66
	2000	0.0300	0.0222	0.124	13.89	201	0.67
Carbon	0	0.0085	0.1311	0.188	2.73	6.70	0.70
dioxide	100	0.0110	0.1077	0.205	3.28	9.79	0.705
(CO₂)	200	0.0135	0.0914	0.217	3.79	13.35	0.71
	400	0.0183	0.0701	0.238	4.73	21.7	0.71
	600	0.0230	0.0569	0.255	5.57	31.5	0.715
	800	0.0276	0.0479	0.269	6.34	42.6	0.72
	1000	0.0318	0.0413	0.280	7.05	54.9	0.72
	1500	0.0417	0.0308	0.300	8.64	90.3	0.72
	2000	0.0506	0.0245	0.313	10.09	132.4	0.72
	2500	0.0586	0.0204	0.321	11.41	180.1	0.72
	3000	0.0660	0.0174	0.326	12.63	233	0.72
	3500	0.0729	0.0152	0.331	13.79	291	0.72

Table B.5　(Continued)

Gas	T, °F	k, Btu/hr ft °F	ρ, lb/ft³	c_p, Btu/lb °F	$\mu \times 10^{7a}$, lb,sec/ft²	$\nu \times 10^5$, ft²/sec	Pr
Helium	−400	0.0204	0.0915	1.242	1.017	3.68	0.74
(He)	−200	0.0536	0.0211	1.242	2.62	40.0	0.70
	−100	0.0680	0.0152	1.242	3.27	69.3	0.70
	0	0.0784	0.0119	1.242	3.76	102.8	0.70
	100	0.0888	0.00978	1.242	4.24	139.6	0.69
	200	0.0977	0.00829	1.242	4.74	187	0.70
	400	0.114	0.00637	1.242	5.74	290	0.72
	600	0.130	0.00517	1.242	6.49	405	0.72
	800	0.145	0.00439	1.242	7.27	532	0.72
	1000	0.159	0.00376	1.242	7.97	683	0.72
	1200	0.172	0.00330	1.242	8.66	841	0.72
Hydrogen	−200	0.054	0.01062	2.92	1.08	32.6	0.68
(H₂)	−100	0.076	0.00767	3.23	1.39	58.1	0.69
	0	0.093	0.00600	3.34	1.66	88.7	0.69
	100	0.109	0.00493	3.42	1.90	123.8	0.69
	200	0.123	0.00418	3.45	2.12	163.2	0.69
	400	0.147	0.00321	3.47	2.54	254	0.69
	600	0.169	0.00261	3.48	2.91	359	0.69
	800	0.190	0.00219	3.49	3.25	477	0.69
	1000	0.211	0.00189	3.52	3.60	608	0.69
	1500	0.262	0.00141	3.62	4.34	989	0.69
	2000	0.314	0.00122	3.75	5.03	1440	0.695
	2500	0.367	0.00093	3.90	5.67	1952	0.70
	3000	0.418	0.00080	4.03	6.26	2523	0.70
	3500	0.472	0.00070	4.15	6.89	3179	0.70
Nitrogen	−200	0.0088	0.1476	0.251	2.10	4.57	0.69
(N₂)	−100	0.0114	0.1066	0.249	2.73	8.25	0.69
	0	0.0137	0.0834	0.249	3.29	12.67	0.69
	100	0.0158	0.0685	0.249	3.79	17.77	0.69
	200	0.0177	0.0582	0.249	4.25	23.5	0.69
	400	0.0214	0.0446	0.252	5.09	36.6	0.69
	600	0.0251	0.0362	0.256	5.85	51.9	0.69
	800	0.0286	0.0305	0.262	6.54	69.1	0.69
	1000	0.0322	0.0263	0.269	7.20	88.1	0.695
	1500	0.0409	0.0196	0.283	8.71	143.1	0.70
	2000	0.0490	0.0156	0.293	10.01	208	0.70
	2500	0.0566	0.0130	0.301	11.40	283	0.70
	3000	0.0636	0.0111	0.306	12.61	366	0.70
	3500	0.0701	0.0097	0.310	13.75	456	0.70
Oxygen	−200	0.0082	0.1686	0.212	2.30	4.38	0.69
(O₂)	−100	0.0111	0.1218	0.215	3.08	8.14	0.69
	0	0.0138	0.0953	0.217	3.79	12.77	0.69

Table B.5 (Continued)

Gas	T, °F	k, Btu/hr ft °F	ρ, lb/ft^3	c_p, Btu/lb °F	$\mu = 10^{7a}$, lb/sec/ft^2	$\nu \times 10^5$, ft^2/sec	Pr
	100	0.0163	0.0783	0.220	4.42	18.17	0.69
	200	0.0187	0.0664	0.223	5.00	24.2	0.69
	400	0.0233	0.0510	0.231	6.05	38.2	0.69
	600	0.0278	0.0414	0.238	6.99	54.3	0.70
	800	0.0321	0.0348	0.246	7.87	72.7	0.70
	1000	0.0363	0.0300	0.252	8.69	93.2	0.70
	1500	0.0458	0.0224	0.263	10.53	151.4	0.70
	2000	0.0543	0.0178	0.270	12.20	220	0.70
	2500	0.0624	0.0148	0.276	13.76	299	0.70
	3000	0.0703	0.0127	0.281	15.23	387	0.705
	3500	0.0779	0.0111	0.286	16.63	483	0.705
Steam	212	0.0145	0.0372	0.451	2.70	23.4	0.96
(H_2O)	300	0.0171	0.0328	0.456	3.11	30.3	0.95
	400	0.0200	0.0288	0.462	3.51	39.5	0.94
	500	0.0228	0.0258	0.470	3.93	49.0	0.94
	600	0.0257	0.0233	0.477	4.41	61.0	0.94
	700	0.0288	0.0213	0.485	4.83	72.5	0.93
	800	0.0321	0.0196	0.494	5.28	85.5	0.92
	900	0.0355	0.0181	0.50	5.62	98.7	0.91
	1000	0.0388	0.0169	0.51	5.96	113	0.91
	1200	0.0457	0.0149	0.53	6.65	144	0.88
	1400	0.053	0.0133	0.55	7.34	178	0.87
	1600	0.061	0.0120	0.56	8.02	214	0.87
	1800	0.068	0.0109	0.58	8.73	258	0.87
	2000	0.076	0.0100	0.60	9.41	303	0.86
	2500	0.096	0.0083	0.64	11.12	430	0.86
	3000	0.114	0.0071	0.67	12.42	575	0.86

[a] This table and subsequent ones are to be read as $\mu \times 10^7 = 2.16$, that is, $\mu = 2.16 \times 10^{-7}$.
From various sources, See, for example, R. A. Svehla, *Estimated Viscosities and Thermal Conductivities of Gases at High Temperatures*, NASA TR R-132, 1962; *JANAF Thermochemical Data*, The Dow Chemical Co., Thermal Laboratory, Midland, Michigan; J. Hilsenrath et al., *Tables of Thermal Properties of Gases*, U.S. National Bureau of Standards Circular 564, Washington, D.C., 1955.

Table B.6 Thermal Properties of Liquid Metals

Liquid Metal Melting point Boiling point	T, °F	k, Btu/hr ft°F	ρ, lb/ft³	c_p, Btu/lb°F	$\mu \times 10^7$, lb, sec/ft²	$\nu \times 10^5$, ft²/sec	Pr
Lead	700	9.6	658	0.038	506	0.247	0.023
621°F M.P.	800	10.1	654	0.037	443	0.218	0.019
3159°F B.P.	1000	11.1	645	0.037	351	0.175	0.014
	1200	11.9	637	0.037	305	0.154	0.011
Lithium	400	26.6	31.6	1.0	118	1.2	0.051
357°F M.P.	600	26.9	31.0	1.0	95	0.99	0.041
2430°F B.P.	800	27.2	30.5	1.0	80	0.85	0.034
	1000	27.5	29.4	1.0	70	0.77	0.029
	1200	27.9	27.6	1.0	62	0.72	0.026
Mercury	50	4.6	847	0.033	336	0.128	0.028
−39°F M.P.	100	5.0	843	0.033	296	0.113	0.023
674°F B.P.	200	5.4	834	0.033	252	0.097	0.018
	400	6.2	817	0.033	211	0.083	0.013
	600	6.9	801	0.033	179	0.072	0.010
Potassium	200	26.6	51.3	0.19	120	0.75	0.010
147°F M.P.	400	25.4	49.7	0.19	59	0.38	0.0051
1400°F B.P.	600	24.1	47.9	0.19	45	0.30	0.0041
	800	22.8	46.3	0.18	38	0.26	0.0035
	1000	21.1	43.0	0.18	33	0.25	0.0033
	1200	19.5	42.9	0.18	29	0.22	0.0031
Sodium	400	46.4	56.3	0.32	92	0.53	0.0073
208°F M.P.	600	43.3	54.6	0.31	68	0.40	0.0056
1621°F B.P.	800	40.5	52.0	0.31	52	0.32	0.0046
	1000	37.6	50.9	0.30	46	0.29	0.0043
	1200	35.5	49.3	0.30	41	0.27	0.0040
	1400	33.7	47.8	0.30	36	0.24	0.0037
	1600	32.2	46.5	0.30	32	0.22	0.0035

Compiled from *Properties of Inorganic Energy Conversion and Heat-Transfer Fluids for Space Applications*, W. D. Weatherford, Jr., et al., WADD Technical Report 61–96, November 1961; *Liquid-Metals Handbook*, R. N. Lyon (ed.), published by the Atomic Energy Commission and Department of the Navy, Washington, D.C., 1952; *Metallic Elements and Their Alloys*, T.P.R.C. Data Book, Vol. 1, Purdue Research Foundation, 1966.

Table B.7 Thermal Properties of Dielectric Liquids

Saturated Liquid Melting point Boiling point	T, °F	k, Btu/hr ft°F	ρ, lb/ft³	c_p, Btu/lb°F	$\mu \times 10^7$, lb$_f$sec/ft²	$\nu \times 10^5$, ft²/sec	Pr
Ammonia	−60	0.316	44.0	1.07	64.5	0.472	2.53
— 108°F M.P.	−40	0.316	43.3	1.07	58.6	0.436	2.30
— 28°F B.P.	−20	0.316	42.3	1.07	54.7	0.416	2.15
	0	0.316	41.5	1.08	52.3	0.406	2.07
	20	0.313	40.5	1.10	50.8	0.403	2.07
	40	0.309	39.7	1.12	49.0	0.398	2.06
	60	0.303	38.6	1.14	46.9	0.391	2.04
	80	0.296	37.6	1.17	44.3	0.379	2.03
	100	0.286	36.6	1.19	41.8	0.367	2.01
	120	0.277	35.5	1.22	39.0	0.354	1.99
Carbon dioxide	−60	0.048	72.6	0.44	28.8	0.128	3.07
— 110°F Subl.	−40	0.058	69.8	0.45	27.5	0.127	2.48
	−20	0.065	66.9	0.47	26.0	0.125	2.18
	0	0.066	63.9	0.49	24.5	0.123	2.10
	20	0.063	60.3	0.54	22.6	0.121	2.25
	40	0.059	56.1	0.65	19.9	0.114	2.54
	60	0.053	50.9	0.96	16.2	0.103	3.41
	80	0.044	42.0	2.6	11.5	0.088	7.87
Engine oil,	40	0.085	56.0	0.43	555000	3190	32600
unused	60	0.084	55.5	0.44	235000	1360	14200
(extra heavy oil	80	0.0835	55.1	0.45	107000	625	6700
as used in	100	0.083	54.8	0.47	51000	300	3350
aircraft,	120	0.082	54.3	0.48	23900	142	1630
SAE 50)	140	0.0815	53.9	0.49	13500	80.7	940
	160	0.081	53.4	0.50	8700	52.5	623
	180	0.080	53.0	0.51	6100	37.1	452
	200	0.0795	52.7	0.52	4300	26.3	345
	250	0.078	51.8	0.55	2080	12.9	170
	300	0.076	50.8	0.58	1140	7.2	100

Table B.7 (Continued)

Saturated Liquid Melting point Boiling point	T, °F	k, Btu/hr ft °F	ρ, lb/ft³	c_p, Btu/lb °F	$\mu \times 10^7$ lb$_f$sec/ft²	$\nu \times 10^5$, ft²/sec	Pr
Freon 12							
CCl₂F₂	−60	0.039	96.7	0.21	100.5	0.299	5.4
−252°F M.P.	−40	0.040	94.8	0.21	88.0	0.271	4.8
−21.6°F B.P.	−20	0.040	92.9	0.215	78.0	0.248	4.3
	0	0.041	91.0	0.218	70.0	0.234	3.9
	20	0.042	89.0	0.221	64.5	0.226	3.7
	40	0.042	85.7	0.224	60.0	0.217	3.5
	60	0.042	83.4	0.229	56.3	0.211	3.5
	80	0.041	81.1	0.234	53.2	0.208	3.5
	100	0.040	78.7	0.238	50.7	0.205	3.5
	120	0.039	76.0	0.243	48.4		
Hydrogen	−435	0.053	4.6	1.6	5.0	0.35	1.9
−434.5°F M.P.	−430	0.060	4.5	1.8	4.1	0.29	1.5
−423.0°F B.P.	−425	0.066	4.3	2.1	3.0	0.226	1.2
Oxygen	−350	0.11	80	0.4	120	0.49	4.8
−361°F M.P.	−300	0.088	72	0.4	41	0.16	2.2
−297°F B.P.							

Table B.7 (Continued)

Saturated Liquid
Melting point
Boiling point

Water
32°F M.P.
212°F B.P.

$T,$ °F	$k,$ Btu/hr ft °F	$\rho,$ lb/ft³	$c_p,$ Btu/lb °F	$\mu \times 10^7,$ lb,sec/ft²	$\nu \times 10^5,$ ft²/sec	Pr
32	0.319	62.4	1.009	374	1.93	13.35
40	0.325	62.4	1.005	324	1.67	11.35
50	0.332	62.4	1.002	274	1.41	9.40
60	0.340	62.3	1.000	234	1.21	7.88
70	0.347	62.3	0.998	205	1.057	6.78
80	0.353	62.2	0.998	180	0.930	5.85
90	0.359	62.1	0.997	160	0.828	5.12
100	0.364	62.0	0.997	143	0.741	4.53
120	0.372	61.7	0.997	118	0.615	3.64
140	0.378	61.4	0.998	98	0.513	3.01
160	0.384	61.0	1.000	84	0.442	2.53
180	0.389	60.6	1.002	73	0.388	2.16
200	0.392	60.1	1.004	64	0.342	1.90
250	0.396	58.8	1.012	48	0.262	1.43
300	0.395	57.3	1.026	39	0.219	1.17
400	0.381	53.7	1.067	29	0.174	1.0
600	0.292	49.0	1.362	18	0.118	1.0

Interpolated from data compiled in *Heat and Mass Transfer*, E. R. G. Eckert and R. M. Drake, McGraw-Hill Book Co., New York, 1959, and *Introduction to Heat Transfer*, A. I. Brown and S. M. Marco, McGraw-Hill Book Co., New York, 1958.

Table B.8 Standard Atmosphere

Attitude, ft	Mean free path, ft	Pressure, atm	Temp., °F	Density ratio, ρ/ρ_0
−1000	2.11×10^{-7}	1.0367	62.6	1.0296
0	2.18×10^{-7}	1.0000	59.0	1.0000
1000	2.24×10^{-7}	0.9644	55.4	0.9711
2000	2.31×10^{-7}	0.9298	51.9	0.9428
3000	2.38×10^{-7}	0.8963	48.3	0.9151
4000	2.45×10^{-7}	0.8637	44.7	0.8881
6000	2.60×10^{-7}	0.8014	37.6	0.8359
8000	2.77×10^{-7}	0.7429	30.5	0.7861
10000	2.95×10^{-7}	0.6878	23.4	0.7386
15000	3.46×10^{-7}	0.5646	5.5	0.6295
20000	4.08×10^{-7}	0.4599	−12.3	0.5332
30000	5.81×10^{-7}	0.2975	−47.8	0.3747
40000	8.81×10^{-7}	0.1858	−69.7	0.2471
60000	2.29×10^{-6}	0.0714	−69.7	0.0949
80000	6.03×10^{-6}	0.0276	−62.0	0.0361
100,000	1.56×10^{-5}	0.0110	−51.1	0.0140
150,000	1.50×10^{-4}	1.34×10^{-3}	19.4	1.45×10^{-3}
200,000	9.81×10^{-4}	1.95×10^{-4}	−2.7	2.22×10^{-4}
300,000	1.12×10^{-1}	1.25×10^{-6}	−126.8	1.95×10^{-6}
400,000	$1.38 \times 10^{+1}$	2.11×10^{-8}	233.9	1.52×10^{-8}
600,000	$4.47 \times 10^{+2}$	1.98×10^{-9}	1647.2	4.38×10^{-10}

From *U.S. Standard Atmosphere*. Washington, D.C.: U.S. Government Printing Office, 1962.

Table B.9 Volume Expansion Coefficients
$$\beta = -(1/\rho)(\partial\rho/\partial T)$$

Fluid	$\beta \times 10^3 [1/°R]$
Gases, ideal	$1000/T$
Liquids	
Ammonia, 68°F	1.36
Bismuth, 1000°F	7.0
Carbon dioxide, 68°F	3.67
Engine oil, extra heavy, 68°F	0.39
Freon 12, −20°F	1.03
Hydrogen, −423°F	8.4
Mercury, 68°F	0.10
Oxygen, −300°F	1.11
Sodium, 200°F	0.15
Water, 40°F	0.045
50°F	0.070
60°F	0.10
70°F	0.13

Table B.9 (Continued)

80°F	0.15
90°F	0.18
100°F	0.20
120°F	0.24
140°F	0.29
160°F	0.33
180°F	0.37
200°F	0.41
250°F	0.50
300°F	0.58
400°F	0.72
500°F	1.18

From E. R. G. Eckert and R. M. Drake, *Heat and Mass Transfer,* New York: McGraw-Hill, 1959; and A. I. Brown and S. M. Marco, *Introduction to Heat Transfer,* 3rd Ed. New York: McGraw-Hill, 1958.

Table B.10 Surface Tensions

Substance	Temperature, °F	Surface Tension, $\sigma_t \times 10^3$, lb_f/ft
Acetone	32	1.80
	68	1.62
	104	1.45
Ammonia	52	1.60
	100	1.24
Bismuth	520	25.8
Ethyl alcohol	32	1.65
	85	1.50
Lead	625	30.8
Mercury	32	33
	140	32
Methyl alcohol	68	1.55
	122	1.34
Potassium	133	28.1
Sodium	194	20
	212	14
	482	13.6

Table B.10   (Continued)

Substance	Temperature, °F	Surface Tension, $\sigma_t \times 10^3$, lb$_f$/ft
Water	32	5.1
	100	4.8
	200	4.2
	300	3.3
	400	2.5
	600	0.6

From the *Handbook of Chemistry and Physics*, 47th Ed., Cleveland, Ohio: The Chemical Rubber Co., 1967.

Table B.11   Thermodynamic Properties of Saturated Steam

T, °F	P, lb$_f$/in.^{2}	$\hat{V} = 1/\rho$, ft^3/lb	$\hat{h}_{fg}$, Btu/lb
32	0.0886	3305.7	1075.1
40	0.1217	2445.1	1070.5
50	0.1780	1704.9	1064.8
60	0.2561	1208.1	1059.1
70	0.3628	868.9	1053.9
80	0.5067	633.7	1047.8
90	0.6980	468.4	1042.1
100	0.9487	350.8	1036.4
110	1.274	265.7	1030.9
120	1.692	203.47	1025.3
130	2.221	157.57	1019.5
140	2.887	123.18	1013.7
150	3.716	97.20	1007.8
160	4.739	77.39	1002.0
170	5.990	62.14	996.1
180	7.510	50.28	990.2
190	9.336	41.01	984.1
200	11.525	33.67	977.8
212	14.696	26.83	970.3

From *Steam Tables*, 3rd Ed., Windsor, Conn.: Combustion Engineering, 1940.

Table B.12   Mass Diffusivities

B.12a Schmidt number for a dilute mixture with air; latent heat^a and boiling point^a at 1 atm

Gas	Sc	$\hat{h}_{fg}$, Btu/lb	T_{BP}, °R
Ammonia, NH_3	0.61	589	432
Benzene, C_6H_6	1.79	170	637
Carbon dioxide, CO_2	1.00	171	350
Carbon monoxide, CO	0.77	91	146
Chlorine, Cl_2	1.42	124	429
Ethanol, CH_3CH_2OH	1.32	367	633
Heptane, C_7H_{16}	1.99	146	669
Hydrogen, H_2	0.22	195	36.6
Methanol, CH_3OH	0.98	473	608
Nitric oxide, NO	0.86	200	218
Oxygen, O_2	0.74	92	163
Sulfur dioxide, SO_2	1.24	171	473
Water vapor, H_2O	0.61	970	672

B.12b Calculated temperature variations (following Chapter 7) in air at 1 atm

Temperature, °F	Diffusivity, ft²/hr						
	H_2O	CO_2	CO	C_7H_{16}	H_2	NO	SO_2
−200	0.28	0.15	0.21	0.07	0.81	0.19	0.12
−100	0.50	0.29	0.38	0.14	1.45	0.34	0.23
0	0.76	0.45	0.59	0.23	2.23	0.53	0.36
100	1.06	0.65	0.83	0.33	3.14	0.76	0.52
200	1.40	0.87	1.11	0.44	4.15	1.00	0.71
400	2.19	1.38	1.74	0.71	6.50	1.59	1.14
600	3.09	1.97	2.47	1.02	9.22	2.26	1.65
800	4.11	2.65	3.30	1.37	12.3	3.03	2.24
1000	5.26	3.40	4.22	1.76	15.7	3.87	2.87
1500	8.60	5.57	6.89	2.90	25.6	6.32	4.73
2000	12.5	8.12	10.0	4.24	37.3	9.21	6.93
2500	17.0	11.0	13.7	5.76	50.8	12.5	9.43
3000	22.1	14.3	17.7	7.46	65.8	16.2	12.2
3500	27.6	17.9	22.2	9.34	82.2	20.3	15.3

B.12c Schmidt number at 68°F[b] for dilute solution in water

Solute	Sc	M
Hydrogen	196	2.016
Hydrogen chloride	381	36.47
Nitric acid	390	63.02
Oxygen	558	32.00
Carbon dioxide	559	44.01
Ammonia	570	17.03
Sulfuric acid	580	98.08
Nitrogen	613	28.02
Nitrous oxide	665	44.02
Sodium hydroxide	665	40.00
Hydrogen sulfide	712	34.08
Sodium chloride	745	58.45
Methanol	785	32.04
Chlorine	824	70.90
Bromine	840	159.83
Ethanol	1005	46.07
Propanol	1150	60.09
Phenol	1200	94.10
Butanol	1310	74.12
Glycerol	1400	92.09
Sucrose	2230	342.30

[a] With the Clausius-Clapyeron relation one may estimate vapor pressure as follows:

$$P_{vap} \doteq \exp\left\{-\frac{M\hat{h}_{fg}}{\mathcal{R}}\left(\frac{1}{T}-\frac{1}{T_{BP}}\right)\right\}[atm]$$

Tables of vapor pressure can be found, for example, in the *Handbook of Chemistry and Physics*. See Table B.11 for water vapor at low temperatures.

[b] For other temperatures use $Sc/Sc_{68°F} = (\mu^2/\rho T)/(\mu^2/\rho T)_{528°R}$ where μ and ρ are for water and T is absolute temperature. For chemically similar solutes but other molecular weights use $Sc_2/Sc_1 \doteq (M_2/M_1)^{0.4}$.

From various sources. See, for example, R. A. Svehla, *Estimated Viscosities and Thermal Conductivities of Gases at High Temperatures*, NASA TR R-132, 1962; the *Handbook of Chemistry and Physics*, 47th Ed., New York: The Chemical Rubber Co., 1967; and D. B. Spalding, *Convective Mass Transfer*, New York: McGraw-Hill, 1963.

Table B.13 Selected Atomic Weights[a]

Aluminum	Al	26.98	Molybdenum	Mo	95.94
Antimony	Sb	121.76	Neodymium	Nd	144.24
Argon	Ar	39.95	Neon	Ne	20.18
Arsenic	As	74.92	Nickel	Ni	58.71
Barium	Ba	137.34	Niobium	Nb	92.91
Beryllium	Be	9.012	Nitrogen	N	14.007
Bismuth	Bi	208.98	Oxygen	O	15.999
Boron	B	10.81	Palladium	Pd	106.4
Bromine	Br	79.90	Phosphorus	P	30.97
Cadmium	Cd	112.40	Platinum	Pt	195.09
Calcium	Ca	40.08	Plutonium	Pu	242
Carbon	C	12.01	Potassium	K	39.10
Cesium	Cs	132.91	Radium	Ra	226
Chlorine	Cl	35.45	Radon	Rn	222
Chromium	Cr	51.996	Rhenium	Re	186.2
Cobalt	Co	58.93	Rhodium	Rh	102.90
Copper	Cu	63.55	Rubidium	Rb	85.47
Fluorine	F	18.998	Selenium	Se	78.96
Gadolinium	Gd	157.25	Silicon	Si	28.09
Gallium	Ga	69.72	Silver	Ag	107.87
Germanium	Ge	72.59	Sodium	Na	22.99
Gold	Au	196.97	Strontium	Sr	87.62
Hafnium	Hf	178.5	Sulfur	S	32.06
Helium	He	4.003	Tantalum	Ta	180.95
Hydrogen	H	1.008	Tellurium	Te	127.60
Indium	In	114.82	Thallium	Tl	204.37
Iodine	I	126.90	Thorium	Th	232.04
Iron	Fe	55.85	Tin	Sn	118.69
Krypton	Kr	83.8	Titanium	Ti	47.90
Lead	Pb	207.19	Tungsten	W	183.85
Lithium	Li	6.939	Uranium	U	238.03
Magnesium	Mg	24.31	Xenon	Xe	131.30
Manganese	Mn	54.94	Zinc	Zn	65.37
Mercury	Hg	200.59	Zirconium	Zr	91.22

[a] Based on the isotope $C^{12} = 12$
Selected from the *Handbook of Chemistry and Physics*, 50th Ed., 1969.

Table B.14   Spectral and Total Absorptances of Metals

Spectral absorptance, normal incidence, $1 < \lambda < 25\ \mu$

$$\alpha(\lambda, T_w) \doteq A\left[\frac{(1+\lambda^2/\lambda_{12}^2)^{1/2}-1}{\lambda^2/2\lambda_{12}^2}\right]^{1/2} + \frac{B}{C+\lambda^2}$$

Total absorptance, normal incidence, $600°R < T_e < 4000°R$

$$\alpha(T_w, T_e) \doteq A\left[\frac{(1+0.078C_2^2/\lambda_{12}^2 T_e^2)^{1/2}-1}{0.039C_2^2/\lambda_{12}^2 T_e^2}\right]^{1/2} + \frac{18.6(BT_e^2/C_2^2)}{1+18.6(CT_e^2/C_2^2)}$$

where $C_2 = hc/k = 25{,}896\ \mu°R$

| Metal | Parameters ($T_w = 550°R$) | | | |
	A	B	C	λ_{12}
Aluminum foil	0.0165	0.23	8.9	14
Cadmium, 99.99%, rolled plate	0.054	2.15	3.2	9
Chromium, polished electroplate	0.076	1.58	3.9	3
Columbium, 99.99%, rolled plate	0.15	0.29	~ 0	1
Copper, 99.99%, polished	0.018	0.077	3.2	45
Gold, 99.99%, polished	0.020	0.056	1.4	45
Indium, 99.99%, scraped	0.060	0.24	1.3	6
Inconel X, rolled plate	0.44	0.036	~ 0	1
Lead, 99.99%, scraped	0.16	0.39	1.1	4
Manganese, 99.99%, polished	0.19	4.8	11	8
Molybdenum, 99.99%	0.033	0.36	~ 0	7
Nickel, 99.99%, polished	0.029	0.83	2.4	5
Platinum, 99.99%, cold rolled	0.038	0.42	~ 0	4
Rhodium, polished electroplate	0.06	1.27	10	6
Silver, polished electroplate	0.011	0.16	11	70
Stainless steel, 303, lapped	− 0.71	~ 0	~ 0	~ 0.125
Tin, 99.99%, rolled plate	0.052	0.56	0.8	7
Titanium, polished electroplate	0.13	2.9	8.3	12
Titanium, 99%, lapped	0.09	6.5	15	12
Tungsten, 99.99%, lapped	0.05	0.49	0.3	3
Vanadium, 99.99%, rolled plate	0.17	0.66	0.93	1
Zinc, 99.9%	0.036	0.26	~ 0	8
Zirconium, 99.99%, rolled plate	0.64	4	35	1

From *Advances in Thermophysical Properties at Extreme Temperatures and Pressures*, Am. Soc. Mech. Engrs., New York: 1965, pp. 189–199.

Table B.15   Total Emittances and Solar Absorptances of Surfaces

	Total Normal Emittance[a]	Extraterrestrial Solar Absorptance
Alumina, flame sprayed	0.80 (− 10°F)	0.28
Aluminum foil, as received	0.04 (70°F)	
Aluminum foil, bright dipped	0.025 (70°F)	0.10
Aluminum, vacuum deposited on duPont mylar	0.025 (70°F)	0.10
Aluminum alloy, 6061, as received	0.03 (70°F)	0.37
Aluminum alloy, 75S-T6 weathered 20,000 hr on a DC6 aircraft	0.16 (150°F)	0.54
Aluminum, hard anodized, 6061-T6, 35 amp/ft² at 45 volts in 20°F sulfuric acid solution, 1 mil thick	0.84 (− 10°F)	0.92
Aluminum, soft anodized *Reflectal* aluminum alloy, 5 amp/ft² for 2 hr in 40°F, 10% H_2SO_4 solution	0.79 (− 10°F)	0.23
Aluminum, 7075-T6, sandblasted with 60 mesh silicon carbide grit	0.30 (70°F)	0.55
Aluminized silicone resin paint Dow Corning XP-310	0.20 (200°F) 0.22 (800°F)	0.27
Beryllium	0.18 (300°F) 0.21 (700°F) 0.30 (1100°F)	0.77
Beryllium, anodized	0.90 (300°F) 0.88 (700°F) 0.82 (1100°F)	
Black paint		
Parson's optical black	0.95 (− 10°F)	0.975
Black silicone high heat National Lead Co. 46H47	0.93 (− 10°F to 1400°F)	0.94
Black epoxy paint, cat-a-lac Finch Paint and Chemical Co 463-1-8	0.89 (− 10°F)	0.95
Black enamel paint, Rinshed-Mason, heated 1000 hr at 700°F in air	0.81 (200°F) 0.80 (800°F)	
Chromium plate heated 50 hr at 1100°F	0.12 (200°F) 0.15 (750°F) 0.15 (95°F)	0.78
Copper, electroplated and black oxidized in Ebanol C	0.03 (70°F) 0.16 (95°F)	0.47 0.91
Glass, second surface mirror		
Aluminized	0.83 (− 10°F)	0.13
Silvered	0.83 (− 10°F)	0.13
Gold, coated on stainless steel, heated in air at 1000°F	0.09 (200°F) 0.14 (750°F)	
Gold, coated on 3M Tape Y9814	0.025 (70°F)	0.21
Graphite, crushed on sodium silicate	0.91 (− 10°F)	0.96

Table B.15   (Continued)

	Total Normal Emittance	Extraterrestrial Solar Absorptance
Inconel X, oxidized 4 hr at 1825°F, 10 hr 1300°F in air	0.71 (−10°F) 0.81 (200°F) 0.79 (800°F)	0.90
Magnesium-thorium alloy	0.07 (200°F) 0.06 (500°F)	
Magnesium, Dow 7 coating	0.36 (700°F)	
Mylar, duPont film aluminized on second surface		
0.00025 in. thick	0.37 (70°F)	0.17
0.001 in. thick	0.63 (70°F)	0.17
0.003 in. thick	0.81 (70°F)	0.24
Nickel, electroplated	0.03 (70°F)	0.22
Nickel, Tabor solar absorber electro-oxidized nickel on copper		
110–30	0.05 (95°F)	0.85
125–30	0.11 (95°F)	0.85
Platinum-coated stainless steel	0.13 (200°F) 0.15 (750°F)	
Annealed in air 300 hr at 700°F	0.11 (200°F) 0.13 (800°F)	
Silica, Corning Glass 7940M sintered powdered fused silica	0.84 (95°F)	0.08
Silica, second surface mirror		
Aluminized	0.83 (70°F)	0.14
Silvered	0.83 (70°F)	0.07
Silicon solar cell		
boron doped, no cover glass	0.32 (95°F)	0.94
0.006 mil quartz cover, blue filter	0.81 (70°F)	0.81
Silver		
Plated on nickel on stainless steel	0.06 (200°F) 0.08 (750°F)	
Heated 300 hr at 700°F	0.11 (200°F) 0.13 (800°F)	
Silver Chromatone paint	0.24 (70°F)	0.20
Stainless steel		
Type 312 heated 300 hr at 500°F	0.27 (200°F) 0.32 (800°F)	

Table B.15   (Continued)

	Total Normal Emittance	Extraterrestrial Solar Absorptance
Type 301 with Armco black oxide	0.75 (−10°F)	0.89
Type 410 heated to 1300°F in air	0.13 (95°F)	0.76
Type 303 sandblasted heavily with 80 mesh aluminum oxide grit	0.42 (200°F)	0.68
Steel blackened by Ebanol S treatment dipped 15 min in 286°F boiling solution	0.10 (95°F)	0.85
Titanium		
75A	0.10 (200°F)	
	0.19 (800°F)	
75A oxidized 300 hr at 850°F in air	0.21 (95°F)	0.80
	0.25 (800°F)	
C-110M oxidized 100 hr at 800°F in air	0.16 (95°F)	0.52
C-110M oxidized 300 hr at 850°F in air	0.20 (95°F)	0.77
Evaporated 80 to 100 μ thick and oxidized 3 hr at 750°F in air	0.14 (95°F)	0.75
Anodized	0.73 (−10°F)	0.51
White acrylic resin paint Sherwin-Williams M49WC8-CA-10144	0.92 (200°F)	
	0.87 (400°F)	
White epoxy paint, Cat-a-lac Finch Paint and Chemical Co. 483-1-8	0.88 (−10°F)	0.25
White potassium zirconium silicate inorganic spacecraft coating	0.89 (70°F)	0.13
Zinc blackened by Tabor solar collector electrochemical treatment 120-20	0.12 (95°F)	0.89

[a]For high emittance materials these values are about 0.03 higher than hemispherical values, while for low emittance metals the values are about 25% lower than hemispherical emittance. See, for example, *Progress in Astronautics and Aeronautics*, Vol. 11, pp. 427–446, 1963; *Solar Energy*, Vol. VI, No. 1, pp. 1–8, 1962; and Report 60–93 of the Department of Engineering, UCLA, 1960.

Table B.16   Theoretical Dimensions of Commercial Pipes (inches)

Nominal Size (~ I.D.)		5	10	Schedule 40	80	160	XX Strong
$\frac{1}{8}$	O.D.	0.405	0.405	0.405	0.405		
	Wall	0.035	0.049	0.068	0.095		
	I.D.	0.335	0.307	0.269	0.215		
$\frac{1}{4}$	O.D.	0.540	0.540	0.540	0.540		
	Wall	0.049	0.065	0.088	0.119		
	I.D.	0.442	0.410	0.364	0.302		
$\frac{3}{8}$	O.D.	0.675	0.675	0.675	0.675		
	Wall	0.049	0.065	0.091	0.126		
	I.D.	0.577	0.545	0.493	0.423		
$\frac{1}{2}$	O.D.	0.840	0.840	0.840	0.840	0.840	0.840
	Wall	0.065	0.083	0.109	0.147	0.187	0.294
	I.D.	0.710	0.674	0.622	0.546	0.466	0.252
$\frac{3}{4}$	O.D.	1.050	1.050	1.050	1.050	1.050	1.050
	Wall	0.065	0.083	0.113	0.154	0.218	0.308
	I.D.	0.920	0.884	0.824	0.742	0.614	0.434
1	O.D.	1.315	1.315	1.315	1.315	1.315	1.315
	Wall	0.065	0.109	0.133	0.179	0.250	0.358
	I.D.	1.185	1.097	1.049	0.957	0.815	0.599
$1\frac{1}{4}$	O.D.	1.660	1.660	1.660	1.660	1.660	1.660
	Wall	0.065	0.109	0.140	0.191	0.250	0.382
	I.D.	1.530	1.442	1.380	1.278	1.160	0.896
$1\frac{1}{2}$	O.D.	1.900	1.900	1.900	1.900	1.900	1.900
	Wall	0.065	0.109	0.145	0.200	0.281	0.400
	I.D.	1.770	1.682	1.610	1.500	1.338	1.100
2	O.D.	2.375	2.375	2.375	2.375	2.375	2.375
	Wall	0.065	0.109	0.154	0.218	0.343	0.436
	I.D.	2.245	2.157	2.067	1.939	1.689	1.503
$2\frac{1}{2}$	O.D.	2.875	2.875	2.875	2.875	2.875	2.875
	Wall	0.083	0.120	0.203	0.276	0.375	0.552
	I.D.	2.709	2.635	2.469	2.323	2.125	1.771
3	O.D.	3.500	3.500	3.500	3.500	3.500	3.500
	Wall	0.083	0.120	0.216	0.300	0.438	0.600
	I.D.	3.334	3.260	3.068	2.900	2.624	2.300
$3\frac{1}{2}$	O.D.	4.000	4.000	4.000	4.000	−	4.000
	Wall	0.083	0.120	0.226	0.318	−	0.636
	I.D.	3.834	3.760	3.548	3.364	−	2.728
4	O.D.	4.500	4.500	4.500	4.500	4.500	4.500
	Wall	0.083	0.120	0.237	0.337	0.531	0.674
	I.D.	4.334	4.260	4.026	3.826	3.438	3.152
5	O.D.	5.563	5.563	5.563	5.563	5.563	5.563
	Wall	0.109	0.134	0.258	0.375	0.625	0.750
	I.D.	5.345	5.295	5.047	4.813	4.313	4.063
6	O.D.	6.625	6.625	6.625	6.625	6.625	6.625
	Wall	0.109	0.134	0.280	0.432	0.718	0.864
	I.D.	6.407	6.357	6.065	5.761	5.189	4.897

Table B.17 Combustion Data

| Fuel Substance | M | Heats of Combustion | | Stoichiometric Combustion (no excess air) | | | | | | | |
| | | Higher Heat Liquid H_2O, Btu/lb | Lower Heat Vapor H_2O, Btu/lb | Required Air lb/lb$_{fuel}$ | | | Major Flue Gases lb/lb$_{fuel}$ | | | |
				O_2	N_2	Air	CO_2	H_2O	N_2	SO_2
C	12.01	14,093	14,093	2.66	8.86	11.53	3.66	—	8.86	
H_2	2.016	61,100	51,623	7.94	26.41	34.34	—	8.94	26.41	
CO	28.01	4,347	4,347	0.57	1.90	2.47	1.57	—	1.90	
CH_4	16.041	23,879	21,520	3.99	13.28	17.27	2.74	2.25	13.28	
C_2H_6	30.067	22,320	20,432	3.73	12.39	16.12	2.93	1.80	12.39	
C_3H_8	44.092	21,661	19,944	3.63	12.07	15.70	2.99	1.63	12.07	
C_4H_{10}	58.118	21,308	19,680	3.58	11.91	15.49	3.03	1.55	11.91	
C_5H_{12}	72.144	21,091	19,517	3.55	11.81	15.35	3.05	1.50	11.81	
C_2H_4	28.051	21,644	20,295	3.42	11.39	14.81	3.14	1.29	11.39	
C_3H_6	42.077	21,041	19,691	3.42	11.39	14.81	3.14	1.29	11.39	
C_4H_8	56.102	20,840	19,496	3.42	11.39	14.81	3.14	1.29	11.39	
C_5H_{10}	70.128	20,712	19,363	3.42	11.39	14.81	3.14	1.29	11.39	
C_6H_6	78.107	18,210	17,480	3.07	10.22	13.30	3.38	0.69	10.22	
C_7H_8	92.132	18,440	17,620	3.13	10.40	13.53	3.34	0.78	10.40	
C_8H_{10}	106.158	18,650	17,760	3.17	10.53	13.70	3.32	0.85	10.53	
C_2H_2	26.036	21,500	20,776	3.07	10.22	13.30	3.38	0.69	10.22	
$C_{10}H_8$	128.162	17,298	16,708	3.00	9.97	12.96	3.43	0.56	9.97	
CH_3OH	32.041	10,259	9,078	1.50	4.98	6.48	1.37	1.13	4.98	
C_2H_5OH	46.067	13,161	11,929	2.08	6.93	9.02	1.92	1.17	6.93	
NH_3	17.031	9,668	8,001	1.41	4.69	6.10	—	1.59	4.69	
S	32.06	3,983	3,983	1.00	3.29	4.29	—	—	3.29	2.00
H_2S	34.076	7,100	6,545	1.41	4.69	6.10	—	0.53	4.69	1.88

From *Fuel Flue Gases*, New York: The American Gas Association, 1940.

Table B.18   Henry's Constant for Dilute Aqueous Solutions (at Moderate Pressures)

Solute	$P_{A,s}/x_{A,s}$ atm					
	60°F	80°F	100°F	120°F	140°F	160°F
NH_3	26.0	30.2	–	–	–	–
Cl_2	480	620	770	880	950	980
SO_2	290	440	620	840	1,090	1,370
CO_2	1,260	1,710	2,220	2,790	3,500	–
O_2	37,000	45,500	53,000	59,000	63,000	67,000
H_2	66,000	72,000	76,000	77,000	77,000	77,000
CO	49,000	60,000	70,000	78,000	83,000	86,000
Air	61,000	74,000	86,000	96,000	104,000	110,000
N_2	74,000	89,000	104,000	–	–	–

From D. B. Spalding, *Convective Mass Transfer*. New York: McGraw-Hill, 1963.

PRINCIPAL SYMBOLS

A area

A_c cross-sectional area available for flow

Bi Biot number

C flow capacity (mass-flow-rate-specific-heat product)

C_{He} Henry's constant

C_R rotary mass-flow-rate-specific-heat product for a rotary regener-
           ator

c molar concentration

c_f skin friction coefficient

c_i molar concentration of species i

c_p heat capacity at constant pressure

c_v heat capacity at constant volume

c speed of light

D diameter

D_h hydraulic diameter

$\mathcal{D}_{ij}$ binary diffusion coefficient

$\mathcal{D}_{im}$ effective binary diffusion coefficient of species i in a multi-
           component mixture

d differential operator; also molecular diameter

Ec Eckert number

F	force: also a function
F_{i-j}	shape factor
f	friction factor
Gr	Grashof number
g	gravitational acceleration
H	elevation or height
He	Henry number
h	convective heat transfer coefficient (also h_1, h_2, h_c)
$\hat{h}$	enthalpy
$\hat{h}_{fg}$	latent heat of phase change (gas-liquid)
$\hat{h}_i{}^\circ$	heat of formation of species i
$\hbar$	Planck's constant
I	radiant intensity; also electric current
I^+	intensity of the radiosity
J	number flux
Ja	Jakob number
J_i	molar diffusion flux of species i relative to mass average velocity
$J_i{}^*$	molar diffusion flux of species i relative to mole average velocity
j_i	mass diffusion flux of species i relative to mass average velocity
K	mass transfer coefficient
K_G	gas side mass transfer coefficient
K_{Goa}	overall mass transfer coefficient
K_L	liquid side mass transfer coefficient
$\mathscr{K}$	von Karman's constant
k	thermal conductivity
k''	heterogeneous reaction rate constant
$\hbar$	Boltzmann's constant
L	length
l	mean free path
M	mean molecular weight of mixture
M_i	molecular weight of species i
$\dot{M}_G$	gas stream molar flow rate
$\dot{M}_L$	liquid stream molar flow rate
$\dot{m}$	mass flow rate
$\dot{m}_G$	gas stream mass flow rate
$\dot{m}_L$	liquid stream mass flow rate
m_i	mass fraction of species i
m	mass of a molecule
N	molar flux
$\mathcal{N}$	number density
N_{Av}	Avogadro's number
N_i	molar flux of species i
N_{tu}	number of transfer units
Nu	Nusselt number
n	mass flux

$\mathscr{n}$	number fraction
n_i	mass flux of species i
P	pressure
Pr	Prandtl number
$\mathscr{P}$	perimeter
p	momentum
ppm	parts per million by volume (mole fraction $\times$ 10^6)
$\dot{Q}$	rate of heat transfer
$\hat{Q}$	heat transferred to or from a system per unit mass
$\dot{Q}_V$	volumetric heat source
q	heat flux (also q_x, q_y, q_z, q_r)
q^+	radiosity
q^-	irradiation
R	tube radius; also thermal or electrical resistance
R_C	flow capacity ratio
Re	Reynolds number
$\dot{R}_i$	mole rate of production of species i
R_i	gas constant for species i
R_R	rotary regenerator flow capacity ratio
$\mathscr{R}$	universal gas constant
r	radial distance in both cylindrical and spherical coordinates
r_f	recovery factor
$\dot{r}_i$	mass rate of production of species i
Sc	Schmidt number
St	Stanton number
T	temperature
t	time
U	overall heat transfer coefficient
$\hat{u}$	internal energy
u	component of velocity in x-direction
V	velocity; also volume
$\hat{V}$	specific volume
$\mathbf{v}$	mass average velocity
$\mathbf{v}_i$	velocity of species i
v_τ	friction velocity
$\mathbf{v}^*$	molar average velocity
W	width of a surface
$\hat{W}$	work done on or by a system per unit mass
$\mathscr{W}$	molar content of a system
w	mass content of a system
x	rectangular coordinate
x_i	mole fraction of species i
y	rectangular coordinate
y^+	dimensionless distance from wall
y_i	mole fraction of species i in the gas phase

z	rectangular coordinate
α	thermal diffusivity; also absorptance
β	thermal coefficient of volume expansion
Γ	flow rate per unit width
γ	ratio of specific heats
Δ	finite increment
δ	film thickness
ϵ	effectiveness; emittance; eddy diffusivity; also Lennard-Jones parameter
ϵ_v	void fraction
η	fin effectiveness
θ	angle in cylindrical or spherical coordinates, radians
λ	wavelength, microns
μ	viscosity; also an abbreviation for microns, 10^{-6} meter
ν	kinematic viscosity; also wavenumber
ν_f	frequency
ρ	density; also reflectance
ρ_i	mass concentration of species i, lb/ft³; also reflectance of surface i
σ	Stefan-Boltzmann constant; also Lennard-Jones parameter
σ_c	collision cross section
σ_t	surface tension, lb$_f$/ft
τ	shear stress or momentum flux
Φ	dissipation function; also heat flux potential
ϕ	angle; intermolecular potential; also neutron flux
Ω	collision integral; also angular rotation rate

Overscores

$\hat{}$	per unit mass
$-$	time smoothed
$\cdot$	rate

Underscores

$\sim, -$ denote vector quantities

Superscripts

$*$	reduced with respect to some reference quantity
$'$	deviation from time-smoothed value
$+$	outward from the wall, dimensionless
$-$	inward toward the wall

Subscripts

A, B	species in binary system (also 1, 2)
a	ambient condition

b	bulk or *cup mixing* value for enclosed stream
C	cold side
c	convective
e	external condition
f, F	film
G	gas side
H	hot side
i,j	species in multicomponent system
i	node number, interface, internal total
L	liquid side; at x, y, or $z = L$
m	for mass transfer; also mixture
r	for radiation
S	solar
s	at a surface, saturated
t	time or thickness
V	on a volume basis
W	of a wall
w	at a wall
0	evaluated at reference surfaces (also 1, 2, . . .)

INDEX

Page 73 (Line 17) — The mass ... $\mathscr{W}M_1$; thus,

The mass ... $\mathscr{W}M_2$; thus,

Page 116 (Line 11) — $M = 2\pi Dc$ ...

$\dot{M} = 2\pi Dc$ ...

Page 140 (Line 11) — Then Gr is found

Thus Gr is found.

Page 140 (Line 14 — air. Figure 5.1, ...

air. Table 5.1, ...

Page 140(bottom of page) — $q_c = ... = 342$

$q_c = ... = 348$

Page 140(bottom of page) — ... (60) = 54.4 ...

... (60) = 55.3 ...

Page 140(Line 23) — Now, inherent in our method of solution is the assumption that $Pr \doteq Sc$. Therefore we arbitrarily use the mean value for both heat and mass transfer

$$Nu_L \doteq 342$$

Now, inherent in our method of solution is the assumption that $Pr \doteq Sc$.

Page 141(Line 5) — $j_1 = \dfrac{342}{0.61}$ ...

$j_1 = \dfrac{335}{0.61}$ ...

Page 141(Line 6) — $j_1 = 0.395$ lb/hr ft^2

$j_1 = 0.387$ lb/hr ft^2

Page 141(Line 8) — $q_{evap} = (0.395)(1019.5) = 403$ ...

$q_{evap} = (0.387)(1019.5) = 395$ Btu/hr ft^2

Page 141(Line 11) — $q_{conv} = 403 + 54 = 457$ Btu/hr ft^2

$q_{conv} = 395 + 55 = 450$ Btu/hr ft^2

Page 142(Table 5.1 column 4) — line 2 — $Nu_x = 0.0296 Re^{0.8}$ ...

$Nu_x = 0.0296 Re_x^{0.8}$ ...

Page 145(Table 5.3) (column 3) — ... $D_h = 4A_c/\rho$

... $D_h = 4A_c/\mathscr{P}$

Page 175(Line 3) — $q_P{}^+ = \dfrac{2\pi k^4 T^4}{h^3{}^2} \displaystyle\int_{\zeta=0}^{\infty} \dfrac{\zeta^3\, d\zeta}{\exp(\zeta) - 1}$

$q_P{}^+ = \dfrac{2\pi k^4 T^4}{h^3 c^2} \displaystyle\int_{\zeta=0}^{\infty} \dfrac{\zeta^3\, d\zeta}{\exp(\zeta) - 1}$

Page 177(Line 4) (Eq. (6.66a) —

$$q^+ = \tfrac{1}{4}\bar{v}\mathscr{N}(2kT) = KJ^+(2kT)$$

$q^+ = \tfrac{1}{4}\bar{v}\mathscr{N}(2kT) = J^+(2kT)$

Page 217(Line 4) (bottom) — 10% ... 500°C. ...

10% ... 500°F. ...

Page 246(Eq. (9.25) —

• $U\mathscr{P} =$

$$\cfrac{1}{\dfrac{1}{ch_c} + \dfrac{\delta_{s,c}}{ck_{s,c}} + \dfrac{\ln(D_o/D_i)}{2\pi k_w} + \dfrac{\delta_{s,H}}{\mathscr{P}_H k_{s,H}} + \dfrac{1}{\mathscr{P}_H h_H}}$$

• $U\mathscr{P} =$

$$\cfrac{1}{\dfrac{1}{\mathscr{P} c h_c} + \dfrac{\delta_{s,c}}{\mathscr{P} c k_{s,c}} + \dfrac{\ln(D_o/D_i)}{2\pi k_w} + \dfrac{\delta_{s,H}}{\mathscr{P}_H k_{s,H}} + \dfrac{1}{\mathscr{P}_H h_H}}$$

Page 253(Line 8) — where ... G_G and G_H.

where ... C_C and C_H.

Page 257(Line 3) — Special Case of $R_R \gg N_{tu}$

Special Case of $R_R \gg N_{tu}$

Page 333(Line 2) (6th column) — $\mu = 10^{7a}$,

$\mu \times 10^{7a}$,